Studies in Fuzziness and Soft Computing

Volume 345

Series editor

Janusz Kacprzyk, Systems Research Institute, Polish Academy of Sciences, Warsaw, Poland
e-mail: kacprzyk@ibspan.waw.pl

About this Series

The series "Studies in Fuzziness and Soft Computing" contains publications on various topics in the area of soft computing, which include fuzzy sets, rough sets, neural networks, evolutionary computation, probabilistic and evidential reasoning, multi-valued logic, and related fields. The publications within "Studies in Fuzziness and Soft Computing" are primarily monographs and edited volumes. They cover significant recent developments in the field, both of a foundational and applicable character. An important feature of the series is its short publication time and world-wide distribution. This permits a rapid and broad dissemination of research results.

More information about this series at http://www.springer.com/series/2941

Xiaolu Zhang · Zeshui Xu

Hesitant Fuzzy Methods for Multiple Criteria Decision Analysis

Springer

Xiaolu Zhang
The Collaborative Innovation Center
Jiangxi University of Finance and Economics
Nanchang, Jiangxi
China

Zeshui Xu
Business School
Sichuan University
Chengdu, Sichuan
China

ISSN 1434-9922 ISSN 1860-0808 (electronic)
Studies in Fuzziness and Soft Computing
ISBN 978-3-319-82476-5 ISBN 978-3-319-42001-1 (eBook)
DOI 10.1007/978-3-319-42001-1

Softcover reprint of the hardcover 1st edition 2016

Printed on acid-free paper

This Springer imprint is published by Springer Nature
The registered company is Springer International Publishing AG
The registered company address is: Gewerbestrasse 11, 6330 Cham, Switzerland

Preface

Multiple criteria decision making (MCDM) is a process to make an optimal choice that has the highest degree of satisfaction from a set of alternatives that are characterized in terms of multiple conflicting criteria, which is a usual task in human activities. In the classical MCDM, both the criteria values and the weights of criteria are usually expressed by crisp (no-fuzzy) numbers. To address the classical MCDM problems, lots of prominent decision-making methods have been developed, such as the TOPSIS (*Technique for Order Preference by Similarity to Ideal Solution*) method (Hwang and Yoon 1981), the TODIM (*an acronym in Portuguese of interactive and multi-criteria decision making*) method (Gomes and Lima 1991, 1992), the LINMAP (*linear programming technique for multidimensional analysis of preferences*) method (Srinivasan and Shocker 1973), the QUALIFLEX (*qualitative flexible multiple criteria method*) method (Paelinck 1976, 1977, 1978), and the ELECTRE (*ELimination Et Choix Traduisant la REalité*) method (Roy 1968). The TOPSIS method is a kind of simple and useful decision-making method and is suitable to handle the classical MCDM problems in which the weights of criteria are completely known in advance. The basic idea of TOPSIS method is that the optimal alternative should have the shortest distance from the positive ideal solution (PIS) and have the farthest distance from the negative ideal solution (NIS), simultaneously. The TODIM method is a discrete multiple criteria decision analysis method based on prospect theory (Kahneman and Tversky 1979) and has been proven to be a valuable tool for solving the classical MCDM problems in case of considering the decision maker's psychological behavior. The QUALIFLEX method is one of outranking methods and is suitable to deal with the decision-making problems where the number of criteria markedly exceeds the number of alternatives. The most characteristic of the QUALIFLEX method is the correct treatment of cardinal and ordinal information (Rebai et al. 2006). The LINMAP method is a practical and useful approach to address the classical MCDM problems in which the weights of criteria and the PIS are unknown in advance. In the LINMAP approach, the decision maker is required to not only provide the ratings of alternatives with respect to each criterion, but also simultaneously give the incomplete preference relations on pair-wise comparisons of alternatives.

Because of the inherent vagueness of human preferences as well as the objects being fuzzy and uncertain, the criteria values and/or weights of criteria involved in the MCDM problems are not always expressed in crisp numbers, but some are more suitable to be denoted by fuzzy numbers and their extensions, such as interval numbers, triangular fuzzy numbers (TFNs), trapezoidal fuzzy numbers (TrFNs), linguistic variables, intuitionistic fuzzy numbers (IFNs), and hesitant fuzzy elements (HFEs). This book mainly investigates the MCDM or multiple criteria group decision-making (MCGDM) problems in which the criteria values and/or the weights of criteria are expressed as HFEs (Xia and Xu 2011). Hesitant fuzzy set (HFS) was originally introduced by Torra (2010) as an extension of fuzzy set (Zadeh 1975). The HFE is the basic element of HFS, which allows the membership of an element to a given set to be represented by a few different values, and is a useful tool to describe and deal with uncertain information in the process of MCDM, especially for the practical MCDM problems in the case of considering the decision maker's hesitation. For example, the individual decision maker may have hesitancy among 0.1, 0.3, and 0.4 when he/she provides the assessment of an alternative with a criterion. Then, this decision maker can employ the HFE $H\{0.1,0.3,0.4\}$ to express his/her assessment provoked by hesitation. In this case, the HFE is used to express the individual decision maker's hesitation. On the other hand, Xu and Xia (2011) gave an example to illustrate how to employ the HFE to express the decision organization's hesitation: A decision organization including several experts is authorized to estimate the degree to which an alternative should satisfy a criterion. Suppose that there are three cases: Some experts provide 0.3, some provide 0.5, and the others provide 0.6, and these three parts cannot persuade each other. Thus, the degree that the alternative satisfies the criterion can be expressed by a HFE $H\{0.3,0.5,0.6\}$. It is noted that the HFE $H\{0.3,0.5,0.6\}$ can describe the above situation more objectively than the IFN $I(0.3,0.4)$ or the interval number [0.3,0.6], because the degree to which the alternative satisfies the criterion is not the convex combination of 0.3 and 0.4, or the interval between 0.3 and 0.6, but just three possible values (Xu and Xia 2011). Obviously, the use of hesitant fuzzy assessments makes the decision makers or the experts' judgments more reliable and informative in decision-making process. The HFEs have also been successfully applied in the fields of decision making (Xia and Xu 2011; Farhadinia 2013; Rodríguez et al. 2014; Zhu and Xu 2014) and clustering analysis (Zhang and Xu 2015c, 2015d), etc.

How to make a scientific decision with hesitant fuzzy information is an interesting and important research topic. In real-life world, one may encounter the hesitant fuzzy decision-making problems with different situations, such as the cases that the weights of criteria and/or experts are completely unknown or known in advance, the cases that the PIS and NIS are known or unknown in advance, the case that the decision maker's psychological behavior should be taken into account, and the case that the number of criteria markedly exceeds the number of alternatives, and the hesitant fuzzy decision-making problems under different situations need different decision-making methods to solve. For this purpose, the authors have recently developed lots of distinct hesitant fuzzy decision-making methods to deal with different MCDM or MCGDM problems effectively, in which the decision data are expressed by hesitant fuzzy information. For example, Xu and Zhang (2013) developed hesitant fuzzy TOPSIS decision analysis method to address the hesitant fuzzy MCDM problems in which the PIS and NIS are known in advance; Zhang and Xu (2014a) proposed the hesitant fuzzy TODIM decision analysis approach to solve the hesitant fuzzy MCDM problems in the case of considering the decision maker's psychological behavior; Zhang and Xu (2015a) put forward the hesitant fuzzy QUALIFLEX decision analysis method to handle the hesitant fuzzy MCDM problems in which the number of criteria markedly exceeds the number of alternatives; Zhang and Xu (2014b) and Zhang et al. (2015b), respectively, developed the hesitant fuzzy LINMAP decision analysis methods to solve the hesitant fuzzy MCGDM problems in which the weights of criteria and the PIS are unknown in advance; we developed the consensus model-based hesitant fuzzy group decision analysis method to investigate the hesitant fuzzy MCGDM problems where the weights of criteria are known in advance but the weights of experts are partially known or completely unknown; Zhang et al. (2015a) proposed the deviation modeling models-based heterogeneous hesitant fuzzy group decision analysis method to solve the MCGDM problems in which the criteria values are expressed as real numbers, interval numbers, linguistic variables, IFNs, HFEs, and hesitant fuzzy linguistic term sets (HFLTSs). It is easily noticed that these aforementioned decision analysis methods possess distinct characteristics and can also be suitable to solve different decision-making problems under various circumstances.

This book provides a thorough and systematic introduction to the above results, which is organized into six chapters that deal with six different but related issues and are listed below:

Chapter 1 introduces a maximizing deviation model-based hesitant fuzzy TOPSIS method for solving the MCDM problems in which the criteria values are expressed by HFEs and the weights of criteria are completely unknown or partially known. There are two key issues being addressed in this approach. The first one is to establish an optimization model on the basis of the idea of the maximizing deviation method, which is mainly used to determine the weights of criteria. According to the idea of the TOPSIS, the second one is to calculate the revised closeness index of each alternative to the hesitant fuzzy positive ideal solution, which can be used to determine the ranking orders of alternatives. On the other hand, this decision-making method is further extended to address the MCDM problems with incomplete weight information in which the criteria values are expressed by interval-valued hesitant fuzzy elements (IVHFEs).

Chapter 2 presents a ranking functions-based hesitant fuzzy TODIM approach to solve the hesitant fuzzy MCDM problems in the case of considering the decision maker's psychological behavior. Firstly, two novel ranking functions are introduced to compare the magnitudes of HFEs and IVHFEs, which are more reasonable and effective compared with the existing ranking functions. Then, a prospect value function is constructed for measuring the dominance degree of each alternative over the others based on novel ranking functions and distance measures. By aggregating these prospect values, the overall prospect value of each alternative is obtained and the corresponding ranking of alternatives is determined. Finally, a practical decision-making problem that concerns the evaluation and ranking of the service quality among domestic airlines is used to illustrate the validity and applicability of the proposed method. On the other hand, as an extension of HFE, hesitant trapezoidal fuzzy number (HTrFN) developed by Zhang et al. (2016) is suitable to tackle the imprecise and ambiguous information in complex decision-making problems and is well enough to represent the uncertainty and vagueness of comparative linguistic expressions. To handle the MCGDM problems in which the decision data are expressed as comparative linguistic expressions based on HTrFNs, this chapter also introduces a hesitant trapezoidal fuzzy TODIM approach developed by Zhang and Liu (2016).

Chapter 3 develops a hesitant fuzzy QUALIFLEX with a signed distance-based comparison method to handle the MCDM problems in which both the assessments of alternatives on criteria and the weights of criteria are expressed by HFEs. We propose a novel concept of hesitancy index for the HFE to measure the degree of hesitancy of the decision maker or the decision organization. By taking the hesitancy indices into account, we present a signed distance-based ranking method to compare the magnitudes of HFEs. Using the signed distance-based comparison approach, we define the concordance/discordance index, the weighted concordance/discordance index, and the comprehensive concordance/discordance index. By investigating all possible permutations of alternatives with respect to the level of concordance/discordance of the complete preference order, the optimal ranking orders of alternatives can be obtained. An application study of the proposed method on green supplier selection is conducted. The study indicates that the proposed method does not require the complicated computation procedures but still yields a reasonable and credible solution. Finally, we extend this technique to manage the heterogeneous information including real numbers, interval numbers, TFNs, IFNs, and HFEs.

Chapter 4 puts forward a hesitant fuzzy LINMAP group decision method with interval programming models for solving the MCGDM problems in which the criteria values are represented by HFEs and the pair-wise comparison judgments over alternatives are taken as interval numbers. The main advantages of this method are that (1) it can sufficiently consider the experts' hesitancy in expressing their assessment information on criteria values by using HFEs and (2) it can take into account the uncertainty of preference information over alternatives by using interval numbers. On the other hand, this chapter also develops a hesitant fuzzy programming model-based LINMAP method for handling the MCDM problems with incomplete and/or inconsistent weight information in which the ratings of alternatives with each criterion are taken as HFEs and the incomplete judgments on pair-wise comparisons of alternatives with hesitant degrees are also represented by HFEs. This proposed approach makes several contributions to the literature and practices. Firstly, the concept of hesitant fuzzy programming model in which both the objective function and some constraints' coefficients take the form of HFEs is defined. Secondly, an effective approach is presented to solve the derived model. Thirdly, a bi-objective programming model is constructed to address the issues of incomplete and inconsistent weights of the criteria under hesitant fuzzy environments.

Chapter 5 investigates the MCGDM problem where the criteria values take the form of HFEs and the weights of criteria are known in advance but the weights of experts are partially known or completely unknown. To solve such a decision-making problem, we propose a consensus model-based hesitant fuzzy group decision-making method, which is motivated by the literature (Zhang and Xu 2014c, 2015b). The proposed method first defines the consensus index from the perspectives of both the ranking and the magnitude of decision information, derives the experts' weights on the basis of the idea of the maximizing consensus, and then utilizes the extended TOPSIS to rank all alternatives. The prominent characteristic of the developed approach is that it can not only take into account the experts' weights and reduce the influence of unjust arguments on the decision results, but also take full advantage of the decision information from both the perspective of ranking and the angle of the sizes of values.

Chapter 6 proposes a deviation modeling method to deal with the heterogeneous MCGDM problems with incomplete weight information. There are three key issues being addressed in this approach. To determine the optimal weights of criteria for each expert, the first one is to construct a maximizing deviation optimal model under heterogeneous fuzzy environment. Borrowing the idea of TOPSIS, the second one is to calculate the relative closeness indices of the alternatives for each expert. The third one is to establish a minimizing deviation optimal model based on the idea that the opinion of the individual expert should be consistent with that of the group to the greatest extent, which is used to determine the weights of experts and identify the optimal alternative. The proposed approach is applied to solve the selection problem of Strategic Freight Forwarder of China Southern Airlines, and a comparison analysis with a similar approach is conducted to demonstrate the advantages of the proposed method.

This book can be used as a reference for researchers and practitioners working in the fields of fuzzy mathematics, operations research, information science, management science and engineering, etc. It can also be used as a textbook for postgraduate and senior undergraduate students.

This work was supported by the National Natural Science Foundation of China (Nos. 71661010 and 71571123), the Major Program of the National Social Science Foundation of China (No. 15ZDC021), and the Natural Science Foundation of Jiangxi Province of China (No. 20161BAB211020).

Nanchang, China Xiaolu Zhang
Chengdu, China Zeshui Xu
March 2016

References

Farhadinia, B.: A novel method of ranking hesitant fuzzy values for multiple attribute decision-making problems. Int. J. Intell. Syst. **28**, 752–767 (2013)

Gomes, L.F.A.M., Lima, M.M.P.P.: TODIM: basics and application to multicriteria ranking of projects with environmental impacts. Found. Comput. Decis. Sci. **16**, 113–127 (1991)

Gomes, L.F.A.M., Lima, M.M.P.P.: From modeling individual preferences to multicriteria ranking of discrete alternatives: a look at prospect theory and the additive difference model. Found. Comput. Decis. Sci. **17**, 171–184 (1992)

Hwang, C.L., Yoon, K.: Multiple Attribute Decision Making Methods and Applications. Springer, Berlin (1981)

Kahneman, D., Tversky, A.: Prospect theory: an analysis of decision under risk. Econometrica, **47**, 263–291 (1979)

Paelinck, J.H.P.: Qualitative multiple criteria analysis, environmental protection and multiregional development. Pap. Reg. Sci. **36**, 59–76 (1976)

Paelinck, J.H.P.: Qualitative multicriteria analysis: an application to airport location. Environ. Plann. A **9**, 883–895 (1977)

Paelinck, J.H.P.: Qualiflex: a flexible multiple-criteria method. Econ. Lett. **1**, 193–197 (1978)

Rebai, A., Aouni, B., Martel, J.-M.: A multi-attribute method for choosing among potential alternatives with ordinal evaluation. Eur. J. Oper. Res. **174**, 360–373 (2006)

Rodríguez, R.M., Martínez, L., Torra, V., Xu, Z.S., Herrera, F.: Hesitant fuzzy sets: state of the art and future directions. Int. J. Intell. Syst. **29**, 495–524 (2014)

Roy, B.: Classement et choix en présence de points de vue multiples. RAIRO-Operations Research-Recherche Opérationnelle **2**, 57–75 (1968)

Srinivasan, V., Shocker, A.D.: Linear programming techniques for multidimensional analysis of preferences. Psychometrika **38**, 337–369 (1973)

Torra, V.: Hesitant fuzzy sets. Int. J. Intell. Syst. **25**, 529–539 (2010)

Xia, M.M., Xu, Z.S.: Hesitant fuzzy information aggregation in decision making. Int. J. Approximate Reasoning **52**, 395–407 (2011)

Xu, Z.S., Xia, M.M.: Distance and similarity measures for hesitant fuzzy sets. Inf. Sci. **181**, 2128–2138 (2011)

Xu, Z.S., Zhang, X.L.: Hesitant fuzzy multi-attribute decision making based on TOPSIS with incomplete weight information. Knowl. Based Syst. **52**, 53–64 (2013)

Zadeh, L.A.: The concept of a linguistic variable and its application to approximate reasoning. Inf. Sci. Part I, II, III (8,9), 199–249, 301–357, 143–180 (1975)

Zhang, X.L., Liu M.F.: Hesitant trapezoidal fuzzy TODIM approach with a closeness index-based ranking method for qualitative group decision making. Technique Report (2016)

Zhang, X.L., Xu, Z.S.: The TODIM analysis approach based on novel measured functions under hesitant fuzzy environment. Knowl. Based Syst. **61**, 48–58 (2014a)

Zhang, X.L., Xu, Z.S.: Interval programming method for hesitant fuzzy multi-attribute group decision making with incomplete preference over alternatives. Comput. Ind. Eng. **75**, 217–229 (2014b)

Zhang, X.L., Xu, Z.S.: Deriving experts' weights based on consistency maximization in intuitionistic fuzzy group decision making. J. Intell. Fuzzy Syst. **27**, 221–233 (2014c)

Zhang, X.L., Xu, Z.S.: Hesitant fuzzy QUALIFLEX approach with a signed distance-based comparison method for multiple criteria decision analysis. Expert Syst. Appl. **42**, 873–884 (2015a)

Zhang, X.L., Xu, Z.S.: Soft computing based on maximizing consensus and fuzzy TOPSIS approach to interval-valued intuitionistic fuzzy group decision making. Appl. Soft Comput. **26**, 42–56 (2015b)

Zhang, X.L., Xu, Z.S.: Hesitant fuzzy agglomerative hierarchical clustering algorithms. Int. J. Syst. Sci. **46**, 562–576 (2015c)

Zhang, X.L., Xu, Z.S.: Novel distance and similarity measures on hesitant fuzzy sets with applications to clustering analysis. J. Intell. Fuzzy Syst. **22**, 2279–2296 (2015d)
Zhang, X.L., Xu, Z.S., Wang, H.: Heterogeneous multiple criteria group decision making with incomplete weight information: a deviation modeling approach. Inf. Fusion **25**, 49–62 (2015a)
Zhang, X.L., Xu, Z.S., Xing X.M.: Hesitant fuzzy programming techniques for multidimensional analysis of hesitant fuzzy preferences. OR Spectr. (2015b). doi:10.1007/s00291-015-0420-0
Zhang, X.L., Xu, Z.S., Liu M.F.: Hesitant trapezoidal fuzzy QUALIFLEX method and its application in the evaluation of green supply chain initiatives. Sustainability, accepted (2016)
Zhu, B., Xu, Z.S.: Consistency measures for hesitant fuzzy linguistic preference relations. IEEE Trans. Fuzzy Syst. **22**(1), 35–45 (2014)

Contents

Chapter 1
Hesitant Fuzzy Multiple Criteria Decision Analysis Based on TOPSIS

Abstract HFE which allows the membership degree of an element to a set represented by several possible values is a powerful tool to describe and deal with uncertain data. This chapter develops the decision making approach based on TOPSIS and the maximizing deviation model for solving MCDM problems in which the evaluation information provided by the decision maker is expressed in HFEs and the information about criteria weights is incomplete. There are two key issues being addressed in this approach. The first one is to establish an optimization model based on the maximizing deviation method, which can be used to determine the weights of criteria. The second one is to calculate the revised closeness index of each alternative to the hesitant fuzzy PIS. The considered alternatives are ranked according to the revised closeness indices of alternatives and the most desirable one is selected. An important advantage of this proposed method is its ability to relieve the influence of subjectivity of the decision maker concerning the weights of criteria and at the same time to remain the original decision information sufficiently. Additionally, the extended results in the interval-valued hesitant fuzzy situations are also pointed out.

The TOPSIS method originally developed by Hwang and Yoon (1981) is a kind of simple and useful decision making method to handle the MCDM problems with crisp data. The basic idea of TOPSIS method is that the optimal alternative should have the shortest distance from the PIS and have the farthest distance from the NIS. The TOPSIS method has been successfully applied in various fields and a state-of the-art survey of TOPSIS applications can be found in the paper (Behzadian et al. 2012). Owing to the increasing complexity of decision making environment, the fuzzy sets and the generalizations of fuzzy sets are usually used by decision makers to express their imprecise and uncertain preference information (Chen and Chang 2015; Chen and Hong 2014). For this reason, lots of papers have recently been devoted to fuzzy extensions of the TOPSIS method in the literature, such as the fuzzy TOPSIS method (Chen 2000), interval-valued TOPSIS method (Chen and Tsao 2008), intuitionistic fuzzy TOPSIS method (Boran et al. 2009), hesitant fuzzy linguistic TOPSIS method (Beg and Rashid 2013), Pythagorean fuzzy TOPSIS method (Zhang and Xu 2014a), etc., have been developed. However, these

X. Zhang and Z. Xu, *Hesitant Fuzzy Methods for Multiple Criteria Decision Analysis*, Studies in Fuzziness and Soft Computing 345,
DOI 10.1007/978-3-319-42001-1_1

TOPSIS-based methods suffer from two limitations: (1) in many practical MCDM problems, the weights of criteria are usually completely unknown or partially known, while these methods under the hypothesis that the weights of criteria are completely known in advance fail to address the MCDM problems with incomplete weight information; (2) these existing methods fail to deal with the HFE and/or IVHFE decision data in the decision making processes.

To overcome these limitations, Xu and Zhang (2013) developed a maximizing deviation model-based hesitant fuzzy TOPSIS method for solving the MCDM problems in which the criteria values are expressed in HFEs and the weights of criteria are completely unknown or partially known. In addition, Xu and Zhang (2013) also extended this developed method to address the interval-valued hesitant fuzzy MCDM problems with incomplete weight information.

1.1 Hesitant Fuzzy Information

HFS originally developed by Torra (2010) can perfectly describe the real-world situations in which the decision makers have hesitancy in providing their preferences over objects in the process of decision making. The definition of HFS is introduced as follows.

Definition 1.1 (Torra 2010). Let T be a reference set, a HFS H on T is defined in terms of a function $h_H(t)$ when applied to T returns a subset of [0, 1].

Xia and Xu (2011) provided the mathematical expression of HFS as below:

$$H = \{ \langle t, h_H(t) \rangle | t \in T \},$$

where $h_H(t)$ is a set of some different values in [0, 1] and represents the possible membership degrees of the element $t \in T$ to H.

The HFS H is called the empty HFS if $H = \{\langle t, h_H(t) = H\{0\} \rangle | t \in T\}$, while the HFS H is regarded as a full HFS if $H = \{\langle t, h_H(t) = H\{1\} \rangle | t \in T\}$ (Torra 2010). Xia and Xu (2011) called $h_H(t)$ a HFE denoted by $h = H\{\gamma^{\lambda} | \lambda = 1, 2, \ldots, \#h\}$. Similarly, the HFE h is called an empty HFE if $h = H\{0\}$, and the HFE h is called a full HFE if $h = H\{1\}$.

Example 1.1 Let $T = \{t_1, t_2, t_3\}$, and let $h_H(t_1) = H\{0.5, 0.4, 0.7\}$, $h_H(t_2) = H\{0.4, 0.6\}$, and $h_H(t_3) = H\{0.6, 0.5, 0.4, 0.8\}$ be three HFEs of t_i $(i = 1, 2, 3)$ to a set H. Then, the set H is called a HFS which is denoted as follows:

$$H = \left\{ \begin{array}{c} \langle t_1, H\{0.5, 0.4, 0.7\} \rangle, \langle t_2, H\{0.4, 0.6\} \rangle, \\ \langle t_3, H\{0.6, 0.5, 0.4, 0.8\} \rangle \end{array} \right\}.$$

Remark 1.1 The symbol $\#h$ in this book is used to represent the number of the elements in a HFE h. The HFS H can be seen as a fuzzy set if there is only one possible value in each HFE $h_H(t)$. Bedregal et al. (2014) regarded the finite and nonempty HFS as the typical HFS, and the elements in typical HFS as the typical

HFEs. From now on, this book will only focus on the study of the typical HFEs. For convenience, this book still employs the HFEs to denote the typical HFEs.

Definition 1.2 (Torra 2010; Xia and Xu 2011). Given three HFEs represented by h, h_1 and h_2, respectively, the basic operations of HFEs are defined as follows:

(1) Multiplication: $\theta h = H\left\{(1-(1-\gamma)^{\theta})|\gamma \in h\right\} (\theta > 0)$;
(2) Complement: $h^c = H\{(1-\gamma)|\gamma \in h\}$;
(3) $\oplus$-union: $h_1 \oplus h_2 = H\{(\gamma_1 + \gamma_2 - \gamma_1\gamma_2)|\gamma_1 \in h_1, \gamma_2 \in h_2\}$;
(4) $\otimes$-intersection: $h_1 \otimes h_2 = H\{\gamma_1\gamma_2|\gamma_1 \in h_1, \gamma_2 \in h_2\}$.

From the definition of HFE, it is noted that the number of values in different HFEs may be different and the values in each HFE are usually out of order. Thus, in real-world decision making processes, the values in all HFEs should be arranged in an increasing/decreasing order for convenience. Suppose that the values of a HFE h are arranged in an increasing order, and let γ^{λ} be the λth smallest value in h. In order to accurately calculate the distance between two HFEs, it is necessary to ensure that the numbers of values of these two HFEs are the same. For two HFEs having different lengths, Xu and Xia (2011a) advised the decision maker to extend the shorter HFE until both of them have the same length and gave the following extension regulations (Xu and Xia's extension method).

For two HFEs h_1, h_2 with $\#h_1 \neq \#h_2$, if $\#h_1 < \#h_2$, then h_1 should be extended by adding the value in it until it has the same length with h_2; if $\#h_1 > \#h_2$, then h_2 should be extended by adding the value in it until it has the same length with h_1. The added value mainly depends on the decision maker's risk preference. Optimists anticipate the desirable outcomes and usually add the maximum value, while pessimists expect unfavorable outcomes and may add the minimal value.

Although Xu and Xia's extension rule is reasonable, it fails to consider the real-world situation in which the decision maker is risk-neutral in the decision process. To this end, Xu and Zhang (2013) introduced a new extension method with a parameter θ.

Definition 1.3 (Xu and Zhang 2013). Assume a HFE $h = H\{\gamma^{\lambda}|\lambda = 1, 2, \ldots, \#h\}$, and stipulate that γ^+ and γ^- are the maximum and minimum values in the HFE h, respectively; then $\gamma^* = \theta\gamma^+ + (1-\theta)\gamma^-$ is called an extension value, where $\theta\,(0 \leq \theta \leq 1)$ is the parameter determined by the decision maker according his/her risk preference.

According to Definition 1.3, different extension values can be selected on the basis of the decision maker's different risk preferences. If $\theta = 1$, then the extension value $\gamma^* = \gamma^+$, which indicates that the decision maker is risk-seeking in the decision making process; while if $\theta = 0$, then $\gamma^* = \gamma^-$, which means that the decision maker is risk-averse. It is clear that Xu and Xia's extension rule is consistent with the proposed extension rule if $\theta = 1$ or $\theta = 0$. Moreover, if the decision maker is risk-neutral, the extension value $\gamma^* = (\gamma^+ + \gamma^-)/2$, i.e., $\theta = 1/2$, is selected.

Example 1.2 For two HFEs $h_1 = H\{0.5, 0.4, 0.7\}$ and $h_2 = H\{0.4, 0.6\}$, it is easy to see $\#h_1(=3) > \#h_2(=2)$. According to Definition 1.3, we should extend h_2 until it has the same length with h_1. Then, we get:

(1) if the decision maker is risk-seeking, then h_2 can be extended as $h_2 = H\{0.4, 0.6, 0.6\}$;
(2) if the decision maker is risk-neutral, then h_2 can be extended as $h_2 = H\{0.4, 0.5, 0.6\}$;
(3) if the decision maker is risk-averse, then h_2 can be extended as $h_2 = H\{0.4, 0.4, 0.6\}$.

It is easy to see from Example 1.2 that the extension method proposed by Xu and Zhang (2013) can reveal all risk preferences of the decision makers (i.e., risk-averse, risk-neutral and risk-seeking), and Xu and Xia's extension method is the special case of the proposed extension method.

Definition 1.4 (Zhang and Xu 2014a). Let $h_1 = H\{\gamma_1^\lambda | \lambda = 1, 2, \ldots, \#h_1\}$ and $h_2 = H\{\gamma_2^\lambda | \lambda = 1, 2, \ldots, \#h_2\}$ be two HFEs with $\#h_1 = \#h_2 = \#h, \gamma_1^\lambda$ be the λth smallest value in h_1 and γ_2^λ be the λth smallest value in h_2. Then, we stipulate:

$$h_1 \leq h_2 \text{ if and only if } \gamma_1^\lambda \leq \gamma_2^\lambda \ (\lambda = 1, 2, \ldots, \#h).$$

Definition 1.5 (Xu and Xia 2011b). For any two HFEs $h_i = H\{\gamma_i^\lambda | \lambda = 1, 2, \ldots, \#h_i\}$ $(i = 1, 2)$ with $\#h_1 = \#h_2 = \#h$, the hesitant fuzzy Hamming distance between them is defined as follows:

$$d_H(h_1, h_2) = \frac{1}{\#h} \sum_{\lambda=1}^{\#h} \left| \gamma_1^\lambda - \gamma_2^\lambda \right| \tag{1.1}$$

and the hesitant fuzzy Euclidean distance between them is defined as:

$$d_E(h_1, h_2) = \sqrt{\frac{1}{\#h} \sum_{\lambda=1}^{\#h} \left(\gamma_1^\lambda - \gamma_2^\lambda \right)^2} \tag{1.2}$$

1.2 Description of the Classical TOPSIS Method

Consider a MCDM problem which is based on m alternatives $\boldsymbol{A} = \{A_1, A_2, \ldots, A_m\}$ and n criteria $\boldsymbol{C} = \{C_1, C_2, \ldots, C_n\}$. The weighting vector of criteria is denoted by $\boldsymbol{w} = (w_1, w_2, \ldots, w_n)^T$ where w_j is the weight of the criterion C_j, satisfying the normalization condition: $\sum_{j=1}^{n} w_j = 1$ and $w_j \geq 0 \, (j \in \{1, 2, \ldots, n\})$. Let x_{ij} be the criterion value of the alternative A_i with respect to the criterion C_j. Usually, such a MCDM problem is characterized in the form of a decision matrix $\Re = (x_{ij})_{m \times n}$. The MCDM problem in which the criteria values and the weights of criteria are expressed by crisp numbers is called the classical MCDM problem. The classical TOPSIS method is a simple and useful method to address the classical MCDM problem. Its algorithm (Algorithm 1.1) consists of the following steps (Hwang and Yoon 1981; Dymova et al. 2013):

- Normalize the decision matrix:

$$\bar{x}_{ij} = \frac{x_{ij}}{\sqrt{\sum_{\xi=1}^{m} (x_{\xi j})^2}} \tag{1.3}$$

- Determine the PIS A^+ and the NIS A^-:

$$\begin{aligned} A^+ &= \left(\bar{x}_1^+, \bar{x}_2^+, \ldots, \bar{x}_n^+\right) \\ &= \left((\max_i \bar{x}_{ij} \mid C_j \in J_I) \; or \; (\min_i \bar{x}_{ij} \mid C_j \in J_{II})\right) \end{aligned} \tag{1.4}$$

$$\begin{aligned} A^- &= \left(\bar{x}_1^-, \bar{x}_2^-, \ldots, \bar{x}_n^-\right) \\ &= \left((\min_i \bar{x}_{ij} \mid C_j \in J_I) \; or \; (\max_i \bar{x}_{ij} \mid C_j \in J_{II})\right) \end{aligned} \tag{1.5}$$

  where J_I is a subset of benefit criteria and J_{II} is a subset of cost criteria, and $J_I \cup J_{II} = \boldsymbol{C}$, $J_I \cap J_{II} = \varnothing$.
- Calculate the distances between the potential alternatives and the PIS as well as the NIS, respectively,

$$d(A_i, A^+) = \sqrt{\sum_{j=1}^{n} w_j (\bar{x}_{ij} - \bar{x}_j^+)^2}, \tag{1.6}$$

$$d(A_i, A^-) = \sqrt{\sum_{j=1}^{n} w_j (\bar{x}_{ij} - \bar{x}_j^-)^2}, \tag{1.7}$$

- Compute the relative closeness index of each alternative to the PIS:

$$CI(A_i) = \frac{d(A_i, A^-)}{d(A_i, A^+) + d(A_i, A^-)} \tag{1.8}$$

  It is easily seen that $CI(A_i) \in [0, 1]$.
- Rank the alternatives according to the closeness indices of alternatives: the bigger $CI(A_i)$ the better the alternative A_i.

1.3 The Maximizing Deviation Model-Based Hesitant Fuzzy TOPSIS Approach

The hesitant fuzzy MCDM problem is usually characterized with m alternatives $\{A_1, A_2, \ldots, A_m\}$ and n criteria $\{C_1, C_2, \ldots, C_n\}$, which is expressed by the hesitant fuzzy decision matrix $\Re = (h_{ij})_{m \times n}$. The element h_{ij} is the criterion value of the alternative A_i with the criterion C_j in this MCDM problem, which is expressed in the form of HFE. In many practical MCDM problems, the weights of criteria are usually partially known or completely unknown in advance. To address the hesitant fuzzy MCDM problems with incomplete weight information, Xu and Zhang (2013) employed the main framework of the classical TOPSIS method to develop a maximizing deviation model-based hesitant fuzzy TOPSIS method. We call it the HF-TOPSIS method. The objective of this HF-TOPSIS method is twofold: (1) a maximizing deviation model is developed to determine the weights of criteria; (2) a hesitant fuzzy TOPSIS is presented to rank the alternatives.

1.3.1 The Maximizing Deviation Model to Determine Weights of Criteria

The maximizing deviation method proposed by Wang (1998) was mainly used to determine the weights of criteria for solving the MCDM problems with crisp data. The main idea of the maximizing deviation method is that in a real-world MCDM problem the criterion with a larger deviation value among alternatives should be assigned a larger weight, while the criterion with a small deviation value among alternatives should be signed a smaller weight. That is to say, if the criterion values of all the alternatives have small differences under a criterion, it shows that such a criterion plays a less important role in the priority procedure. Contrarily, if a criterion makes the criterion values of all alternatives have obvious differences, then such a criterion plays a more important role in choosing the best alternative. Thus, from the standpoint of ranking the alternatives, if one criterion has similar criterion values across alternatives, it should be assigned a small weight; otherwise, the criterion which makes larger deviations should be evaluated a bigger weight, in spite of the degree of its own importance. Especially, if all available alternatives score equally with respect to a given criterion, then such a criterion will be judged unimportant (Wang 1998).

Based on the idea of the maximizing deviation method, Xu and Zhang (2013) constructed an optimization model to determine the optimal weights of criteria under hesitant fuzzy environment. Firstly, for the criteria C_j ($j \in \{1, 2, \ldots, n\}$), the deviation values of the alternatives A_ξ ($\xi \in \{1, 2, \ldots, m\}$) to all other alternatives can be defined as below:

$$\begin{aligned} D_{\xi j}(w_j) &= w_j \sum_{\zeta=1}^{m} d_E(h_{\xi j}, h_{\zeta j}) \\ &= w_j \sum_{\zeta=1}^{m} \sqrt{\frac{1}{\#h} \sum_{\lambda=1}^{\#h} (\gamma_{\xi j}^{\lambda} - \gamma_{\zeta j}^{\lambda})^2}, \quad j \in \{1, 2, \ldots, n\} \end{aligned} \tag{1.9}$$

The deviation values of all alternatives to all other alternatives for the criterion $C_j\ (j \in \{1,2,\ldots,n\})$ are obtained by the following equations:

$$\begin{aligned} D_j(w_j) &= \sum_{\xi=1}^{m} D_{\xi j}(w_j) \\ &= \sum_{\xi=1}^{m}\sum_{\zeta=1}^{m} w_j \sqrt{\frac{1}{\#h}\sum_{\lambda=1}^{\#h}(\gamma_{\xi j}^{\lambda}-\gamma_{\zeta j}^{\lambda})^2},\quad j \in \{1,2,\ldots,n\} \end{aligned} \tag{1.10}$$

Then, if the weights of criteria are completely unknown in advance, a non-linear programming model (MOD-1.1) which maximizes all deviation values for all criteria is established to select the weight vector $\boldsymbol{w}$ (Xu and Zhang 2013):

$$\begin{cases} \max \quad D(\boldsymbol{w}) = \sum\limits_{j=1}^{n}\sum\limits_{\xi=1}^{m}\sum\limits_{\zeta=1}^{m} w_j \sqrt{\frac{1}{\#h}\sum\limits_{\lambda=1}^{\#h}(\gamma_{\xi j}^{\lambda}-\gamma_{\zeta j}^{\lambda})^2} \\ s.t. \quad w_j \geq 0,\quad j=1,2,\ldots,n,\quad \sum\limits_{j=1}^{n}(w_j)^2 = 1 \end{cases} \tag{MOD − 1.1}$$

The Lagrange function of the optimization model (MOD-1.1) is obtained as below:

$$Q(\boldsymbol{w},\eta) = \sum_{j=1}^{n}\sum_{\xi=1}^{m}\sum_{\zeta=1}^{m} w_j \sqrt{\frac{1}{\#h}\sum_{\lambda=1}^{\#h}(\gamma_{\xi j}^{\lambda}-\gamma_{\zeta j}^{\lambda})^2} + \frac{\eta}{2}\left(\sum_{j=1}^{n}(w_j)^2 - 1\right) \tag{1.11}$$

where η is a real number, denoting the Lagrange multiplier variable.

Then, the partial derivatives of Q are computed as:

$$\begin{cases} \frac{\partial Q}{\partial w_j} = \sum\limits_{\xi=1}^{m}\sum\limits_{\zeta=1}^{m} \sqrt{\frac{1}{\#h}\sum\limits_{\lambda=1}^{\#h}(\gamma_{\xi j}^{\lambda}-\gamma_{\zeta j}^{\lambda})^2} + \eta w_j = 0 \\ \frac{\partial Q}{\partial \eta} = \frac{1}{2}\left(\sum\limits_{j=1}^{n}(w_j)^2 - 1\right) = 0 \end{cases} \tag{1.12}$$

It follows from Eq. (1.12) that

$$w_j = -\frac{\sum_{\xi=1}^{m}\sum_{\zeta=1}^{m}\sqrt{\frac{1}{\#h}\sum_{\lambda=1}^{\#h}(\gamma_{\xi j}^{\lambda}-\gamma_{\zeta j}^{\lambda})^2}}{\eta},\quad j=1,2,\ldots,n \tag{1.13}$$

Putting Eq. (1.13) into Eq. (1.12), we have

$$\eta = -\sqrt{\sum_{j=1}^{n}\left(\sum_{\xi=1}^{m}\sum_{\zeta=1}^{m}\sqrt{\frac{1}{\#h}\sum_{\lambda=1}^{\#h}\left(\gamma_{\xi j}^{\lambda}-\gamma_{\zeta j}^{\lambda}\right)^{2}}\right)^{2}} \tag{1.14}$$

By combining Eqs. (1.13) and (1.14), we can get

$$w_j = \frac{\sum_{\xi=1}^{m}\sum_{\zeta=1}^{m}\sqrt{\frac{1}{\#h}\sum_{\lambda=1}^{\#h}\left(\gamma_{\xi j}^{\lambda}-\gamma_{\zeta j}^{\lambda}\right)^{2}}}{\sqrt{\sum_{j=1}^{n}\left(\sum_{\xi=1}^{m}\sum_{\zeta=1}^{m}\sqrt{\frac{1}{\#h}\sum_{\lambda=1}^{\#h}\left(\gamma_{\xi j}^{\lambda}-\gamma_{\zeta j}^{\lambda}\right)^{2}}\right)^{2}}} \tag{1.15}$$

For the sake of simplicity, let

$$\aleph_j = \sum_{\xi=1}^{m}\sum_{\zeta=1}^{m}\sqrt{\frac{1}{\#h}\sum_{\lambda=1}^{\#h}\left(\gamma_{\xi j}^{\lambda}-\gamma_{\zeta j}^{\lambda}\right)^{2}},\quad j=1,2,\ldots,n,$$

and Eq. (1.15) can be rewritten as:

$$w_j = \frac{\aleph_j}{\sqrt{\sum_{j=1}^{n}\left(\aleph_j\right)^{2}}},\quad j=1,2,\ldots,n \tag{1.16}$$

It can be verified easily from Eq. (1.16) that $w_j\,(j=1,2,\ldots,n)$ are positive such that they satisfy the constrained conditions in the model (MOD-1.1) and the solution is unique.

By normalizing $w_j\,(j=1,2,\ldots,n)$, it is easy to get

$$w_j^{*} = \frac{w_j}{\sum_{j=1}^{n} w_j} = \frac{\aleph_j}{\sum_{j=1}^{n}\aleph_j} = \frac{\sum_{\xi=1}^{m}\sum_{\zeta=1}^{m}\sqrt{\frac{1}{\#h}\sum_{\lambda=1}^{\#h}\left(\gamma_{\xi j}^{\lambda}-\gamma_{\zeta j}^{\lambda}\right)^{2}}}{\sum_{j=1}^{n}\sum_{\xi=1}^{m}\sum_{\zeta=1}^{m}\sqrt{\frac{1}{\#h}\sum_{\lambda=1}^{\#h}\left(\gamma_{\xi j}^{\lambda}-\gamma_{\zeta j}^{\lambda}\right)^{2}}} \tag{1.17}$$

On the other hand, the weights of criteria in several real-world decision making problems are not completely unknown but partially known in advance. The incomplete weight information usually possesses several different structure forms. Several works (Park and Kim 1997; Chen 2014) have reported the structure forms of criteria weights which are roughly divided into the following five forms.

Let $\mathbf{\Delta_0} = \{(w_1, w_2, \ldots, w_n) | w_j \geq 0\,(j = 1, 2, \ldots, n),\ \sum_{j=1}^{n} w_j = 1\}$, then

(1) A weak ranking:

$$\mathbf{\Delta_1} = \left\{(w_1, w_2, \ldots, w_n) \in \mathbf{\Delta_0} \middle| w_{j_1} \geq w_{j_2} \quad \text{for all } j_1 \in \Lambda_{(1)1} \text{ and } j_2 \in \Lambda_{(2)1}\right\}$$

where $\Lambda_{(1)1}$ and $\Lambda_{(2)1}$ are two disjoint subsets of the subscript index set $\{1, 2, \ldots, n\}$ of all criteria.

(2) A strict ranking:

$$\Delta_2 = \left\{(w_1, w_2, \ldots, w_n) \in \Delta_0 \middle| \begin{array}{c} \tau_{j_1j_2}^L \leq w_{j_1} - w_{j_2} \leq \tau_{j_1j_2}^U \\ \textit{for all } j_1 \in \Lambda_{(1)2} \textit{ and } j_2 \in \Lambda_{(2)2} \end{array}\right\}$$

where $\tau_{j_1j_2}^L$ and $\tau_{j_1j_2}^U$ are the constants that satisfy the condition: $0 < \tau_{j_1j_2}^L < \tau_{j_1j_2}^U$, $\Lambda_{(1)2}$ and $\Lambda_{(2)2}$ are two disjoint subsets of $\{1, 2, \ldots, n\}$.

(3) A ranking of differences:

$$\Delta_3 = \left\{(w_1, w_2, \ldots, w_n) \in \Delta_0 \middle| \begin{array}{c} w_{j_1} - w_{j_2} \geq w_{j_3} - w_{j_4} \\ \textit{for all } j_1 \in \Lambda_{(1)3}, j_2 \in \Lambda_{(2)3}, j_3 \in \Lambda_{(3)3} \textit{ and } j_4 \in \Lambda_{(4)3} \end{array}\right\}$$

where $\Lambda_{(1)3}, \Lambda_{(2)3}, \Lambda_{(3)3}$ and $\Lambda_{(4)3}$ are four disjoint subsets of $\{1, 2, \ldots, n\}$.

(4) A ranking with multiples:

$$\Delta_4 = \left\{(w_1, w_2, \ldots, w_n) \in \mathbf{\Delta_0} \middle| w_{j_1} \geq \tau_{j_1j_2} \cdot w_{j_2} \quad \text{for all } j_1 \in \Lambda_{(1)4} \text{ and } j_2 \in \Lambda_{(2)4}\right\}$$

where $\tau_{j_1j_2}$ is a constant that satisfies the condition: $\tau_{j_1j_2} > 0$, $\Lambda_{(1)4}$ and $\Lambda_{(2)4}$ are two disjoint subsets of $\{1, 2, \ldots, n\}$.

(5) An interval form:

$$\mathbf{\Delta_5} = \left\{(w_1, w_2, \ldots, w_n) \in \mathbf{\Delta_0} \middle| \tau_{j_1}^L \leq w_{j_1} \leq \tau_{j_1}^U \quad \text{for all } j_1 \in \Lambda_{(1)5}\right\}$$

where $\tau_{j_1}^L$ and $\tau_{j_1}^U$ are two constants that satisfy the condition: $0 < \tau_{j_1}^L < \tau_{j_1}^U$, and $\Lambda_{(1)5}$ is a subset of $\{1, 2, \ldots, n\}$.

In general, the preference information structure of criteria importance may consist of several sets of the above basic sets or may contain all the five basic sets, which depends on the characteristic and need of the real-world decision making problems. Let $\mathbf{\Delta}$ denote a set of the known information on the criteria weights, and $\Delta = \mathbf{\Delta_1} \cup \mathbf{\Delta_2} \cup \mathbf{\Delta_3} \cup \mathbf{\Delta_4} \cup \mathbf{\Delta_5}$.

For this case, Xu and Zhang (2013) constructed the following constrained optimization model to determine the optimal weights of criteria:

$$\begin{cases} \max \quad D(\boldsymbol{w}) = \sum_{j=1}^{n} \sum_{\xi=1}^{m} \sum_{\zeta=1}^{m} w_j \sqrt{\frac{1}{\#h} \sum_{\lambda=1}^{\#h} \left(\gamma_{\xi j}^{\lambda} - \gamma_{\zeta j}^{\lambda}\right)^2} \\ s.t. \quad \boldsymbol{w} \in \boldsymbol{\Delta} \end{cases} \tag{MOD-1.2}$$

where $\boldsymbol{\Delta}$ is also a set of constraint conditions that the weight value w_j should satisfy according to the requirements in real situations.

The model (MOD-1.2) is a linear programming model that can be executed using the LINGO 11.0 or MATLAB 7.4.0 mathematics software package. By solving this model, the optimal solution, namely $\boldsymbol{w} = (w_1, w_2, \ldots, w_n)^T$, can be obtained.

1.3.2 The Hesitant Fuzzy TOPSIS Model to Rank the Alternatives

After obtaining the weights of criteria on the basis of the maximizing deviation model, Xu and Zhang (2013) further extended the classical TOPSIS method to take HFEs into account and utilized the distance measures of HFEs to obtain the final ranking of alternatives.

Firstly, the hesitant fuzzy PIS can be denoted by A^+ which is determined by the following formula:

$$A^+ = \left(h_1^+, h_2^+, \ldots, h_n^+\right) = \left(\begin{array}{l} \left(H\{\max_i \gamma_{ij}^1, \max_i \gamma_{ij}^2, \ldots, \max_i \gamma_{ij}^{\#h_{ij}}\} \mid C_j \in J_I \right) \\ or \left(H\{\min_i \gamma_{ij}^1, \min_i \gamma_{ij}^2, \ldots, \min_i \gamma_{ij}^{\#h_{ij}}\} \mid C_j \in J_{II} \right) \end{array} \right) \tag{1.18}$$

Usually, the hesitant fuzzy PIS A^+ does not exist in the real-life MCDM problems. In other words, the hesitant fuzzy PIS A^+ is not the feasible alternative, namely, $A^+ \notin \boldsymbol{A}$. Otherwise, the hesitant fuzzy PIS A^+ is the optimal alternative of the MCDM problem.

Then, the distance between the alternatives $A_i\,(i = 1, 2, \ldots, m)$ and the hesitant fuzzy PIS A^+ is calculated by using the hesitant fuzzy Euclidean distance as follows:

$$d^h(A_i, A^+) = \sum_{j=1}^{n} w_j \sqrt{\frac{1}{\#h_{ij}} \sum\nolimits_{\lambda=1}^{\#h_{ij}} \left(\gamma_{ij}^{\lambda} - (\gamma_j^{\lambda})^+\right)^2} \tag{1.19}$$

Usually, the smaller the $d^h(A_i, A^+)$ is, the better the alternative A_i is; and let

$$d^h_{\min}(A_i, A^+) = \min_{1 \le i \le m} d^h(A_i, A^+) \tag{1.20}$$

However, the alternative with the closest distance to hesitant fuzzy PIS may be not the farthest from hesitant fuzzy NIS. We denote the hesitant fuzzy NIS by A^-, which is determined by the following formula:

$$A^- = \left(h_1^-, h_2^-, \ldots, h_n^-\right) = \left(\begin{array}{c} \left(H\{\min_i \gamma_{ij}^1, \min_i \gamma_{ij}^2, \ldots, \min_i \gamma_{ij}^{\#h_{ij}}\} \mid C_j \in J_I \right) \\ or \left(H\{\max_i \gamma_{ij}^1, \max_i \gamma_{ij}^2, \ldots, \max_i \gamma_{ij}^{\#h_{ij}}\} \mid C_j \in J_{II} \right) \end{array} \right) \tag{1.21}$$

It is easily seen from Eq. (1.21) that the obtained value of hesitant fuzzy NIS under each criterion is minimal value among all the alternatives. Usually, in the practical MCDM process, the hesitant fuzzy NIS A^- is an unfeasible alternative, namely, $A^- \notin \mathbf{A}$. Otherwise, the hesitant fuzzy NIS A^- is the worst alternative of the MCDM problem, which should be directly deleted in the decision making process.

Using Eq. (1.2), the distance between the alternatives A_i $(i = 1, 2, \ldots, m)$ and the hesitant fuzzy NIS A^- can be obtained as follows:

$$d^h(A_i, A^-) = \sum_{j=1}^{n} w_j \sqrt{\frac{1}{\#h_{ij}} \sum\nolimits_{\lambda=1}^{\#h_{ij}} \left(\gamma_{ij}^{\lambda} - (\gamma_j^{\lambda})^-\right)^2} \tag{1.22}$$

In general, the bigger the $d^h(A_i, A^-)$ is, the better the alternative A_i is; and let

$$d^h_{\max}(A_i, A^-) = \max_{1 \le i \le m} d^h(A_i, A^-) \tag{1.23}$$

According to the idea of the classical TOPSIS method, we need to calculate the closeness indices of the alternatives A_i $(i = 1, 2, \ldots, m)$ to the hesitant fuzzy PIS A^+ as below:

$$CI^h(A_i) = \frac{d^h(A_i, A^-)}{d^h(A_i, A^+) + d^h(A_i, A^-)}, \tag{1.24}$$

On the basis of the closeness index $CI^h(A_i)$, the ranking orders of alternatives can be determined and the optimal alternative is selected. However, Hadi-Vencheh and Mirjaberi (2014) showed that in some situations the relative closeness indices of alternatives cannot achieve the aim that the optimal solution should have the shortest distance from the PIS and the farthest distance from the NIS, simultaneously. Thus, they advised us to use the following formula instead of the relative closeness index [i.e., Eq. (1.24)] to rank the alternatives:

$$\mathscr{F}^h(A_i) = \frac{d^h(A_i, A^-)}{d^h_{\max}(A_i, A^-)} - \frac{d^h(A_i, A^+)}{d^h_{\min}(A_i, A^+)} \tag{1.25}$$

which is called the revised closeness index used to measure the extent to which the alternative A_i closes to the hesitant fuzzy PIS A^+ and is far away from the hesitant fuzzy NIS A^-, simultaneously.

It can be easily seen that $\mathscr{F}^h(A_i) \leq 0\,(i = 1, 2, \ldots, m)$ and the bigger $\mathscr{F}^h(A_i)$ the better the alternative A_i. If there exists an alternative A^* satisfying the conditions that $d^h(A^*, A^-) = d^h_{\max}(A_i, A^-)$ and $d^h(A^*, A^+) = d^h_{\min}(A_i, A^+)$, simultaneously; then $\mathscr{F}^h(A_i) = 0$ and obviously, the alternative A^* is the best alternative that is closest to the hesitant fuzzy PIS A^+ and farthest away from the hesitant fuzzy NIS A^-, simultaneously. Therefore, according to the revised closeness index $\mathscr{F}^h(A_i)$, we can determine the ranking orders of all alternatives and select the best one from a set of feasible alternatives.

Now, a practical algorithm is introduced to solve the hesitant fuzzy MCDM problems in which the weights of criteria are partially known or completely unknown. The practical algorithm (Algorithm 1.2) involves the following steps (Xu and Zhang 2013):

Step 1. For a hesitant fuzzy MCDM problem, we construct the hesitant fuzzy decision matrix $\Re = (h_{ij})_{m \times n}$, where the argument h_{ij} is HFE, and indicates the rating of the alternative $A_i \in \boldsymbol{A}$ with respect to the criterion $C_j \in \boldsymbol{C}$.

Step 2. If the information about the criteria weights is completely unknown, then we can obtain the criteria weights by using Eq. (1.17); if the information about the criteria weights is partially known, then we solve the model (MOD-1.2) to obtain the criteria weights.

Step 3. Utilize Eqs. (1.18) and (1.21) to determine the corresponding hesitant fuzzy PIS A^+ and the hesitant fuzzy NIS A^-, respectively.

Step 4. Utilize Eqs. (1.19) and (1.22) to calculate the distance measures between the alternative A_i and the hesitant fuzzy PIS A^+ as well as the hesitant fuzzy NIS A^-, respectively.

Step 5. Utilize Eq. (1.25) to calculate the revised closeness indices $\mathscr{F}^h(A_i)\,(i = 1, 2, \ldots, m)$ of the alternatives $A_i\,(i = 1, 2, \ldots, m)$.

Step 6. Rank all alternatives according to the revised closeness indices $\mathscr{F}^h(A_i)\,(i = 1, 2, \ldots, m)$ and select the optimal alternative.

1.4 Numerical Example and Comparison Analysis

Xu and Zhang (2013) gave an energy police selection problem [adopted from the paper Xu and Xia (2011a)] to demonstrate the applicability and the implementation process of the proposed method. Meanwhile, the comparison analysis of the computational results was also conducted to show its superiority.

1.4.1 Problem Description and Decision Making Process

Energy is an indispensable factor for the social and economic development of societies. The selection of energy policy plays an important role in economic development and environment. Suppose that there are five energy projects $A_i\,(i = 1, 2, 3, 4, 5)$, and they are evaluated based on the following four criteria: technological (C_1); environmental (C_2); socio-political (C_3); economic (C_4). The decision maker needs to select an optimal energy project from the five potential energy projects based on these four criteria. The ratings of the energy projects $A_i\,(i = 1, 2, 3, 4, 5)$ with respect to the criteria $C_j\,(j = 1, 2, 3, 4)$ provided by the decision maker are contained in Table 1.1 (Xu and Zhang 2013).

It is easily noted that the numbers of values in different HFEs in Table 1.1 are different. In order to accurately calculate the distance between two HFEs, we should extend the shorter one until both of them have the same length. According to Definition 1.3, we here assume that the decision maker is pessimistic and further normalize the decision data in Table 1.1 by adding the minimal values. The normalized results are displayed in Table 1.2 (Xu and Zhang 2013).

Then, we utilize the maximizing deviation model-based hesitant fuzzy TOPSIS approach to solve the above decision making problem which involves the following two cases (Xu and Zhang 2013).

Table 1.1 Hesitant fuzzy decision matrix

	C_1	C_2
A_1	H{0.3, 0.4, 0.5}	H{0.1, 0.7, 0.8, 0.9}
A_2	H{0.3, 0.5}	H{0.2, 0.5, 0.6, 0.7, 0.9}
A_3	H{0.6, 0.7}	H{0.6, 0.9}
A_4	H{0.3, 0.4, 0.7, 0.8}	H{0.2, 0.4, 0.7}
A_5	H{0.1, 0.3, 0.6, 0.7, 0.9}	H{0.4, 0.6, 0.7, 0.8}
	C_3	C_4
A_1	H{0.2, 0.4, 0.5}	H{0.3, 0.5, 0.6, 0.9}
A_2	H{0.1, 0.5, 0.6, 0.8}	H{0.3, 0.4, 0.7}
A_3	H{0.3, 0.5, 0.7}	H{0.4, 0.6}
A_4	H{0.1, 0.8}	H{0.6, 0.8, 0.9}
A_5	H{0.7, 0.8, 0.9}	H{0.3, 0.6, 0.7, 0.9}

Table 1.2 Hesitant fuzzy normalized decision matrix

	C_1	C_2
A_1	H{0.3, 0.3, 0.3, 0.4, 0.5}	H{0.1, 0.1, 0.7, 0.8, 0.9}
A_2	H{0.3, 0.3, 0.3, 0.3, 0.5}	H{0.2, 0.5, 0.6, 0.7, 0.9}
A_3	H{0.6, 0.6, 0.6, 0.6, 0.7}	H{0.6, 0.6, 0.6, 0.6, 0.9}
A_4	H{0.3, 0.3, 0.4, 0.7, 0.8}	H{0.2, 0.2, 0.2, 0.4, 0.7}
A_5	H{0.1, 0.3, 0.6, 0.7, 0.9}	H{0.4, 0.4, 0.6, 0.7, 0.8}
	C_3	C_4
A_1	H{0.2, 0.2, 0.2, 0.4, 0.5}	H{0.3, 0.3, 0.5, 0.6, 0.9}
A_2	H{0.1, 0.1, 0.5, 0.6, 0.8}	H{0.3, 0.3, 0.3, 0.4, 0.7}
A_3	H{0.3, 0.3, 0.3, 0.5, 0.7}	H{0.4, 0.4, 0.4, 0.4, 0.6}
A_4	H{0.1, 0.1, 0.1, 0.1, 0.8}	H{0.6, 0.6, 0.6, 0.8, 0.9}
A_5	H{0.7, 0.7, 0.7, 0.8, 0.9}	H{0.3, 0.3, 0.6, 0.7, 0.9}

Case 1.1 The weights of criteria are assumed to be completely unknown in advance, and we get the best alternative according to the following steps:

In Steps 1–2, we utilize Eq. (1.17) to get the optimal weight vector:

$$w = (0.2341, 0.2474, 0.3181, 0.2004)^T.$$

In Step 3, we use Eqs. (1.18) and (1.21) to determine the hesitant fuzzy PIS A^+ and the hesitant fuzzy NIS A^-, respectively:

$$A^+ = \begin{pmatrix} H\{0.6,0.6,0.6,0.7,0.9\}, H\{0.6,0.6,0.7,0.8,0.9\}, \\ H\{0.7,0.7,0.7,0.8,0.9\}, H\{0.6,0.6,0.6,0.8,0.9\} \end{pmatrix},$$

$$A^- = \begin{pmatrix} H\{0.1,0.3,0.3,0.3,0.5\}, H\{0.1,0.1,0.2,0.4,0.7\}, \\ H\{0.1,0.1,0.1,0.1,0.5\}, H\{0.3,0.3,0.3,0.4,0.6\} \end{pmatrix}.$$

In Step 4, we employ Eqs. (1.19) and (1.22) to calculate the distances between the alternative A_i and the hesitant fuzzy PIS A^+ as well as the hesitant fuzzy NIS A^-, respectively. The results are shown in Table 1.3 (Xu and Zhang 2013).

In Step 5, we utilize Eq. (1.25) to calculate the revised closeness index $\mathscr{F}^h(A_i)$ of the alternative A_i, and the results are also listed in Table 1.3. In Step 6, according to $\mathscr{F}^h(A_i)$, we can obtain the ranking of all alternatives which is shown in Table 1.3, and the most desirable alternative is A_5.

Table 1.3 The results obtained by the Algorithm 1.2 under Case 1.1

	$d^h(A_i, A^+)$	$d^h(A_i, A^-)$	$\mathscr{F}^h(A_i)$	Ranking
A_1	0.3555	0.1976	−1.5975	4
A_2	0.3277	0.2384	−1.3328	3
A_3	0.2418	0.2905	−0.6986	2
A_4	0.3865	0.2040	−1.7628	5
$\mathbf{A_5}$	**0.1702**	**0.4023**	**0.0000**	**1**

Table 1.4 The results obtained by Algorithm 1.2 under Case 1.2

	$d^h(A_i,A^+)$	$d^h(A_i,A^-)$	$\mathscr{F}^h(A_i)$	Ranking
A_1	0.3527	0.1910	−1.6821	4
A_2	0.3344	0.2313	−1.4720	3
A_3	0.2615	0.2645	−0.9465	2
A_4	0.3865	0.2200	−1.8171	5
$\boldsymbol{A_5}$	**0.1641**	**0.4088**	**0.0000**	**1**

Case 1.2 The information about the criteria weights is partially known and the known weight information is given as follows:

$$\Delta = \left\{ \begin{array}{c} 0.15 \le w_1 \le 0.2, 0.16 \le w_2 \le 0.18, 0.3 \le w_3 \le 0.35, \\ 0.3 \le w_4 \le 0.45, \sum_{j=1}^{4} w_j = 1 \end{array} \right\}.$$

For Case 1.2, we can obtain the best alternative according to the following steps:

In Steps 1–2, we utilize the model (MOD-1.2) to construct the single-objective model as follows:

$$\begin{cases} \max & D(\boldsymbol{w}) = 4.4467w_1 + 4.6999w_2 + 6.0431w_3 + 3.8068w_4 \\ s.t. & 0.15 \le w_1 \le 0.2, 0.16 \le w_2 \le 0.18, \\ & 0.3 \le w_3 \le 0.35, 0.3 \le w_4 \le 0.45, \\ & \sum_{j=1}^{4} w_j = 1 \end{cases}$$

By solving this model, we get the optimal weight vector $\boldsymbol{w} = (0.17, 0.18, 0.35, 0.3)^T$.

In Step 3, please see Step 3 in Case 1.1. In Step 4, we also employ Eqs. (1.19) and (1.22) to calculate the distances between the alternative A_i and the hesitant fuzzy PIS A^+ as well as the hesitant fuzzy NIS A^-, respectively. The results are shown in Table 1.4 (Xu and Zhang 2013).

In Step 5, we utilize Eq. (1.25) to calculate the revised closeness index $\mathscr{F}^h(A_i)$ of the alternative A_i, and the results are also listed in Table 1.4. In Step 6, according to $\mathscr{F}^h(A_i)$, we can obtain the ranking of all alternatives which is shown in Table 1.4 and thus the most desirable alternative is A_5.

1.4.2 Comparison Analysis

Recently, Nan et al. (2008) proposed an intuitionistic fuzzy TOPSIS method for solving the MCDM problems with IFNs. We call it the IF-TOPSIS method. Xu and Zhang (2013) made a comparison with the IF-TOPSIS method which is the closest to the proposed approach because the HFEs' envelopes are IFNs.

Definition 1.6 (Atanassov 1986). Let a set T be a universe of discourse. An intuitionistic fuzzy set (IFS) I is an object having the form $I = \{\langle t, \mu_I(t), v_I(t) \rangle | t \in T\}$, where the function $\mu_I : T \to [0, 1]$ defines the degree of membership and

$v_I : T \to [0,1]$ defines the degree of non-membership of the element $t \in T$ to I, respectively, and for every $t \in T$, it holds that $0 \le \mu_I(t) + v_I(t) \le 1$.

The $\pi_I(t) = 1 - \mu_I(t) - v_I(t)$ is called the degree of indeterminacy of $t \in T$ to I. For simplicity, Xu (2007) called $\chi = I(\mu_\chi, v_\chi)$ an IFN, where $\mu_\chi \in [0,1]$, $v_\chi \in [0,1]$ and $\mu_\chi + v_\chi \le 1$. In the practical decision making process, people usually use the IFNs instead of the IFSs to express their assessments.

The Hamming distance between IFNs is introduced as follows (Szmidt and Kacprzyk 2000):

$$d(\chi_1, \chi_2) = \frac{1}{2}\left(\left|\mu_{\chi_1} - \mu_{\chi_2}\right| + \left|v_{\chi_1} - v_{\chi_2}\right| + \left|\pi_{\chi_1} - \pi_{\chi_2}\right|\right) \tag{1.26}$$

and the Euclidean distance between IFNs is defined as below (Szmidt and Kacprzyk 2000):

$$d(\chi_1, \chi_2) = \sqrt{\frac{1}{2}\left(\left(\mu_{\chi_1} - \mu_{\chi_2}\right)^2 + \left(v_{\chi_1} - v_{\chi_2}\right)^2 + \left(\pi_{\chi_1} - \pi_{\chi_2}\right)^2\right)} \tag{1.27}$$

In addition, Torra (2010) indicated that the envelope of a HFE is an IFN, which is shown as follows.

Definition 1.7 (Torra 2010). Given a HFE $h = H\{\gamma^\lambda | \lambda = 1, 2, \ldots, \#h\}$, its envelope is an IFN $I_{env}(h)$ which can be represented as $I_{env}(h) = I(\gamma^-, 1 - \gamma^+)$ with $\gamma^- = \min\{\gamma | \gamma \in h\}$ and $\gamma^+ = \max\{\gamma | \gamma \in h\}$.

According to Definition 1.7, the hesitant fuzzy data of the energy police selection problem can be transformed into the intuitionistic fuzzy data, which are listed in Table 1.5 (Xu and Zhang 2013). Moreover, because Nan et al. (2008)'s method needs to obtain weights of criteria in advance, we assume that the weight vector of criteria is given by the decision maker in advance as $w = (0.2341, 0.2474, 0.3181, 0.2004)^T$.

With Nan et al. (2008)'s method, we first need to determine the intuitionistic fuzzy PIS A^+ and the intuitionistic fuzzy NIS A^-, respectively:

$$A^+ = \{I(0.6, 0.3), I(0.6, 0.1), I(0.7, 0.1), I(0.6, 0.1)\},$$
$$A^- = \{I(0.1, 0.1), I(0.1, 0.1), I(0.1, 0.2), I(0.3, 0.1)\}.$$

Table 1.5 Intuitionistic fuzzy decision matrix

	C_1	C_2	C_3	C_4
A_1	I(0.3, 0.5)	I(0.1, 0.1)	I(0.2, 0.5)	I(0.3, 0.1)
A_2	I(0.3, 0.5)	I(0.2, 0.1)	I(0.1, 0.2)	I(0.3, 0.3)
A_3	I(0.6, 0.3)	I(0.6, 0.1)	I(0.3, 0.3)	I(0.4, 0.4)
A_4	I(0.3, 0.2)	I(0.2, 0.3)	I(0.1, 0.2)	I(0.6, 0.1)
A_5	I(0.1, 0.1)	I(0.4, 0.2)	I(0.7, 0.1)	I(0.3, 0.1)

Then, according to the Euclidean distance of IFNs introduced in Eq. (1.27), we can calculate the distances between the alternative A_i and the intuitionistic fuzzy PIS A^+ as well as the intuitionistic fuzzy NIS A^-, respectively:

$$d^i(A_1,A^+) = 0.3261,\ d^i(A_1,A^-) = 0.2138,$$
$$d^i(A_2,A^+) = 0.3372,\ d^i(A_2,A^-) = 0.1521,$$
$$d^i(A_3,A^+) = 0.1044,\ d^i(A_3,A^-) = 0.4209,$$
$$d^i(A_4,A^+) = 0.3715,\ d^i(A_4,A^-) = 0.1035,$$
$$d^i(A_5,A^+) = 0.2335,\ d^i(A_5,A^-) = 0.2615.$$

Furthermore, according to the following formula:

$$CI^i(A_i) = \frac{d^i(A_i,A^-)}{d^i(A_i,A^-) + d^i(A_i,A^+)}$$

it is easy to obtain the relative closeness index of the alternative A_i to the intuitionistic fuzzy PIS A^+ as follows:

$$CI^i(A_1) = 0.3960, CI^i(A_2) = 0.3108, CI^i(A_3) = 0.7942,$$
$$CI^i(A_4) = 0.2458, CI^i(A_5) = 0.5283.$$

Finally, on the basis of the relative closeness indices $CI^i(A_i)\ (i = 1,2,3,4,5)$, the ranking of the alternatives $A_i\ (i = 1,2,3,4,5)$ is obtained as: $A_3 \succ A_5 \succ A_1 \succ A_2 \succ A_4$. Thus, the most desirable alternative is A_3.

To provide a better view of the comparison results, we put the results of the ranking of alternatives obtained by the HF-TOPSIS method and the IF-TOPSIS method into Fig. 1.1 (Xu and Zhang 2013).

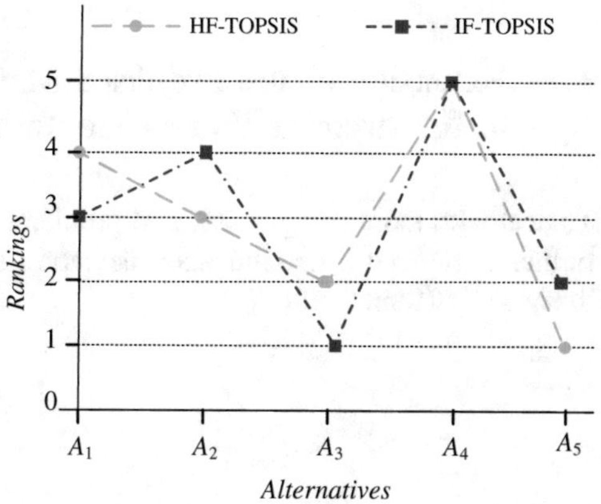

Fig. 1.1 The pictorial representation of the HF-TOPSIS and IF-TOPSIS rankings

It is noticed from Fig. 1.1 that the ranking order of alternatives obtained by Nan et al. (2008)'s method is remarkably different from that obtained by the HF-TOPSIS method. The differences are the ranking orders between A_3 and A_5, and between A_1 and A_2, i.e., $A_3 \succ A_5$ and $A_1 \succ A_2$ for the former while $A_5 \succ A_3$ and $A_2 \succ A_1$ for the latter. Namely, the ranking orders of these pairs of alternatives are just converse. The main reason is that the HF-TOPSIS method considers the hesitant fuzzy information which is represented by several possible values, not by a margin of error (as in IFNs), while if adopting the IF-TOPSIS method, it needs to transform HFEs into IFNs, which gives rise to a difference in the accuracy of data in the two types, and it will have an effect on the final decision results. Thus, it is not hard to see that the HF-TOPSIS method has some desirable advantages over Nan et al. (2008)'s method as follows:

(1) The HF-TOPSIS method, by extending the TOPSIS method to take into account the hesitant fuzzy assessments which are well-suited to handle the ambiguity and impreciseness inherent in the MCDM problems, does not need to transform HFEs into IFNs but directly deals with these problems, and thus obtains better final decision results. In particular, when we meet some situations where the decision data are represented by several possible values, our approach shows its great superiority in handling those decision making problems with hesitant fuzzy information.
(2) The HF-TOPSIS method utilizes the maximizing deviation method to objectively determine the weights of criteria, which is more reasonable; while Nan et al. (2008)'s method needs the decision maker to provide the weights of criteria in advance, which is subjective and sometime cannot yield the persuasive results.

In addition, compared with the aggregating operators-based decision making methods (Xia and Xu 2011; Zhu et al. 2012; Zhang 2013), this technique is based on the revised closeness index of each alternative to determine the ranking order of all alternatives, which avoids producing the loss of too much information in the process of information aggregation.

1.5 Extension of the Proposed Method in the Interval-Valued Hesitant Fuzzy Situations

To deal with more complex MCDM problems with IVHFEs, Xu and Zhang (2013) further developed a maximization deviation model-based interval-valued hesitant fuzzy TOPSIS method.

1.5.1 Interval-Valued Hesitant Fuzzy Information

In some complex real-world decision processes, it is somewhat difficult for the decision makers to assign exact values for the membership degrees of certain elements to a given set, but a range of values belonging to [0, 1] may be assigned. To cope with such a situation, Chen et al. (2013) put forward the concept of interval-valued hesitant fuzzy set (IVHFS) which permits the membership of an element to a given set having a few different interval values. The basic term of IVHFS is introduced as follows.

Definition 1.8 (Chen et al. 2013). Let T be a reference set, and $D\,[0,1]$ be the set of all closed subintervals of [0,1]. An IVHFS on T is defined as:

$$\tilde{H} = \{\langle t, \tilde{h}_{\tilde{H}}(t)\rangle | t \in T\},$$

where $\tilde{h}_{\tilde{H}}(t) : T \to D\,[0,1]$ denotes all possible interval membership degrees of the element $t \in T$ to the set $\tilde{H}$.

Usually, $\tilde{h}_{\tilde{H}}(t)$ is called an interval-valued hesitant fuzzy element (IVHFE) denoted by $\tilde{h} = \tilde{H}\{\tilde{\gamma}^1, \tilde{\gamma}^2, \ldots, \tilde{\gamma}^{\#\tilde{h}}\}$ where $\tilde{\gamma}^\lambda\,(\lambda = 1, 2, \ldots, \#\tilde{h})$ is an interval number denoted by $\tilde{\gamma}^\lambda = [\tilde{\gamma}^{\lambda L}, \tilde{\gamma}^{\lambda U}] \in D\,[0,1]$. The IVHFE is reduced to the HFE if $\tilde{\gamma}^{\lambda L} = \tilde{\gamma}^{\lambda U}\,(\lambda = 1, 2, \ldots, \#\tilde{h})$. We employ the symbol $\#\tilde{h}$ to represent the number of the elements in an IVHFE $\tilde{h}$ in the rest of this book without explicitly mentioning it.

Definition 1.9 (Chen et al. 2013). Let $\tilde{h}, \tilde{h}_1$ and $\tilde{h}_2$ be three IVHFEs, the basic operations of IVHFEs are defined as follows:

(1) $\otimes$-intersection: $\tilde{h}_1 \otimes \tilde{h}_2 = \tilde{H}\{[\tilde{\gamma}_1^L \cdot \tilde{\gamma}_2^L, \tilde{\gamma}_1^U \cdot \tilde{\gamma}_2^U] | \tilde{\gamma}_1 \in \tilde{h}_1, \tilde{\gamma}_2 \in \tilde{h}_2\}$;

(2) $\oplus$-union: $\tilde{h}_1 \oplus \tilde{h}_2 = \tilde{H}\left\{\left[\begin{array}{c}\tilde{\gamma}_1^L + \tilde{\gamma}_2^L - \tilde{\gamma}_1^L \cdot \tilde{\gamma}_2^L, \\ \tilde{\gamma}_1^U + \tilde{\gamma}_2^U - \tilde{\gamma}_1^U \cdot \tilde{\gamma}_2^U\end{array}\right] | \tilde{\gamma}_1 \in \tilde{h}_1, \tilde{\gamma}_2 \in \tilde{h}_2\right\}$;

(3) Multiplication: $\theta\tilde{h} = \tilde{H}\left\{[1-(1-\tilde{\gamma}^L)^\theta, 1-(1-\tilde{\gamma}^U)^\theta] | \tilde{\gamma} \in \tilde{h}\right\}$, $\theta > 0$;

(4) Complement: $\tilde{h}^c = \tilde{H}\{[1-\tilde{\gamma}^U, 1-\tilde{\gamma}^L] | \tilde{\gamma} \in \tilde{h}\}$.

According to the definition of IVHFE, it is noted that the numbers of interval values in different IVHFEs are usually different and the interval values in an IVHFE are often out of order. For convenience, let $\#\tilde{h}_1$ and $\#\tilde{h}_2$ be the numbers of interval numbers in the IVHFEs $\tilde{h}_1$ and $\tilde{h}_2$, respectively, and arrange the interval numbers in an increasing or decreasing order according to the possibility degree formula-based ranking method of interval numbers (Xu and Da 2002). In order to accurately calculate the distance between $\tilde{h}_1$ and $\tilde{h}_2$ with $\#\tilde{h}_1 \neq \#\tilde{h}_2$, we should also extend the shorter one until both of them have the same length. Similar to Definition 1.3, an extension rule for IVHFEs is introduced as follows.

Definition 1.10 (Xu and Zhang 2013). For an IVHFE $\tilde{h} = \tilde{H}\{\tilde{\gamma}^1, \tilde{\gamma}^2, \ldots, \tilde{\gamma}^{\#\tilde{h}}\}$, let $\tilde{\gamma}^+$ and $\tilde{\gamma}^-$ be the maximum and the minimum interval values in the IVHFE $\tilde{h}$, respectively, then $\tilde{\gamma}^* = \theta\tilde{\gamma}^+ + (1-\theta)\tilde{\gamma}^-$ is called an extension interval value, where $\theta\,(0 \le \theta \le 1)$ is the parameter determined by the decision maker according to his/her risk preference.

Therefore, for two IVHFEs $\tilde{h}_1$ and $\tilde{h}_2$ with $\#\tilde{h}_1 \neq \#\tilde{h}_2$, we can add the different interval values to the shorter IVHFE using the parameter θ according to the decision maker's risk preference. In general, if the decision maker is risk-neutral, then we can add the extension interval value $\tilde{\gamma}^* = \frac{1}{2}(\tilde{\gamma}^+ + \tilde{\gamma}^-)$, i.e., $\theta = \frac{1}{2}$; if the decision maker is risk-seeking, then we can add the extension interval value $\tilde{\gamma}^* = \tilde{\gamma}^+$, i.e., $\theta = 1$; while the decision maker is risk-averse, we can add the extension interval value $\tilde{\gamma}^* = \tilde{\gamma}^-$, i.e., $\theta = 0$.

Example 1.3 (Xu and Zhang 2013). For two IVHFEs $\tilde{h}_1 = \tilde{H}\{[0.1, 0.2], [0.2, 0.3], [0.1, 0.3]\}$ and $\tilde{h}_2 = \tilde{H}\{[0.3, 0.4], [0.4, 0.5]\}$, it is easy to know $\#\tilde{h}_1(=3) > \#\tilde{h}_2(=2)$. In order to accurately calculate the distance between $\tilde{h}_1$ and $\tilde{h}_2$, we should extend $\tilde{h}_2$ until it has the same length with $\tilde{h}_1$. According to Definition 1.10, we can get:

(a) if the decision maker is the risk-seeking, then $\tilde{h}_2$ can be extended as $\tilde{h}_2 = \tilde{H}\{[0.3, 0.4], [0.4, 0.5], [0.4, 0.5]\}$;
(b) if the decision maker is the risk-averse, then $\tilde{h}_2$ can be extended as $\tilde{h}_2 = \tilde{H}\{[0.3, 0.4], [0.3, 0.4], [0.4, 0.5]\}$;
(c) if the decision maker is the risk-neutral, then $\tilde{h}_2$ can be extended as $\tilde{h}_2 = \tilde{H}\{[0.3, 0.4], [0.35, 0.45], [0.4, 0.5]\}$.

Chen et al. (2013) gave two different distance measures for IVHFEs:

- The interval-valued hesitant fuzzy Hamming distance:

$$d_H(\tilde{h}_1, \tilde{h}_2) = \frac{1}{2\#\tilde{h}} \sum_{\lambda=1}^{\#\tilde{h}} \begin{pmatrix} \left|(\tilde{\gamma}_1^L)^\lambda - (\tilde{\gamma}_2^L)^\lambda\right| \\ + \left|(\tilde{\gamma}_1^U)^\lambda - (\tilde{\gamma}_2^U)^\lambda\right| \end{pmatrix} \tag{1.28}$$

- The interval-valued hesitant fuzzy Euclidean distance:

$$d_E(\tilde{h}_1, \tilde{h}_2) = \sqrt{\frac{1}{2\#\tilde{h}} \sum_{\lambda=1}^{\#\tilde{h}} \begin{pmatrix} \left((\tilde{\gamma}_1^L)^\lambda - (\tilde{\gamma}_2^L)^\lambda\right)^2 \\ + \left((\tilde{\gamma}_1^U)^\lambda - (\tilde{\gamma}_2^U)^\lambda\right)^2 \end{pmatrix}} \tag{1.29}$$

where $\#\tilde{h}_1 = \#\tilde{h}_2 = \#\tilde{h}$, $\tilde{\gamma}_1^\lambda = [\tilde{\gamma}_1^{\lambda L}, \tilde{\gamma}_1^{\lambda U}]$ and $\tilde{\gamma}_2^\lambda = [\tilde{\gamma}_2^{\lambda L}, \tilde{\gamma}_2^{\lambda U}]$ are the λth smallest interval values in $\tilde{h}_1$ and $\tilde{h}_2$, respectively, which will be used thereafter.

1.5.2 The Maximizing Deviation Model-Based Interval-Valued Hesitant Fuzzy TOPSIS Approach

Consider an interval-valued hesitant fuzzy MCDM problem which is based on m alternatives $\boldsymbol{A} = \{A_1, A_2, \ldots, A_m\}$ and n criteria $\boldsymbol{C} = \{C_1, C_2, \ldots, C_n\}$. The rating of the alternative A_i with respect to the criterion C_j is expressed by an IVHFE, i.e., $x_{ij} = \tilde{h}_{ij}$. Then, such a MCDM problem is expressed by the interval-valued hesitant fuzzy decision matrix $\Re = (\tilde{h}_{ij})_{m\times n}$. To address this MCDM problem in which the weights of criteria are completely unknown or partially known, Xu and Zhang (2013) proposed a maximizing deviation model-based interval-valued hesitant fuzzy TOPSIS method.

Firstly, motivated by the HF-TOPSIS method, a non-linear programming model which maximizes all deviation values for all criteria is established to identify the weight vector $\boldsymbol{w}$ as follows (Xu and Zhang 2013):

$$\begin{cases} \max \quad \sum_{j=1}^{n}\sum_{\xi=1}^{m}\sum_{\zeta=1}^{m} w_j \sqrt{\frac{1}{2\#\tilde{h}}\sum_{\lambda=1}^{\#\tilde{h}}\left(\begin{array}{c}\left((\tilde{\gamma}_{\xi j}^{L})^\lambda - (\tilde{\gamma}_{\zeta j}^{L})^\lambda\right)^2 \\ +\left((\tilde{\gamma}_{\xi j}^{U})^\lambda - (\tilde{\gamma}_{\zeta j}^{U})^\lambda\right)^2\end{array}\right)} \\ s.t. \quad w_j \geq 0, \quad j = 1, 2, \ldots, n, \quad \sum_{j=1}^{n}(w_j)^2 = 1. \end{cases} \qquad (\text{MOD}-1.3)$$

To solve the above model, let

$$Q(\boldsymbol{w}, \eta) = \sum_{j=1}^{n}\sum_{\xi=1}^{m}\sum_{\zeta=1}^{m} w_j \sqrt{\frac{1}{2\#\tilde{h}}\sum_{\lambda=1}^{\#\tilde{h}}\left(\begin{array}{c}\left((\tilde{\gamma}_{\xi j}^{L})^\lambda - (\tilde{\gamma}_{\zeta j}^{L})^\lambda\right)^2 \\ +\left((\tilde{\gamma}_{\xi j}^{U})^\lambda - (\tilde{\gamma}_{\zeta j}^{U})^\lambda\right)^2\end{array}\right)} + \frac{\eta}{2}\left(\sum_{j=1}^{n} w_j^2 - 1\right) \tag{1.30}$$

which indicates the Lagrange function of the optimization model (MOD-1.3) and the parameter η is a real number denoting the Lagrange multiplier variable.

Then, the partial derivatives of Q are computed as:

$$\begin{cases} \frac{\partial Q}{\partial w_j} = \sum_{\xi=1}^{m}\sum_{\zeta=1}^{m}\sqrt{\frac{1}{2\#\tilde{h}}\sum_{\lambda=1}^{\#\tilde{h}}\left(\begin{array}{c}\left((\tilde{\gamma}_{\xi j}^{L})^{\lambda}-(\tilde{\gamma}_{\zeta j}^{L})^{\lambda}\right)^2\\ +\left((\tilde{\gamma}_{\xi j}^{U})^{\lambda}-(\tilde{\gamma}_{\zeta j}^{U})^{\lambda}\right)^2\end{array}\right)} + \eta w_j = 0 \\ \frac{\partial Q}{\partial \eta} = \frac{1}{2}\left(\sum_{j=1}^{n} w_j^2 - 1\right) = 0 \end{cases} \tag{1.31}$$

By solving Eq. (1.31), we can get

$$w_j = \frac{\sum_{\xi=1}^{m}\sum_{\zeta=1}^{m}\sqrt{\frac{1}{2\#\tilde{h}}\sum_{\lambda=1}^{\#\tilde{h}}\left(\begin{array}{c}\left((\tilde{\gamma}_{\xi j}^{L})^{\lambda}-(\tilde{\gamma}_{\zeta j}^{L})^{\lambda}\right)^2\\ +\left((\tilde{\gamma}_{\xi j}^{U})^{\lambda}-(\tilde{\gamma}_{\zeta j}^{U})^{\lambda}\right)^2\end{array}\right)}}{\sqrt{\sum_{j=1}^{n}\left(\sum_{\xi=1}^{m}\sum_{\zeta=1}^{m}\sqrt{\frac{1}{2\#\tilde{h}}\sum_{\lambda=1}^{\#\tilde{h}}\left(\begin{array}{c}\left((\tilde{\gamma}_{\xi j}^{L})^{\lambda}-(\tilde{\gamma}_{\zeta j}^{L})^{\lambda}\right)^2\\ +\left((\tilde{\gamma}_{\xi j}^{U})^{\lambda}-(\tilde{\gamma}_{\zeta j}^{U})^{\lambda}\right)^2\end{array}\right)}\right)^2}} \tag{1.32}$$

For convenience, let

$$\aleph_j = \sum_{\xi=1}^{m}\sum_{\zeta=1}^{m}\sqrt{\frac{1}{2\#\tilde{h}}\sum_{\lambda=1}^{\#\tilde{h}}\left(\begin{array}{c}\left((\tilde{\gamma}_{\xi j}^{L})^{\lambda}-(\tilde{\gamma}_{\zeta j}^{L})^{\lambda}\right)^2\\ +\left((\tilde{\gamma}_{\xi j}^{U})^{\lambda}-(\tilde{\gamma}_{\zeta j}^{U})^{\lambda}\right)^2\end{array}\right)},$$

and make the sum of w_j $(j = 1, 2, \ldots, n)$ into a unit by normalizing w_j, we get

$$w_j^* = \frac{w_j}{\sum_{j=1}^{n} w_j} = \frac{\aleph_j}{\sum_{j=1}^{n}\aleph_j} = \frac{\sum_{\xi=1}^{m}\sum_{\zeta=1}^{m}\sqrt{\frac{1}{2\#\tilde{h}}\sum_{\lambda=1}^{\#\tilde{h}}\left(\begin{array}{c}\left((\tilde{\gamma}_{\xi j}^{L})^{\lambda}-(\tilde{\gamma}_{\zeta j}^{L})^{\lambda}\right)^2\\ +\left((\tilde{\gamma}_{\xi j}^{U})^{\lambda}-(\tilde{\gamma}_{\zeta j}^{U})^{\lambda}\right)^2\end{array}\right)}}{\sum_{j=1}^{n}\sum_{\xi=1}^{m}\sum_{\zeta=1}^{m}\sqrt{\frac{1}{2\#\tilde{h}}\sum_{\lambda=1}^{\#\tilde{h}}\left(\begin{array}{c}\left((\tilde{\gamma}_{\xi j}^{L})^{\lambda}-(\tilde{\gamma}_{\zeta j}^{L})^{\lambda}\right)^2\\ +\left((\tilde{\gamma}_{\xi j}^{U})^{\lambda}-(\tilde{\gamma}_{\zeta j}^{U})^{\lambda}\right)^2\end{array}\right)}} \tag{1.33}$$

As we have remarked earlier, there are actual situations that the information about the weighting vector is not completely unknown but partially known. Let $\mathbf{\Delta}$ be a set of constraint conditions that the weight value w_j should satisfy according to the requirements in the real situations, Xu and Zhang (2013) constructed the following constrained optimization model:

$$\begin{cases} \max \sum_{j=1}^{n}\sum_{\xi=1}^{m}\sum_{\zeta=1}^{m}\left(w_j\sqrt{\frac{1}{2\#\tilde{h}}\sum_{\lambda=1}^{\#\tilde{h}}\left(\begin{array}{c}\left((\tilde{\gamma}_{\xi j}^{L})^{\lambda}-(\tilde{\gamma}_{\zeta j}^{L})^{\lambda}\right)^2\\ +\left((\tilde{\gamma}_{\xi j}^{U})^{\lambda}-(\tilde{\gamma}_{\zeta j}^{U})^{\lambda}\right)^2\end{array}\right)}\right) \\ s.t. \quad \boldsymbol{w}\in\boldsymbol{\Delta} \end{cases} \quad (\text{MOD}-1.4)$$

The solution to this model (MOD-1.4) could be found by using the MATLAB 7.4.0 or LINGO software package.

Once the weights of criteria are obtained, it is necessary to rank the alternatives and select the optimal alternative. Xu and Zhang (2013) extended the classical TOPSIS method to take into account interval-valued hesitant fuzzy information and utilized it to obtain the final ranking of the alternatives. Under interval-valued hesitant fuzzy environment, the interval-valued hesitant fuzzy PIS denoted by A^+, and the interval-valued hesitant fuzzy NIS denoted by A^-, respectively, can be defined as follows:

$$A^+ = \left(\tilde{h}_1^+,\tilde{h}_2^+,\ldots,\tilde{h}_n^+\right) = \left(\begin{array}{c}\left(\tilde{H}\{\max_i \tilde{\gamma}_{ij}^1, \max_i \tilde{\gamma}_{ij}^2,\ldots,\max_i \tilde{\gamma}_{ij}^{\#\tilde{h}_{ij}}\}|C_j\in J_I\right)\\ or\left(\tilde{H}\{\min_i \tilde{\gamma}_{ij}^1, \min_i \tilde{\gamma}_{ij}^2,\ldots,\min_i \tilde{\gamma}_{ij}^{\#\tilde{h}_{ij}}\}|C_j\in J_{II}\right)\end{array}\right) \quad (1.34)$$

$$A^- = \left(\tilde{h}_1^-,\tilde{h}_2^-,\ldots,\tilde{h}_n^-\right) = \left(\begin{array}{c}\left(\tilde{H}\{\min_i \tilde{\gamma}_{ij}^1, \min_i \tilde{\gamma}_{ij}^2,\ldots,\min_i \tilde{\gamma}_{ij}^{\#\tilde{h}_{ij}}\}|C_j\in J_I\right)\\ or\left(\tilde{H}\{\max_i \tilde{\gamma}_{ij}^1, \max_i \tilde{\gamma}_{ij}^2,\ldots,\max_i \tilde{\gamma}_{ij}^{\#\tilde{h}_{ij}}\}|C_j\in J_{II}\right)\end{array}\right) \quad (1.35)$$

Then, the distances between each alternative and the interval-valued hesitant fuzzy PIS A^+ as well as the interval-valued hesitant fuzzy NIS A^- can be obtained by the following equations, respectively:

$$d^{ih}(A_i,A^+) = \sum_{j=1}^{n}\left(w_j\sqrt{\frac{1}{2\#\tilde{h}_{ij}}\sum_{\lambda=1}^{\#\tilde{h}_{ij}}\left(\begin{array}{c}\left((\tilde{\gamma}_{ij}^{L})^{\lambda}-(\tilde{\gamma}_{j}^{L+})^{\lambda}\right)^2\\ +\left((\tilde{\gamma}_{ij}^{U})^{\lambda}-(\tilde{\gamma}_{j}^{U+})^{\lambda}\right)^2\end{array}\right)}\right) \quad (1.36)$$

$$d^{ih}(A_i,A^-) = \sum_{j=1}^{n}\left(w_j\sqrt{\frac{1}{2\#\tilde{h}_{ij}}\sum_{\lambda=1}^{\#\tilde{h}_{ij}}\left(\begin{array}{c}\left((\tilde{\gamma}_{ij}^{L})^{\lambda}-(\tilde{\gamma}_{j}^{L-})^{\lambda}\right)^2\\ +\left((\tilde{\gamma}_{ij}^{U})^{\lambda}-(\tilde{\gamma}_{j}^{U-})^{\lambda}\right)^2\end{array}\right)}\right) \quad (1.37)$$

Usually, the smaller the $d^{ih}(A_i, A^+)$ the better the alternative A_i, and let

$$d^{ih}_{\min}(A_i, A^+) = \min_{1 \le i \le m} d^{ih}(A_i, A^+) \tag{1.38}$$

On the contrary, the bigger the $d^{ih}(A_i, A^-)$ the better the alternative A_i, and let

$$d^{ih}_{\max}(A_i, A^-) = \max_{1 \le i \le m} d^{ih}(A_i, A^-) \tag{1.39}$$

Then, the revised closeness index of the alternative A_i is presented as follows:

$$\mathscr{F}^{ih}(A_i) = \frac{d^{ih}(A_i, A^-)}{d^{ih}_{\max}(A_i, A^-)} - \frac{d^{ih}(A_i, A^+)}{d^{ih}_{\min}(A_i, A^+)} \tag{1.40}$$

which is used to measure the extent to which the alternative A_i closes to the interval-valued hesitant fuzzy PIS A^+ and is far away from the interval-valued hesitant fuzzy NIS A^-, simultaneously.

In general, the bigger the $\mathscr{F}^{ih}(A_i)$ is, the better the alternative A_i is. If there exists an alternative A^* satisfying the conditions that $d^{ih}(A^*, A^-) = d^{ih}_{\max}(A_i, A^-)$ and $d^{ih}(A^*, A^+) = d^{ih}_{\min}(A_i, A^+)$, then $\mathscr{F}^{ih}(A^*) = 0$, and obviously, the alternative A^* is the best alternative that is closest to the interval-valued hesitant fuzzy PIS A^+ and farthest away from the interval-valued hesitant fuzzy NIS A^-, simultaneously. Therefore, according to the revised closeness index $\mathscr{F}^{ih}(A_i)$, we can determine the ranking orders of all alternatives and select the best one from a set of feasible alternatives.

Next, a practical algorithm is presented to address the interval-valued hesitant fuzzy MCDM problems in which the information about criteria weights is incompletely known or completely unknown. The algorithm (Algorithm 1.3) involves the following steps (Xu and Zhang 2013):

Step 1. For an interval-valued hesitant fuzzy MCDM problem, we construct the interval-valued hesitant fuzzy decision matrix $\tilde{\Re} = (\tilde{h}_{ij})_{m \times n}$, where all the arguments $\tilde{h}_{ij}\,(i = 1, 2, \ldots, m;\ j = 1, 2, \ldots, n)$ are IVHFEs.

Step 2. If the weights of criteria are completely unknown in advance, then we employ Eq. (1.33) to determine the optimal weights; while if the information about the criteria weights is partially known, then we employ the model (MOD-1.4) to obtain the optimal weights.

Step 3. Utilize Eqs. (1.34) and (1.35) to determine the interval-valued hesitant fuzzy PIS A^+ and the interval-valued hesitant fuzzy NIS A^-, respectively.

Step 4. Use Eqs. (1.36) and (1.37) to calculate the distances between the alternative A_i and the interval-valued hesitant fuzzy PIS A^+ as well as the interval-valued hesitant fuzzy NIS A^-, respectively.

Step 5. Employ Eq. (1.40) to calculate the revised closeness indices $\mathscr{F}^{ih}(A_i)\,(i = 1, 2, \ldots, m)$ of the alternatives $A_i\,(i = 1, 2, \ldots, m)$.

Step 6. Rank the alternatives $A_i\,(i = 1, 2, \ldots, m)$ according to the revised closeness indices $\mathscr{F}^{ih}(A_i)\,(i = 1, 2, \ldots, m)$ and select the optimal alternative.

1.5.3 Numerical Example with IVHFEs

We assume that all criteria values in the numerical example introduced in Sect. 1.4.1 are expressed in the form of IVHFEs, and these IVHFEs are listed in the interval-valued hesitant fuzzy decision matrix as Table 1.6 (Xu and Zhang 2013). Next, the maximizing deviation model-based interval-valued hesitant fuzzy TOPSIS method is used to address this MCDM problem with IVHFEs.

Because the numbers of values in different IVHFEs in Table 1.6 are different, we should extend the shorter one until all of them have the same length. According to Definition 1.10, we normalize the interval-valued hesitant fuzzy data by adding the minimal values. The normalized results are listed in Table 1.7 (Xu and Zhang 2013).

Then, we proceed to utilize the developed approach to get the optimal alternative, which involves the following two cases (Xu and Zhang 2013):

Table 1.6 Interval-valued hesitant fuzzy decision matrix

	C_1
A_1	$\tilde{H}\{[0.2,0.3],[0.3,0.4],[0.3,0.5]\}$
A_2	$\tilde{H}\{[0.2,0.3],[0.2,0.5]\}$
A_3	$\tilde{H}\{[0.5,0.6],[0.5,0.7]\}$
A_4	$\tilde{H}\{[0.2,0.3],[0.3,0.4],[0.6,0.7],[0.7,0.8]\}$
A_5	$\tilde{H}\{[0.1,0.3],[0.2,0.4],[0.5,0.6],[0.6,0.7],[0.7,0.9]\}$
	C_2
A_1	$\tilde{H}\{[0.1,0.3],[0.6,0.7],[0.7,0.8],[0.7,0.9]\}$
A_2	$\tilde{H}\{[0.1,0.2],[0.4,0.5],[0.4,0.6],[0.6,0.7],[0.7,0.9]\}$
A_3	$\tilde{H}\{[0.5,0.6],[0.7,0.9]\}$
A_4	$\tilde{H}\{[0.1,0.2],[0.3,0.4],[0.6,0.7]\}$
A_5	$\tilde{H}\{[0.3,0.4],[0.5,0.6],[0.6,0.7],[0.6,0.8]\}$
	C_3
A_1	$\tilde{H}\{[0.1,0.2],[0.3,0.4],[0.3,0.5]\}$
A_2	$\tilde{H}\{[0.1,0.3],[0.4,0.4],[0.4,0.6],[0.6,0.8]\}$
A_3	$\tilde{H}\{[0.2,0.3],[0.4,0.5],[0.6,0.7]\}$
A_4	$\tilde{H}\{[0.1,0.3],[0.7,0.8]\}$
A_5	$\tilde{H}\{[0.6,0.7],[0.7,0.8],[0.7,0.9]\}$
	C_4
A_1	$\tilde{H}\{[0.2,0.3],[0.4,0.5],[0.4,0.6],[0.8,0.9]\}$
A_2	$\tilde{H}\{[0.2,0.3],[0.3,0.4],[0.6,0.7]\}$
A_3	$\tilde{H}\{[0.3,0.4],[0.5,0.6]\}$
A_4	$\tilde{H}\{[0.5,0.6],[0.7,0.8],[0.7,0.9]\}$
A_5	$\tilde{H}\{[0.2,0.3],[0.5,0.6],[0.5,0.7],[0.7,0.9]\}$

Case 1.3 In the above interval-valued hesitant fuzzy MCDM problem, the weights of criteria are completely unknown. In this case, we can obtain the optimal alternative according to the following decision making steps:

In Steps 1–2, we utilize Eq. (1.33) to get the optimal weight vector:

$$\boldsymbol{w} = (0.2537, 0.2428, 0.29, 0.2135)^T$$

In Step 3, we employ Eqs. (1.34) and (1.35) to determine the interval-valued hesitant fuzzy PIS A^+ and the interval-valued hesitant fuzzy NIS A^-, respectively:

Table 1.7 Interval-valued hesitant fuzzy normalized decision matrix

	C_1
A_1	$\tilde{H}\{[0.2,0.3],[0.2,0.3],[0.2,0.3],[0.3,0.4],[0.3,0.5]\}$
A_2	$\tilde{H}\{[0.2,0.3],[0.2,0.3],[0.2,0.3],[0.2,0.3],[0.2,0.5]\}$
A_3	$\tilde{H}\{[0.5,0.6],[0.5,0.6],[0.5,0.6],[0.5,0.6],[0.5,0.7]\}$
A_4	$\tilde{H}\{[0.2,0.3],[0.2,0.3],[0.3,0.4],[0.6,0.7],[0.7,0.8]\}$
A_5	$\tilde{H}\{[0.1,0.3],[0.2,0.4],[0.5,0.6],[0.6,0.7],[0.7,0.9]\}$
	C_2
A_1	$\tilde{H}\{[0.1,0.3],[0.1,0.3],[0.6,0.7],[0.7,0.8],[0.7,0.9]\}$
A_2	$\tilde{H}\{[0.1,0.2],[0.4,0.5],[0.4,0.6],[0.6,0.7],[0.7,0.9]\}$
A_3	$\tilde{H}\{[0.5,0.6],[0.5,0.6],[0.5,0.6],[0.5,0.6],[0.7,0.9]\}$
A_4	$\tilde{H}\{[0.1,0.2],[0.1,0.2],[0.1,0.2],[0.3,0.4],[0.6,0.7]\}$
A_5	$\tilde{H}\{[0.3,0.4],[0.3,0.4],[0.5,0.6],[0.6,0.7],[0.6,0.8]\}$
	C_3
A_1	$\tilde{H}\{[0.1,0.2],[0.1,0.2],[0.1,0.2],[0.3,0.4],[0.3,0.5]\}$
A_2	$\tilde{H}\{[0.1,0.3],[0.1,0.3],[0.4,0.4],[0.4,0.6],[0.6,0.8]\}$
A_3	$\tilde{H}\{[0.2,0.3],[0.2,0.3],[0.2,0.3],[0.4,0.5],[0.6,0.7]\}$
A_4	$\tilde{H}\{[0.1,0.3],[0.1,0.3],[0.1,0.3],[0.1,0.3],[0.7,0.8]\}$
A_5	$\tilde{H}\{[0.6,0.7],[0.6,0.7],[0.6,0.7],[0.7,0.8],[0.7,0.9]\}$
	C_4
A_1	$\tilde{H}\{[0.2,0.3],[0.2,0.3],[0.4,0.5],[0.4,0.6],[0.8,0.9]\}$
A_2	$\tilde{H}\{[0.2,0.3],[0.2,0.3],[0.2,0.3],[0.3,0.4],[0.6,0.7]\}$
A_3	$\tilde{H}\{[0.3,0.4],[0.3,0.4],[0.3,0.4],[0.3,0.4],[0.5,0.6]\}$
A_4	$\tilde{H}\{[0.5,0.6],[0.5,0.6],[0.5,0.6],[0.7,0.8],[0.7,0.9]\}$
A_5	$\tilde{H}\{[0.2,0.3],[0.2,0.3],[0.5,0.6],[0.5,0.7],[0.7,0.9]\}$

$$\begin{aligned}A^{+} = {} & (\tilde{H}\{[0.5,0.6],[0.5,0.6],[0.5,0.6],[0.6,0.7],[0.7,0.9]\},\tilde{H}\{[0.5,0.6],[0.5,0.6],\\ & [0.6,0.7],[0.7,0.8],[0.7,0.9]\},\tilde{H}\{[0.6,0.7],[0.6,0.7],[0.6,0.7],[0.7,0.8],\\ & [0.7,0.9]\},\tilde{H}\{[0.5,0.6],[0.5,0.6],[0.5,0.6],[0.7,0.8],[0.8,0.9]\});\\ A^{-} = {} & (\tilde{H}\{[0.1,0.3],[0.2,0.3],[0.2,0.3],[0.2,0.3],[0.2,0.5]\},\tilde{H}\{[0.1,0.2],[0.1,0.2],\\ & [0.1,0.2],[0.3,0.5],[0.6,0.7]\},\tilde{H}\{[0.1,0.2],[0.1,0.2],[0.1,0.2],[0.1,0.3],\\ & [0.3,0.5]\},\tilde{H}\{[0.2,0.3],[0.2,0.3],[0.2,0.3],[0.3,0.4],[0.5,0.6]\}).\end{aligned}$$

In Step 4, we employ Eqs. (1.36) and (1.37) to calculate the distances between the alternative A_i and the interval-valued hesitant fuzzy PIS A^+ as well as the interval-valued hesitant fuzzy NIS A^-, respectively. The results are shown in Table 1.8 (Xu and Zhang 2013).

In Step 5, we utilize Eq. (1.40) to calculate the revised closeness indices $\mathscr{F}^{ih}(A_i)$ $(i = 1,2,3,4,5)$ of the alternatives A_i $(i = 1,2,3,4,5)$, and the results are also listed in Table 1.8. In Step 6, according to $\mathscr{F}^{ih}(A_i)$ $(i = 1,2,3,4,5)$, we can obtain the ranking of all alternatives which is shown in Table 1.8 and the most desirable alternative is A_5.

Case 1.4 The information about the criteria weights is partially known, which is given as follows:

$$\Delta = \left\{ \begin{array}{c} 0.15 \le w_1 \le 0.2, 0.16 \le w_2 \le 0.18, 0.3 \le w_3 \le 0.35, \\ 0.3 \le w_4 \le 0.45, \quad \sum_{j=1}^{4} w_j = 1 \end{array} \right\}$$

In this case, the optimal alternative can be identified by the following steps:

In Steps 1–2, we utilize the model (MOD-1.4) to construct the single-objective model as follows:

$$\begin{cases} \max & D(w) = 4.4063w_1 + 4.2167w_2 + 5.0365w_3 + 3.7098w_4 \\ s.t. & 0.15 \le w_1 \le 0.2, 0.16 \le w_2 \le 0.18 \\ & 0.3 \le w_3 \le 0.35, 0.3 \le w_4 \le 0.45, \\ & \sum_{j=1}^{4} w_j = 1 \end{cases}$$

Table 1.8 The results obtained by Algorithm 1.3 under Case 1.3

	$d^{ih}(A_i, A^+)$	$d^{ih}(A_i, A^-)$	$\mathscr{F}^{ih}(A_i)$	Ranking
A_1	0.3293	0.1760	−1.5694	5
A_2	0.2977	0.1807	−1.3557	3
A_3	0.2329	0.2451	−0.7572	2
A_4	0.3209	0.2156	−1.4004	4
$\mathbf{A_5}$	**0.1580**	**0.3419**	**0.0000**	**1**

Table 1.9 The results obtained by the Algorithm 1.3 under Case 1.4

	$d^{ih}(A_i, A^+)$	$d^{ih}(A_i, A^-)$	$\mathscr{F}^{ih}(A_i)$	Ranking
A_1	0.3401	0.1625	−1.7285	5
A_2	0.3043	0.1749	−1.4625	4
A_3	0.2581	0.2181	−1.0418	2
A_4	0.3139	0.2286	−1.3715	3
$\boldsymbol{A_5}$	**0.1552**	**0.3511**	**0.0000**	**1**

By solving this model, we get the optimal weight vector $\boldsymbol{w} = (0.19, 0.16, 0.35, 0.3)^T$.

In Step 3, please see Step 3 in Case 1.3. In Step 4, we also employ Eqs. (1.36) and (1.37) to calculate the distances between each of the alternatives $A_i\,(i = 1,2,3,4,5)$ and the interval-valued hesitant fuzzy PIS A^+ as well as the interval-valued hesitant fuzzy NIS A^-, respectively. The results are shown in Table 1.9 (Xu and Zhang 2013).

In Step 5, we utilize Eq. (1.40) to calculate the revised closeness indices $\mathscr{F}^{ih}(A_i)\,(i = 1,2,3,4,5)$ of the alternative $A_i\,(i = 1,2,3,4,5)$, and the calculation results are also listed in Table 1.9. In Step 6, according to $\mathscr{F}^{ih}(A_i)\,(i = 1,2,3,4,5)$, we can obtain the ranking of all alternatives which is shown in Table 1.9 and the most desirable alternative is A_5.

1.6 Conclusions

In general, many real-world MCDM problems take place in a complex and uncertain environment and the decision data are usually imprecise and vague. The HFEs or the IVHFEs are adequate for dealing with the decision making situations in which the decision maker has hesitancy in providing his/her judgments for alternatives with respect to each criterion. Based on the idea that the criterion with a larger deviation value among alternatives should be assigned with a larger weight, this chapter introduced a maximizing deviation model developed by Xu and Zhang (2013) to determine the optimal weights of criteria under hesitant fuzzy environment. An important advantage of this proposed method is its ability to relieve the influence of subjectivity of the decision makers and at the same time remain the original decision information sufficiently. Then, a hesitant fuzzy TOPSIS method developed by Xu and Zhang (2013) is presented to rank the alternatives for solving the MCDM problems with hesitant fuzzy information. Compared with the aggregating operators-based decision methods (Xia and Xu 2011; Zhu et al. 2012; Zhang 2013), this technique is based on the revised closeness index of each alternative to determine the ranking order of all alternatives, which avoids producing the loss of too much information in the process of information aggregation. Furthermore, this developed method is extended to tackle appropriately the MCDM problems under interval-valued hesitant fuzzy environment. Finally, the effectiveness and applicability of the proposed method has been illustrated with an energy police selection problem. Apparently, the developed approach can also be applied to the other managerial decision making problems under hesitant fuzzy or interval-valued hesitant fuzzy environment.

References

Atanassov, K. T. (1986). Intuitionistic fuzzy sets. *Fuzzy Sets and Systems, 20*, 87–96.

Bedregal, B., Reiser, R., Bustince, H., Lopez-Molina, C., & Torra, V. (2014). Aggregation functions for typical hesitant fuzzy elements and the action of automorphisms. *Information Sciences, 255*, 82–99.

Beg, I., & Rashid, T. (2013). TOPSIS for hesitant fuzzy linguistic term sets. *International Journal of Intelligent Systems, 28*, 1162–1171.

Behzadian, M., Otaghsara, S. K., Yazdani, M., & Ignatius, J. (2012). A state-of the-art survey of TOPSIS applications. *Expert Systems with Applications, 39*, 13051–13069.

Boran, F. E., Genç, S., Kurt, M., & Akay, D. (2009). A multi-criteria intuitionistic fuzzy group decision making for supplier selection with TOPSIS method. *Expert Systems with Applications, 36*, 11363–11368.

Chen, C. T. (2000). Extensions of the TOPSIS for group decision-making under fuzzy environment. *Fuzzy Sets and Systems, 114*, 1–9.

Chen, T. Y. (2014). Interval-valued intuitionistic fuzzy QUALIFLEX method with a likelihood-based comparison approach for multiple criteria decision analysis. *Information Sciences, 261*, 149–169.

Chen, S. M., & Chang, C.-H. (2015). A novel similarity measure between Atanassov's intuitionistic fuzzy sets based on transformation techniques with applications to pattern recognition. *Information Sciences, 291*, 96–114.

Chen, S. M., & Hong, J. A. (2014). Multicriteria linguistic decision making based on hesitant fuzzy linguistic term sets and the aggregation of fuzzy sets. *Information Sciences, 286*, 63–74.

Chen, T. Y., & Tsao, C.-Y. (2008). The interval-valued fuzzy TOPSIS method and experimental analysis. *Fuzzy Sets and Systems, 159*, 1410–1428.

Chen, N., Xu, Z. S., & Xia, M. M. (2013). Interval-valued hesitant preference relations and their applications to group decision making. *Knowledge-Based Systems, 37*, 528–540.

Dymova, L., Sevastjanov, P., & Tikhonenko, A. (2013). An approach to generalization of fuzzy TOPSIS method. *Information Sciences, 238*, 149–162.

Hadi-Vencheh, A., & Mirjaberi, M. (2014). Fuzzy inferior ratio method for multiple attribute decision making problems. *Information Sciences, 277*, 263–272.

Hwang, C. L., & Yoon, K. (1981). *Multiple attribute decision making methods and applications*. Berlin: Springer.

Nan, J. X., Li, D. F., & Zhang, M. J. (2008). TOPSIS for Multiattribute decision making in IF-set setting. *Operations Research and Management Science in China, 3*, 34–37.

Park, K. S., & Kim, S. H. (1997). Tools for interactive multiattribute decision making with incompletely identified information. *European Journal of Operational Research, 98*, 111–123.

Szmidt, E., & Kacprzyk, J. (2000). Distances between intuitionistic fuzzy sets. *Fuzzy Sets and Systems, 114*, 505–518.

Torra, V. (2010). Hesitant fuzzy sets. *International Journal of Intelligent Systems, 25*, 529–539.

Wang, Y. M. (1998). Using the method of maximizing deviations to make decision for multi-indices. *System Engineering and Electronics, 7*, 31.

Xia, M. M., & Xu, Z. S. (2011). Hesitant fuzzy information aggregation in decision making. *International Journal of Approximate Reasoning, 52*, 395–407.

Xu, Z. S. (2007). Intuitionistic fuzzy aggregation operators. *IEEE Transactions on Fuzzy Systems, 15*, 1179–1187.

Xu, Z. S., & Da, Q. L. (2002). The uncertain OWA operator. *International Journal of Intelligent Systems, 17*, 569–575.

Xu, Z. S., & Xia, M. M. (2011a). Distance and similarity measures for hesitant fuzzy sets. *Information Sciences, 181*, 2128–2138.

Xu, Z. S., & Xia, M. M. (2011b). On distance and correlation measures of hesitant fuzzy information. *International Journal of Intelligent Systems, 26*, 410–425.

Xu, Z. S., & Zhang, X. L. (2013). Hesitant fuzzy multi-attribute decision making based on TOPSIS with incomplete weight information. *Knowledge-Based Systems, 52*, 53–64.

Zhang, Z. M. (2013). Hesitant fuzzy power aggregation operators and their application to multiple attribute group decision making. *Information Sciences, 234*, 150–181.

Zhang, X. L., & Xu, Z. S. (2014a). Extension of TOPSIS to multiple criteria decision making with pythagorean fuzzy sets. *International Journal of Intelligent Systems, 29*, 1061–1078.

Zhang, X. L., & Xu, Z. S. (2014b). The TODIM analysis approach based on novel measured functions under hesitant fuzzy environment. *Knowledge-Based Systems, 61*, 48–58.

Zhu, B., Xu, Z. S., & Xia, M. M. (2012). Hesitant fuzzy geometric Bonferroni means. *Information Sciences, 205*, 72–85.

Chapter 2
Hesitant Fuzzy Multiple Criteria Decision Analysis Based on TODIM

Abstract The TODIM is a valuable technique for solving classical MCDM problems in case of considering the decision maker's psychological behavior. One main goal of this chapter is to introduce the measured functions-based hesitant fuzzy TODIM technique to deal with the behavioral MCDM problem under hesitant fuzzy environments. The main advantages of this technique are that (1) it can handle the MCDM problems in which the ratings of alternatives with respect to each criterion are represented by HFEs or IVHFEs and (2) it can take the decision maker's psychological behavior into account. Another aim of this chapter is to present the hesitant trapezoidal fuzzy TODIM method with a closeness index-based ranking approach to handle MCGDM problems in which decision data is expressed as comparative linguistic expressions based on HTrFNs. This proposed method first transforms comparative linguistic expressions into HTrFNs for carrying out computing with word processes. Then, a closeness index-based ranking method is proposed for comparing the magnitude of HTrFNs. By using such a ranking method, the dominance values of alternatives over others for each expert are calculated. Next, a nonlinear programming model is established to derive the dominance values of alternatives over others for the group and correspondingly the optimal ranking order of alternatives is determined.

The classical TODIM method originally proposed by Gomes and Lima (1991, 1992) is a discrete multiple criteria decision analysis method based on prospect theory (Kahneman and Tversky 1979) and has been proven to be a valuable tool for solving the classical MCDM problems in case of considering the decision maker's psychological behavior. In the classical TODIM approach, the prospect value function is first built to measure the dominance degree of each alternative over the others, which reflects the decision maker's behavioral characteristics such as reference dependence and loss aversion; and then the ranking orders of alternatives can be obtained by calculating the overall prospect value of each alternative.

The classical TODIM method has been extensively applied in various fields of decision making, such as the selection of the destination of natural gas (Gomes et al. 2009), the evaluation of residential properties (Gomes and Rangel 2009), and the oil

X. Zhang and Z. Xu, *Hesitant Fuzzy Methods for Multiple Criteria Decision Analysis*, Studies in Fuzziness and Soft Computing 345,
DOI 10.1007/978-3-319-42001-1_2

spill response problem (Passos et al. 2014), etc. Considering the fact that in some real-world situations the relationships among criteria are interdependent, Gomes et al. (2013) developed a method combining Choquet integral and the classical TODIM to handle the MCDM problems with criteria interactions. Owing to the fact that in many situations crisp data are inadequate or insufficient to model the real-world decision making problems, the fuzzy set and its extensions are more appropriate to model human judgments. This realization has motivated many researchers to extend the classical TODIM method for dealing with the MCDM problems under various fuzzy environments. For instance, considering the decision data assessed by TFNs or TrFNs, Krohling and de Souza (2012) developed a fuzzy extension of TODIM (named F-TODIM) for solving the fuzzy MCDM problems. Fan et al. (2013) proposed another extension of TODIM (named H-TODIM) to deal with the hybrid MCDM problems with three forms of criteria values (crisp numbers, interval numbers and fuzzy numbers). More recently, Lourenzutti and Krohling (2013) also presented a generalization of the TODIM method (named IF-RTODIM) which considers intuitionistic fuzzy information and an underlying random vector.

Although the existing TODIM methods can solve effectively the classical MCDM problems or fuzzy MCDM problems in case of considering the decision maker's psychological behavior, they fail to handle such MCDM problems under hesitant fuzzy environment. The MCDM problems with HFEs and/or IVHFEs have recently received increasing attentions and many corresponding MCDM methods (Farhadinia 2013; Liao and Xu 2013; Xu and Zhang 2013; Zhang 2013) have also been developed, but none of them can be used to solve the hesitant fuzzy MCDM problems in case of considering the decision maker's psychological behavior. To this end, Zhang and Xu (2014a) extended the classical TODIM method to solve the hesitant fuzzy MCDM problems in case of considering the decision maker's psychological behavior. In this approach, two novel ranking functions are developed for comparing the magnitude of HFEs and IVHFEs, which are more reasonable and effective compared with the existing ranking functions. Then, the prospect values of each alternative related to the others are calculated based on novel ranking functions and distance measures. By aggregating these prospect values, the overall prospect value of each alternative is further obtained and the ranking of alternatives is also obtained. Finally, Zhang and Xu (2014a) provided a decision making problem that concerns the evaluation and ranking of the service quality among domestic airlines to illustrate the validity and applicability of this approach.

On the other hand, Zhang et al. (2016) proposed a new concept of HTrFN which is an extension of HFE and is well enough to represent the uncertainty and vagueness of comparative linguistic expressions. The HTrFNs benefited from the superiority of both TrFNs and HFEs can be directly applied in MCDM and MCGDM. To handle the MCGDM problems in which the decision data are expressed by comparative linguistic expressions based on HTrFNs, Zhang and Liu (2016) developed a hesitant trapezoidal fuzzy TODIM method with a closeness index-based ranking approach. This proposed method first transforms comparative linguistic expressions into HTrFNs for carrying out computing with word processes. Then, a closeness index-based ranking method is proposed for comparing the magnitude of HTrFNs. By using the closeness index-based ranking method of

HTrFNs, the gain and loss of each alternative relative to the others are identified. Next, the dominance values of alternatives over others for each expert are calculated. Furthermore, a nonlinear programming model is established to derive the dominance values of alternatives over others for the group and correspondingly the optimal ranking order of alternatives is determined. At length, the proposed method is implemented in an evaluation problem of the service quality of airlines in order to demonstrate its decision making process and its applicability.

2.1 Description of the Classical TODIM Method

The classical TODIM method is to measure the dominance degree of each alternative over the others by establishing a prospect value function based on prospect theory (Kahneman and Tversky 1979). The main advantage of the classical TODIM method is its capability of capturing the decision maker's psychological behavior. It is worth mentioning that the classical TODIM method can only be suitable to deal with the classical MCDM problems in which the criteria values and the weights of criteria are in the format of crisp numbers. The algorithm (Algorithm 2.1) of the classical TODIM approach is introduced as follows (Zhang and Xu 2014a; Fan et al. 2013):

Step 1. Identify the decision matrix $\Re = (x_{ij})_{m\times n}$, and normalize it into $\bar{\Re} = (\bar{x}_{ij})_{m\times n}$ by using Eq. (1.3), where both x_{ij} and $\bar{x}_{ij}$ $(i \in \{1,2,\ldots,m\}, j \in \{1,2,\ldots,n\})$ are crisp numbers.

Step 2. Determine the relative weight w_{jR} of the criterion C_j to the reference criterion C_R according to the following expression:

$$w_{jR} = w_j / w_R, \quad j \in \{1,2,\ldots,n\} \tag{2.1}$$

where w_j is the weight value of the criterion C_j and $w_R = \max_{j=1}^{n}\{w_j\}$.

Step 3. Calculate the prospect values of the alternatives A_ξ $(\xi \in \{1,2,\ldots,m\})$ over the alternatives A_ζ $(\zeta \in \{1,2,\ldots,m\})$ using the following expression:

$$\mathscr{H}(A_\xi, A_\zeta) = \sum_{j=1}^{n} \mathscr{D}_j(A_\xi, A_\zeta), \quad \forall(\xi,\zeta) \tag{2.2}$$

where

$$\mathscr{D}_j(A_\xi, A_\zeta) = \begin{cases} \sqrt{w_{jR}\left(\bar{x}_{\xi j} - \bar{x}_{\zeta j}\right) \Big/ \sum_{j=1}^{n} w_{jR}}, & \text{if } \bar{x}_{\xi j} - \bar{x}_{\zeta j} > 0 \\ 0, & \text{if } \bar{x}_{\xi j} - \bar{x}_{\zeta j} = 0 \\ -\frac{1}{\theta}\sqrt{\left(\sum_{j=1}^{n} w_{jR}\right)\left(\bar{x}_{\zeta j} - \bar{x}_{\xi j}\right) \Big/ w_{jR}}, & \text{if } \bar{x}_{\xi j} - \bar{x}_{\zeta j} < 0 \end{cases} \tag{2.3}$$

The term $\mathscr{D}_j(A_\xi, A_\zeta)$ represents the contribution of the criterion C_j to the function $\mathscr{H}(A_\xi, A_\zeta)$ when comparing the alternative A_ξ with the alternative A_ζ. The parameter θ represents the attenuation factor of the losses, which can be tuned according to the problem at hand. In Eq. (2.3), three cases can occur:

(1) if $\bar{x}_{\xi j} - \bar{x}_{\zeta j} > 0$, then $\mathscr{D}_j(A_\xi, A_\zeta)$ represents a gain;
(2) if $\bar{x}_{\xi j} - \bar{x}_{\zeta j} = 0$, then $\mathscr{D}_j(A_\xi, A_\zeta)$ represents a nil;
(3) if $\bar{x}_{\xi j} - \bar{x}_{\zeta j} < 0$, then $\mathscr{D}_j(A_\xi, A_\zeta)$ represents a loss.

Step 4. Calculate the overall prospect values of the alternatives A_ξ ($\xi \in \{1, 2, \ldots, m\}$) according to the following expression:

$$\mathscr{Q}(A_\xi) = \frac{\sum_{\zeta=1}^{m} \mathscr{H}(A_\xi, A_\zeta) - \min_\xi \left\{ \sum_{\zeta=1}^{m} \mathscr{H}(A_\xi, A_\zeta) \right\}}{\max_\xi \left\{ \sum_{\zeta=1}^{m} \mathscr{H}(A_\xi, A_\zeta) \right\} - \min_\xi \left\{ \sum_{\zeta=1}^{m} \mathscr{H}(A_\xi, A_\zeta) \right\}} \tag{2.4}$$

Step 5. Rank all the alternatives by comparing their overall prospect values $\mathscr{Q}(A_\xi)$ ($\xi \in \{1, 2, \ldots, m\}$).

It is easily noted that the main idea of the classical TODIM method is to construct a prospect value function for measuring the dominance degree of each alternative over the others by comparing the criteria values. That is to say, to address the MCDM problem with hesitant fuzzy data by using the TODIM-based method, it is necessary to develop an effective ranking method for comparing the magnitudes of hesitant fuzzy data. To this end, Zhang and Xu (2014a) proposed novel ranking methods for HFEs and IVHFEs.

2.2 Ranking Functions Related to HFEs and IVHFEs

In practical fuzzy decision making process, the ranking of fuzzy information plays an important role in solving the fuzzy MCDM problems; while the ranking method is essentially based on the ranking function of fuzzy information which maps fuzzy information into crisp number. In general, the ranking methods of fuzzy information can be classified into two categories: the algorithmic ranking approaches and the non-algorithmic ranking approaches. In the non-algorithmic ranking approaches of fuzzy information, the ranking results are achieved in only one step, such as Xia and Xu (2011)'s ranking method, Farhadinia (2013)'s ranking method; while the ranking results obtained by the algorithmic ranking approaches of fuzzy information usually need to perform several steps, for instance, Liao and Xu (2013)'s ranking method.

In the sequel, we first review the existing ranking functions of HFEs, and then present a comparative study of these existing ranking functions. Furthermore, a novel ranking function developed by Zhang and Xu (2014a) is introduced for comparing the magnitude of HFEs, and its extension for comparing the magnitude of IVHFEs is also presented.

2.2.1 The Existing Ranking Functions of HFEs

Firstly, we review Xia and Xu (2011)'s ranking method which is referred to the non-algorithmic ranking approach of HFEs, as follows:

Definition 2.1 (Xia and Xu 2011). For a HFE $h = H\{\gamma^{\lambda}|\lambda = 1, 2, \ldots, \#h\}$, the score function of h is defined as follows:

$$s(h) = \frac{\gamma^1 + \gamma^2 + \cdots + \gamma^{\#h}}{\#h} \tag{2.5}$$

Then, for two HFEs h_1 and h_2, it is obtained that:

(1) if $s(h_1) > s(h_2)$, then h_1 is superior to h_2, denoted by $h_1 \succ_s h_2$;
(2) if $s(h_1) = s(h_2)$, then h_1 is indifferent to h_2, denoted by $h_1 \sim_s h_2$;
(3) if $s(h_1) < s(h_2)$, then h_1 is inferior to h_2, denoted by $h_1 \prec_s h_2$.

Example 2.1 (Zhang and Xu 2014a). Given three HFEs $h_1 = H\{0.3, 0.5\}$, $h_2 = H\{0.4\}$ and $h_3 = H\{0.2, 0.4, 0.6\}$, we need to compare the magnitude of these three HFEs.

According to the score function of HFE, we can obtain:

$$s(h_1) = 0.4, \quad s(h_2) = 0.4, \quad s(h_3) = 0.4.$$

Thus, we get $h_1 \sim_s h_2 \sim_s h_3$ by using Xia and Xu (2011)'s ranking method.

It is easy to notice that there yield indistinguishable HFEs when using the ranking function $s(h)$. In other words, the score-based ranking of HFEs is invalid in this situation and therefore should be improved. This realization has motivated Farhadinia (2013) to propose an improved score function and the corresponding ranking approach for HFEs as follows:

Assumption 2.1 (Farhadinia 2013). (1) The arrangement of elements in a HFE h is in an increasing order; (2) for two HFEs h_1 and h_2 with $\#h_1 \neq \#h_2$, we should extend the shorter one by adding the maximum element until them have the same length.

Remark 2.1 We point out that Assumption 2.1 is one special case of Definition 1.3, i.e., the special situation where the decision maker is optimist or risk-seeking.

Definition 2.2 (Farhadinia 2013). For a HFE $h = H\{\gamma^{\lambda}|\lambda = 1, 2, \ldots, \#h\}$, an improved score function $Is(h)$ of h is defined as follows:

$$Is(h) = \frac{\sum_{\lambda=1}^{\#h} f(\lambda)\gamma^{\lambda}}{\sum_{\lambda=1}^{\#h} f(\lambda)} \tag{2.6}$$

where $f(\lambda)$ $(\lambda = 1, 2, \ldots, \#h)$ is a positive-valued monotonic increasing sequence of index λ.

For convenience, Farhadinia (2013) suggested $f(\lambda) = \lambda$, and Eq. (2.6) is transformed into the following equation:

$$Is(h) = \frac{2\sum_{\lambda=1}^{\#h} \lambda\gamma^{\lambda}}{\#h \times (\#h + 1)} \tag{2.7}$$

Then, under Assumption 2.1, for two HFEs h_1 and h_2, we have:

(1) if $Is(h_1) > Is(h_2)$, then h_1 is superior to h_2, denoted by $h_1 \succ_{Is} h_2$;
(2) if $Is(h_1) = Is(h_2)$, then h_1 is indifferent to h_2, denoted by $h_1 \sim_{Is} h_2$;
(3) if $Is(h_1) < Is(h_2)$, then h_1 is inferior to h_2, denoted by $h_1 \prec_{Is} h_2$.

Example 2.2 (Zhang and Xu 2014a). For three HFEs shown in Example 2.1, according to Assumption 2.1 we extend the shorter one by adding the maximum element until these three HFEs have the same length. Namely, h_1 is extended to $h_1 = H\{0.3, 0.5, 0.5\}$ and h_2 is extended to $h_2 = H\{0.4, 0.4, 0.4\}$. Using the improved score function we can obtain:

$$Is(h_1) = 0.383, \quad Is(h_2) = 0.4, \quad Is(h_3) = 0.467.$$

Thus, according to Farhadinia (2013)'s ranking method, we can get $h_3 \succ_{Is} h_2 \succ_{Is} h_1$.

Obviously, the comparison of results of Examples 2.1 and 2.2 shows that the improved score function $Is(h)$ is more effective in ranking than the score function $s(h)$. However, we also notice that in the definition of $Is(h)$, the elements being an increasing order in h have the increasing importance, which is not consistent with the definition of the HFE (i.e. Definition 1.1) in which the importance weights of all elements in a HFE are the same.

Drawing on the mean-variance model in statistics, Liao and Xu (2013) developed a score-variance model to rank HFEs. The concept of variance function $Var(h)$ of HFEs is introduced as follows:

Definition 2.3 (Liao and Xu 2013). For a HFE $h = H\{\gamma^1, \gamma^2, \ldots, \gamma^{\#h}\}$, the variance function $Var(h)$ of h is defined as follows:

$$Var(h) = \frac{\sqrt{\sum_{\gamma^{\xi}, \gamma^{\zeta} \in h} \left(\gamma^{\xi} - \gamma^{\zeta}\right)^2}}{\#h} \tag{2.8}$$

It is easy to see that $Var(h)$ reflects the variance value among all possible values in h. Based on $s(h)$ and $Var(h)$, a score-variance model is introduced to rank the HFEs:

Definition 2.4 (Liao and Xu 2013). Let h_1 and h_2 be two HFEs, the $s(h_1)$ and $s(h_2)$ be the scores of h_1 and h_2, respectively, $Var(h_1)$ and $Var(h_2)$ be the deviation values of h_1 and h_2, respectively. Then, it is concluded that

(1) if $s(h_1) < s(h_2)$, then $h_1 \prec_s h_2$;

(2) if $s(h_1) = s(h_2)$, then $\begin{cases} Var(h_1) < Var(h_2) \Rightarrow h_1 \succ_{sv} h_2 \\ Var(h_1) = Var(h_2) \Rightarrow h_1 \sim_{sv} h_2 \\ Var(h_1) > Var(h_2) \Rightarrow h_1 \prec_{sv} h_2 \end{cases}$;

(3) if $s(h_1) > s(h_2)$, then $h_1 \succ_s h_2$.

Example 2.3 (Zhang and Xu 2014a). For three HFEs displayed in Example 2.1, by Eqs. (2.5) and (2.8) it is easy to obtain:

$$\begin{aligned} s(h_1) = 0.4, \quad s(h_2) = 0.4, \quad s(h_3) = 0.4, \\ Var(h_1) = 0.1, \quad Var(h_2) = 0, \quad Var(h_3) = 0.1633. \end{aligned}$$

According to Liao and Xu (2013)'s ranking approach, we can obtain the ranking $h_3 \prec_{sv} h_1 \prec_{sv} h_2$, which is different from the result ($h_1 \prec_{Is} h_2 \prec_{Is} h_3$) obtained by Farhadinia (2013)'s ranking approach.

With the help of Examples 2.1, 2.2 and 2.3, it is easily showed that Liao and Xu (2013)'s ranking approach is superior to Xia and Xu (2011)'s ranking approach and Farhadinia (2013)'s ranking approach.

Although Liao and Xu (2013)'s ranking approach can be applied to HFEs and seems to be effective, this approach which is referred to the algorithm ranking approach makes the process of decision making more time-consuming. Because when we utilize this approach to deal with the hesitant fuzzy MCDM problems, the process of decision making is required to be divided into several steps and it is necessary to add other rules for obtaining the best alternative.

Bearing this fact in mind, Zhang and Xu (2014a) developed a novel ranking function which is referred to the non-algorithmic ranking approach for comparing the HFEs.

2.2.2 The Proposed Ranking Functions

Definition 2.5 (Zhang and Xu 2014a). For a HFE $h = H\{\gamma^1, \gamma^2, \ldots, \gamma^{\#h}\}$, a new ranking function $Hr_\delta(h)$ of h is defined as follows:

$$Hr_\delta(h) = \left(\frac{(\gamma^1)^\delta + (\gamma^2)^\delta + \cdots + (\gamma^{\#h})^\delta}{\#h} \right)^{1/\delta} \tag{2.9}$$

where δ ($0 < \delta \leq 1$) is a parameter determined by the decision maker, which can be tuned according to the practical situation.

In particular, if $\delta = 1$, then the new ranking function $Hr_{\delta=1}(h) = (\gamma^1 + \gamma^2 + \cdots + \gamma^{\#h})/\#h$, which is reduced to the score function of HFEs developed by Xia and Xu (2011).

Meanwhile, Zhang and Xu (2014a) also discussed some properties of the new ranking function $Hr_\delta(h)$ as below:

Proposition 2.1 (Zhang and Xu 2014a). For any HFE $h = H\{\gamma^1, \gamma^2, \ldots, \gamma^{\#h}\}$, the proposed ranking function $Hr_\delta(h) \in [0, 1]$.

Proof Let $\gamma^+ = \max_{\lambda=1}^{\#h}\{\gamma^\lambda\}$ and $\gamma^- = \min_{\lambda=1}^{\#h}\{\gamma^\lambda\}$, because $0 < \delta \leq 1$ and $\gamma^\lambda \in [0, 1]$ $(\lambda \in \{1, 2, \ldots, \#h\})$, then

$$\begin{aligned} Hr_\delta(h) &= \left(\left((\gamma^1)^\delta + (\gamma^2)^\delta + \cdots + (\gamma^{\#h})^\delta\right)\Big/\#h\right)^{1/\delta} \\ &\leq \left(\left((\gamma^+)^\delta + (\gamma^+)^\delta + \cdots + (\gamma^+)^\delta\right)\Big/\#h\right)^{1/\delta} = \gamma^+ \leq 1, \end{aligned}$$

$$\begin{aligned} Hr_\delta(h) &= \left(\left((\gamma^1)^\delta + (\gamma^2)^\delta + \cdots + (\gamma^{\#h})^\delta\right)\Big/\#h\right)^{1/\delta} \\ &\geq \left(\left((\gamma^-)^\delta + (\gamma^-)^\delta + \cdots + (\gamma^-)^\delta\right)\Big/\#h\right)^{1/\delta} = \gamma^- \geq 0. \end{aligned}$$

Obviously, $0 \leq Hr_\delta(h) \leq 1$, which completes the proof (Zhang and Xu 2014a). □

Proposition 2.2 (Zhang and Xu 2014a). For a single-valued HFE $h = H\{\gamma\}$, the proposed ranking function $Hr_\delta(h) = \gamma$. In particular, if h is the hesitant empty element, i.e., $h = H\{0\}$, then $Hr_\delta(h) = 0$; if h is the hesitant full element, i.e., $h = H\{1\}$, then $Hr_\delta(h) = 1$.

Proposition 2.3 (Zhang and Xu 2014a). For two HFEs h_1 and h_2 having the same length, and the arrangement of elements in these two HFEs are in increasing orders, if $h_1 \leq h_2$, then $Hr_\delta(h_1) \leq Hr_\delta(h_2)$.

Proof For two HFEs $h_i = H\{\gamma_i^\lambda | \lambda = 1, 2, \ldots, \#h_i\}$ $(i = 1, 2)$, because they have the same length (i.e., $\#h = \#h_1 = \#h_2$) and the arrangement of elements in these two HFEs are in increasing orders. According to Definition 1.4, if $h_1 \leq h_2$, then $\gamma_1^\lambda \leq \gamma_2^\lambda$ $(\lambda \in \{1, 2, \ldots, \#h\})$. Owing to $0 < \delta \leq 1$, then $(\gamma_1^\lambda)^\delta \leq (\gamma_2^\lambda)^\delta$ $(\lambda \in \{1, 2, \ldots, \#h\})$. Apparently, we have

$$\left(\frac{(\gamma_1^1)^\delta + (\gamma_1^2)^\delta + \cdots + (\gamma_1^{\#h})^\delta}{\#h}\right)^{\frac{1}{\delta}} \leq \left(\frac{(\gamma_2^1)^\delta + (\gamma_2^2)^\delta + \cdots + (\gamma_2^{\#h})^\delta}{\#h}\right)^{\frac{1}{\delta}}$$

Namely, $Hr_\delta(h_1) \leq Hr_\delta(h_2)$, which completes the proof (Zhang and Xu 2014a). □

On the basis of the ranking function proposed by Definition 2.5, Zhang and Xu (2014a) introduced the following ranking method for HFEs which is referred to the non-algorithmic ranking approach:

Definition 2.6 (Zhang and Xu 2014a). For two HFEs h_1 and h_2, $Hr_\delta(h_1)$ and $Hr_\delta(h_2)$ are the new ranking functions of h_1 and h_2, respectively, then:

(1) if $Hr_\delta(h_1) > Hr_\delta(h_2)$, then h_1 is superior to h_2, denoted by $h_1 \succ_{Hr} h_2$;
(2) if $Hr_\delta(h_1) = Hr_\delta(h_2)$, then h_1 is indifferent to h_2, denoted by $h_1 \sim_{Hr} h_2$;
(3) if $Hr_\delta(h_1) < Hr_\delta(h_2)$, then h_1 is inferior to h_2, denoted by $h_1 \prec_{Hr} h_2$.

Example 2.4 (Zhang and Xu 2014a). For three HFEs being shown in Example 2.1, according to Definitions 2.5 and 2.6 it is easy to obtain the ranking results of these HFEs which are listed in Table 2.1.

In order to provide a synthetic view of the comparison results, we put all the results of the ranking of alternatives with different approaches into Table 2.2 (Zhang and Xu 2014a).

As shown in Table 2.2, the ranking order of HFEs obtained by Zhang and Xu (2014a)'s ranking method is the same as the result obtained by Liao and Xu (2013)'s ranking approach, but not consistent with the results obtained by Xia and Xu (2011)'s ranking approach and Farhadinia (2013)'s ranking approach. The main reason is that Xia and Xu (2011)'s ranking approach only considers the average value of all elements in HFEs and is just the special case of Zhang and Xu (2014a)'s ranking method, which cannot distinguish these three HFEs; While Farhadinia (2013)'s ranking approach is not only based on Assumption 2.1 which is an especial case of Definition 1.3, but also adopts the index $f(\lambda)$, which is unreasonable. Although Liao and Xu (2013)'s ranking approach is consistent with Zhang and Xu (2014a)'s ranking method, this approach referred to the algorithmic ranking approach makes the process of decision making more time-consuming. Therefore, we can confidently say that Zhang and Xu (2014a)'s ranking method referred to the

Table 2.1 The calculation results obtained by Definitions 2.5 and 2.6

Zhang and Xu (2014a)'s ranking method	h_1	h_2	h_3	The ranking orders
$Hr_{\delta=0.001}(h)$	0.3873	0.4	0.3635	$h_3 \prec_{Hr} h_1 \prec_{Hr} h_2$
$Hr_{\delta=0.01}(h)$	0.3874	0.4	0.3638	$h_3 \prec_{Hr} h_1 \prec_{Hr} h_2$
$Hr_{\delta=0.1}(h)$	0.3886	0.4	0.3672	$h_3 \prec_{Hr} h_1 \prec_{Hr} h_2$

Table 2.2 The comparison results of rankings of HFEs

The ranking approach	Ranking orders of HFEs
Xia and Xu (2011)'s ranking approach	$h_1 \sim_s h_2 \sim_s h_3$
Farhadinia (2013)'s ranking approach	$h_1 \prec_{Is} h_2 \prec_{Is} h_3$
Liao and Xu (2013)'s ranking approach	$h_3 \prec_{sv} h_1 \prec_{sv} h_2$
Zhang and Xu (2014a)'s ranking method	$h_3 \prec_{Hr} h_1 \prec_{Hr} h_2$

non-algorithmic ranking approach is much superior to Xia and Xu (2011)'s ranking approach, Liao and Xu (2013)'s ranking approach and Farhadinia (2013)'s ranking approach, which implies that Zhang and Xu (2014a)'s ranking method can provide a more useful technique than the previous ones to efficiently assist the decision maker.

Analogously, a new ranking method for IVHFEs is introduced as below:

Definition 2.7 (Zhang and Xu 2014a). For an IVHFE $\tilde{h} = \tilde{H}\{[\gamma^{1L}, \gamma^{1U}], [\gamma^{2L}, \gamma^{2U}], \ldots, [\gamma^{\#\tilde{h}L}, \gamma^{\#\tilde{h}U}]\}$, a new ranking function $Ir_\delta(\tilde{h})$ of $\tilde{h}$ is defined as follows:

$$Ir_\delta(\tilde{h}) = \left(\frac{(\gamma^{1L})^\delta + (\gamma^{2L})^\delta + \cdots + (\gamma^{\#\tilde{h}L})^\delta}{\#\tilde{h}}\right)^{1/\delta} + \left(\frac{(\gamma^{1U})^\delta + (\gamma^{2U})^\delta + \cdots + (\gamma^{\#\tilde{h}U})^\delta}{\#\tilde{h}}\right)^{1/\delta} \tag{2.10}$$

where δ $(0 < \delta \leq 1)$ is a parameter determined by the decision maker, which can be tuned according to the real-life decision problem at hand.

Based on the ranking function of IVHFEs, it is easy to give the following ranking method for IVHFEs, which is also referred to the non-algorithmic ranking approach:

Definition 2.8 (Zhang and Xu 2014a). For two IVHFEs $\tilde{h}_1$ and $\tilde{h}_2$, $Ir_\delta(\tilde{h}_1)$ and $Ir_\delta(\tilde{h}_2)$ are the new ranking functions of $\tilde{h}_1$ and $\tilde{h}_2$, respectively, then:

(1) if $Ir_\delta(\tilde{h}_1) > Ir_\delta(\tilde{h}_2)$, then $\tilde{h}_1$ is superior to $\tilde{h}_2$, denoted by $\tilde{h}_1 \succ_{Ir} \tilde{h}_2$;
(2) if $Ir_\delta(\tilde{h}_1) = Ir_\delta(\tilde{h}_2)$, then $\tilde{h}_1$ is indifferent to $\tilde{h}_2$, denoted by $\tilde{h}_1 \sim_{Ir} \tilde{h}_2$;
(3) if $Ir_\delta(\tilde{h}_1) < Ir_\delta(\tilde{h}_2)$, then $\tilde{h}_1$ is inferior to $\tilde{h}_2$, denoted by $\tilde{h}_1 \prec_{Ir} \tilde{h}_2$.

Next, on the basis of the proposed ranking functions and the ranking methods of HFEs and IVHFEs, Zhang and Xu (2014a) extended the classical TODIM method to solve the MCDM problems in which the criteria values are denoted by two different formats (i.e., HFEs or IVHFEs).

2.3 The Ranking Functions-Based Hesitant Fuzzy TODIM Approach

The MCDM problem is to identify the desirable compromise solution from a set of feasible alternatives which are assessed based on a set of conflicting criteria. Let $\boldsymbol{A} = \{A_1, A_2, \ldots, A_m\}$ be a finite alternative set and $\boldsymbol{C} = \{C_1, C_2, \ldots, C_n\}$ be a finite criteria set. The classical MCDM problem can be expressed in a decision matrix $\Re = (x_{ij})_{m \times n}$ whose elements x_{ij} indicate the ratings of the alternatives

$A_i(i \in \{1,2,\ldots,m\})$ with respect to the criteria $C_j(j \in \{1,2,\ldots,n\})$. In this chapter, we extend the decision matrix $\Re$ to the hesitant fuzzy decision matrix by considering the MCDM under hesitant fuzzy environment. Three scenarios are described in the decision matrix $\Re$ as follows (Zhang and Xu 2014a):

(1) All the criteria values in $\Re$ are expressed in the format of HFEs, i.e.,

$$x_{ij} = h_{ij} = H\{\gamma^1, \gamma^2, \ldots, \gamma^{\#h_{ij}}\}_{ij}$$

In this case, x_{ij} should be interpreted that the degree to which the alternative A_i satisfies the criterion C_j is some possible values between 0 and 1.

(2) Consider some practical situations that it is somewhat difficult for the decision maker to assign exact values for the membership degrees of certain elements to a given set, but an interval number may be assigned. In this case, all criteria values in $\Re$ take the form of IVHFEs, i.e.,

$$x_{ij} = \tilde{h}_{ij} = \tilde{H}\{[\tilde{\gamma}^{1L}, \tilde{\gamma}^{1U}], [\tilde{\gamma}^{2L}, \tilde{\gamma}^{2U}], \ldots, [\tilde{\gamma}^{\#\tilde{h}_{ij}L}, \tilde{\gamma}^{\#\tilde{h}_{ij}U}]\}_{ij}$$

The criterion value x_{ij} indicates that the degree to which the alternative A_i satisfies the criterion C_j is several possible intervals belonging to the interval [0, 1].

(3) The criteria values in the decision matrix $\Re$ are denoted by two different formats: HFEs and IVHFEs. In this case, the set of criteria $\boldsymbol{C}$ can be divided into two subsets $\boldsymbol{C}^{HF}$ and $\boldsymbol{C}^{IVHF}$, representing the criteria whose values are in the formats of HFEs and IVHFEs, respectively. Let $\boldsymbol{C}^{HF} = \{C_1, C_2, \ldots, C_{j_1}\}$, $\boldsymbol{C}^{IVHF} = \{C_{j_1+1}, C_{j_1+2}, \ldots, C_n\}$, then $\boldsymbol{C}^{HF} \cup \boldsymbol{C}^{IVHF} = \boldsymbol{C}$ and $\boldsymbol{C}^{HF} \cap \boldsymbol{C}^{IVHF} = \varnothing$. If $C_j \in \boldsymbol{C}^{HF}$, then x_{ij} is a HFE denoted by $x_{ij} = h_{ij} = H\{\gamma^1, \gamma^2, \ldots, \gamma^{\#h_{ij}}\}_{ij}$; while if $C_j \in \boldsymbol{C}^{IVHF}$, then x_{ij} is an IVHFE denoted by $\tilde{H}\{[\tilde{\gamma}^{1L}, \tilde{\gamma}^{1U}], [\tilde{\gamma}^{2L}, \tilde{\gamma}^{2U}], \ldots, [\tilde{\gamma}^{\#\tilde{h}_{ij}L}, \tilde{\gamma}^{\#\tilde{h}_{ij}U}]\}_{ij}$.

It is noticed that Scenarios (1) and (2) are the special cases of Scenario (3), and in this chapter we only take Scenario (3) into account. In practical MCDM problems, there usually exist benefit criteria (the larger the better) and cost criteria (the smaller the better). Meanwhile, the dimensions and measurements of criteria values are often different since the natures of these criteria are different. In order to ensure the compatibility between criteria values, the criteria values must be converted into a compatible scale (or dimensionless indices). Consequently, to eliminate the effect of different physical dimensions and measurements in the final decision results, in this chapter the criteria values of the cost type are transformed into the criteria values of the benefit type by normalizing the hesitant fuzzy decision matrix $\Re = (x_{ij})_{m \times n}$ to yield a corresponding normalized hesitant fuzzy decision matrix $\bar{\Re} = (\bar{x}_{ij})_{m \times n}$, using the following Eqs. (2.11) and (2.12):

(a) If $C_j \in \boldsymbol{C}^{HF}$, then the criterion value x_{ij} is a HFE h_{ij}. It can be normalized as follows (Zhu et al. 2012):

$$\bar{x}_{ij} = \begin{cases} h_{ij}, & \text{if } C_j \in \left(\boldsymbol{C}_I \cap \boldsymbol{C}^{HF}\right) \\ \left(h_{ij}\right)^c, & \text{if } C_j \in \left(\boldsymbol{C}_{II} \cap \boldsymbol{C}^{HF}\right) \end{cases} \tag{2.11}$$

(b) If $C_j \in \boldsymbol{C}^{IVHF}$, then the criterion value x_{ij} is an IVHFE $\tilde{h}_{ij}$. We can normalize it as follows (Zhang and Xu 2014a):

$$\bar{x}_{ij} = \begin{cases} \tilde{h}_{ij}, & \text{if } C_j \in \left(\boldsymbol{C}_I \cap \boldsymbol{C}^{IVHF}\right) \\ \left(\tilde{h}_{ij}\right)^c, & \text{if } C_j \in \left(\boldsymbol{C}_{II} \cap \boldsymbol{C}^{IVHF}\right) \end{cases} \tag{2.12}$$

To solve the aforementioned MCDM problem in case of considering the decision maker's psychological behavior, Zhang and Xu (2014a) developed a novel ranking functions-based hesitant fuzzy TODIM method. We call it the HF-TODIM method for convenience. Similar to the steps of the classical TODIM approach, in the HF-TODIM method, we first need to normalize the original decision matrix by using Eqs. (2.11) and (2.12). Then, we proceed to measure the prospect value of each alternative over the others by constructing the prospect value function based on prospect theory (Kahneman and Tversky 1979).

For this, we first identify a reference criterion and calculate the relative weight of each criterion to the reference criterion. Usually, the criterion with the highest weight can be regarded as the reference criterion and then the relative weight w_{jR} of the criterion C_j to the reference criterion C_R can be obtained by Eq. (2.1). Afterwards, based on the novel ranking functions $Hr_\delta(h)$ and $Ir_\delta(\tilde{h})$, we can compare with the magnitudes of the ratings of alternatives regarding each criterion which are represented by the HFE or IVHFE. Furthermore, analogous to Eq. (2.2), we can calculate the gain and loss of the alternative A_ξ over the alternative A_ζ concerning the criterion C_j.

(1) If $C_j \in \boldsymbol{C}^{HF}$, the criteria values x_{ij} $(i \in \{1,2,\ldots,m\}, j \in \{1,2,\ldots,j_1\})$ are expressed by HFEs, i.e., $x_{ij} = h_{ij}$, then the gain and loss of the alternative A_ξ over the alternative A_ζ concerning the criterion C_j can be obtained by the following expression (Zhang and Xu 2014a):

$$\mathscr{D}_j^h(A_\xi, A_\zeta) = \begin{cases} \sqrt{\frac{w_{jR} d(h_{\xi j}, h_{\zeta j})}{\sum_{j=1}^n w_{jR}}}, & \text{if } Hr_\delta(h_{\xi j}) - Hr_\delta(h_{\zeta j}) > 0 \\ 0, & \text{if } Hr_\delta(h_{\xi j}) - Hr_\delta(h_{\zeta j}) = 0 \\ -\frac{1}{\theta}\sqrt{\frac{\left(\sum_{j=1}^n w_{jR}\right) d(h_{\xi j}, h_{\zeta j})}{w_{jR}}}, & \text{if } Hr_\delta(h_{\xi j}) - Hr_\delta(h_{\zeta j}) < 0 \end{cases} \tag{2.13}$$

where the parameter θ represents the attenuation factor of the losses, and the $d(\tilde{h}_{\xi j}, \tilde{h}_{\zeta j})$ denotes the distance between the HFEs $h_{\xi j}$ and $h_{\zeta j}$ using Eq. (1.2).

(2) If $C_j \in \boldsymbol{C}^{IVHF}$, the criteria values $x_{ij}(i \in \{1,2,\ldots,m\}, j \in \{j_1+1, j_1+2,\ldots,n\})$ are expressed by IVHFEs, i.e., $x_{ij} = \tilde{h}_{ij}$, then the gain and loss of the alternative A_ξ over the alternative A_ζ concerning the criterion C_j is obtained by the following expression (Zhang and Xu 2014a):

$$\mathscr{D}_j^h(A_\xi, A_\zeta) = \begin{cases} \sqrt{\frac{w_{jR} d(\tilde{h}_{\xi j}, \tilde{h}_{\zeta j})}{\sum_{j=1}^n w_{jR}}}, & if\ Ir_\delta(\tilde{h}_{\xi j}) - Ir_\delta(\tilde{h}_{\zeta j}) > 0 \\ 0, & if\ Ir_\delta(\tilde{h}_{\xi j}) - Ir_\delta(\tilde{h}_{\zeta j}) = 0 \\ -\frac{1}{\theta}\sqrt{\frac{\left(\sum_{j=1}^n w_{jR}\right) d(\tilde{h}_{\xi j}, \tilde{h}_{\zeta j})}{w_{jR}}}, & if\ Ir_\delta(\tilde{h}_{\xi j}) - Ir_\delta(\tilde{h}_{\zeta j}) < 0 \end{cases} \tag{2.14}$$

where the $d(\tilde{h}_{\xi j}, \tilde{h}_{\zeta j})$ denotes the distance between the IVHFEs $\tilde{h}_{\xi j}$ and $\tilde{h}_{\zeta j}$ using Eq. (1.29).

It is easily observed from Eq. (2.13) that there exist three cases:

(1) If $Hr_\delta(h_{\xi j}) - Hr_\delta(h_{\zeta j}) > 0$, then $\mathscr{D}_j^h(A_\xi, A_\zeta)$ represents a gain;
(2) If $Hr_\delta(h_{\xi j}) - Hr_\delta(h_{\zeta j}) = 0$, then $\mathscr{D}_j^h(A_\xi, A_\zeta)$ represents a nil;
(3) If $Hr_\delta(h_{\xi j}) - Hr_\delta(h_{\zeta j}) < 0$, then $\mathscr{D}_j^h(A_\xi, A_\zeta)$ represents a loss.

Likewise, there exist three cases in Eq. (2.14):

(1) If $Ir_\delta(\tilde{h}_{\xi j}) - Ir_\delta(\tilde{h}_{\zeta j}) > 0$, then $\mathscr{D}_j^h(A_\xi, A_\zeta)$ represents a gain;
(2) If $Ir_\delta(\tilde{h}_{\xi j}) - Ir_\delta(\tilde{h}_{\zeta j}) = 0$, then $\mathscr{D}_j^h(A_\xi, A_\zeta)$ represents a nil;
(3) If $Ir_\delta(\tilde{h}_{\xi j}) - Ir_\delta(\tilde{h}_{\zeta j}) < 0$, then $\mathscr{D}_j^h(A_\xi, A_\zeta)$ represents a loss.

By aggregating $\mathscr{D}_j^h(A_\xi, A_\zeta)$ with each criterion C_j, the prospect value of the alternative A_ξ over the alternative A_ζ can be obtained as follows (Zhang and Xu 2014a):

$$\mathscr{H}^h(A_\xi, A_\zeta) = \sum_{j=1}^n \mathscr{D}_j^h(A_\xi, A_\zeta), \quad \forall(\xi, \zeta) \tag{2.15}$$

At length, we calculate the overall prospect value of the alternatives A_ξ $(\xi \in \{1,2,\ldots,m\})$ according to the following expression (Zhang and Xu 2014a):

$$\mathscr{Q}^h(A_\xi) = \frac{\sum_{\zeta=1}^m \mathscr{H}^h(A_\xi, A_\zeta) - \min_\xi\left\{\sum_{\zeta=1}^m \mathscr{H}^h(A_\xi, A_\zeta)\right\}}{\max_\xi\left\{\sum_{\zeta=1}^m \mathscr{H}^h(A_\xi, A_\zeta)\right\} - \min_\xi\left\{\sum_{\zeta=1}^m \mathscr{H}^h(A_\xi, A_\zeta)\right\}} \tag{2.16}$$

Obviously, $0 \le \mathcal{Q}^h(A_\xi) \le 1$, and the greater the $\mathcal{Q}^h(A_\xi)$ is, the better the alternative A_ξ will be. Therefore, we can determine the ranking of all alternatives according to the increasing orders of the overall prospect values of the alternatives A_ξ ($\xi \in \{1, 2, \ldots, m\}$), and select the desirable alternative(s) from the alternative set $\boldsymbol{A} = \{A_1, A_2, \ldots, A_m\}$.

Based on the above models and analysis, an algorithm (Algorithm 2.2) for the HF-TODIM approach is presented as follows (Zhang and Xu 2014a):

Step 1. Identify the original decision matrix $\Re = (x_{ij})_{m \times n}$, and obtain the normalized decision matrix $\bar{\Re} = (\bar{x}_{ij})_{m \times n}$ by Eqs. (2.11) and (2.12).

Step 2. Determine the reference criterion C_R, and calculate the relative weight w_{jR} of the criterion C_j to the reference criterion C_R using Eq. (2.2).

Step 3. Calculate the gain and loss of the alternative A_ξ over the alternative A_ζ concerning each criterion C_j using Eqs. (2.13) and (2.14), respectively.

Step 4. Calculate the prospect value of the alternative A_ξ over the alternative A_ζ using Eq. (2.15).

Step 5. Calculate the overall prospect values of the alternatives A_ξ ($\xi \in \{1, 2, \ldots, m\}$) using Eq. (2.16).

Step 6. The ranking of all alternatives is generated according to the increasing orders of the overall prospect values of alternatives and the desirable alternative from $\boldsymbol{A} = \{A_1, A_2, \ldots, A_m\}$ is determined.

Remark 2.2 It is worth pointing out that in the HF-TODIM model, the decision data take the forms of HFEs and IVHFEs. Whereas, in the classical TODIM model (Gomes and Lima 1991, 1992), the F-TODIM model (Krohling and de Souza 2012), the H-TODIM model (Fan et al. 2013) and the IF-RTODIM model (Lourenzutti and Krohling 2013), the decision data are represented by crisp numbers, fuzzy numbers, hybrid types (crisp numbers, intervals numbers and fuzzy numbers) and IFNs, respectively. Apparently, these aforementioned models cannot deal directly with the decision making problems where the decision information takes the forms of HFEs and IVHFEs. On the other hand, the existing hesitant fuzzy MCDM methods (Farhadinia 2013; Liao and Xu 2013; Xu and Zhang 2013; Zhang 2013) are based on the strict assumption regarding the complete rationality of the decision maker and thus fail to capture the decision maker's psychological behavior, while the HF-TODIM approach which is based on prospect theory can fully consider the decision maker's behavior in the hesitant fuzzy MCDM process.

2.4 Illustration Example Based on the Evaluation Problem of Service Quality

To demonstrate the applicability and the implementation process of the HF-TODIM approach, Zhang and Xu (2014a) presented a practical decision making problem that concerns the evaluation and ranking of the service quality among domestic airlines [adapted from Liou et al. (2011), Liao and Xu (2013)].

2.4.1 Description

Due to the development of high-speed railroad, the domestic airline marketing has faced a stronger challenge in Taiwan. More and more airlines have attempted to attract customers by reducing price. Unfortunately, they soon found that there was a no-win situation and only service quality is the critical and fundamental element to survive in this highly competitive domestic market. In order to improve the service quality of domestic airline, the civil aviation administration of Taiwan (CAAT) wants to know which airline is the best in Taiwan and then calls for the others to learn from it. Thus, the CAAT constructs a committee to investigate the four major domestic airlines, which are UNI Air (A_1), Transasia (A_2), Mandarin (A_3), and Daily Air (A_4), and four major criteria are given based on the research of Liou et al. (2011) to evaluate these four domestic airlines. These four main criteria are: Booking and ticketing service (C_1), Check-in and boarding process (C_2), Cabin service (C_3), and Responsiveness (C_4); a detailed description of the four criteria are given in Table 2.3 (Liou et al. 2011).

After the survey about passengers' importance degrees and perceptions for the service criteria done by Liou et al. (2011), they found that the cabin service is

Table 2.3 Criteria for evaluating domestic airlines

Criteria	Description of criteria
Booking and ticketing service C_1	Booking and ticketing service involves conveniences of booking or buying ticket, promptness of booking or buying ticket, courtesy of booking or buying ticket
Check-in and boarding process C_2	Check-in and boarding process consists of convenience check-in, efficient check-in, courtesy of employee, clarity of announcement and so on
Cabin service C_3	Cabin service can be divided into cabin safety demonstration, variety of newspapers and magazines, courtesy of flight attendants, flight attendant willing to help, clean and comfortable interior, in-flight facilities, and captain's announcement
Responsiveness C_4	Responsiveness consists of fair waiting-list call, handing of delayed flight, complaint handing, and missing baggage handling

Table 2.4 Hesitant fuzzy decision matrix

	C_1	C_2
A_1	$\tilde{H}$ {[0.4, 0.55], [0.7, 0.9]}	H {0.6, 0.8}
A_2	$\tilde{H}$ {[0.5, 0.6], [0.8, 0.9]}	H {0.5, 0.8, 0.9}
A_3	$\tilde{H}$ {[0.3, 0.4], [0.5, 0.65], [0.8,0.9]}	H {0.6, 0.7, 0.95}
A_4	$\tilde{H}$ {[0.4, 0.5], [0.6, 0.7]}	H {0.7, 0.9}
	C_3	C_4
A_1	$\tilde{H}$ {[0.35, 0.4], [0.5, 0.65], [0.8, 0.95]}	H {0.4, 0.5, 0.9}
A_2	$\tilde{H}$ {[0.4, 0.5], [0.65, 0.9]}	H {0.4, 0.6, 0.7}
A_3	$\tilde{H}$ {[0.3,0.45], [0.6,0.8]}	H {0.5, 0.8}
A_4	$\tilde{H}$ {[0.2, 0.4], [0.6, 0.7]}	H {0.4, 0.5}

considered the most important factor of service quality, which can be interpreted easily because the cabin service occupies more of a passenger's travelling time than other aspects. Meanwhile, the booking and ticketing service is less important due to the fact that the work is mainly done by computers. Therefore, the weight vector of the criteria is $\boldsymbol{w} = (0.15, 0.25, 0.35, 0.25)^T$, which is consistent with the result of the survey done by Liou et al. (2011). Suppose that the committee gives the criteria values by using HFEs and IVHFEs, and then the hesitant fuzzy decision matrix is presented in Table 2.4 (Zhang and Xu 2014a).

2.4.2 Decision Making Model

In the following, we use the HF-TODIM decision model to solve the decision making problem mentioned in Sect. 2.4.1. The solution process and the computation results are summarized as follows (Zhang and Xu 2014a):

Firstly, we take the criterion C_3 as the reference criterion because the cabin service (C_3) is considered the most important factor of service quality. Thus, the weight of the reference criterion is $w_R = 0.35$. Meanwhile, we take $\theta = 1$, which means that the losses will contribute with their real value to the global value (Gomes and Rangel 2009).

Secondly, based on the novel ranking functions $Hr_\delta(h)$ and $Ir_\delta(\tilde{h})$, we compare with the magnitudes of the ratings of alternatives regarding each criterion which are represented by HFEs or IVHFEs. Here we take $\delta = 0.1$. As a result, the superior-inferior table can be obtained and is listed in Table 2.5 (Zhang and Xu 2014a). The top-left cell ${}_{1/2}\mathrm{I}_1$ in Table 2.5 means that for the booking and ticketing service C_1, UNI Air A_1 is inferior to Transasia A_2, because the criterion value of A_1 for C_1 is equal to $\tilde{h}_{11} = \tilde{H}$ {[0.4, 0.55], [0.7, 0.9]}, and the criterion value of A_2 for C_1 is equal to $\tilde{h}_{21} = \tilde{H}$ {[0.5, 0.6], [0.8, 0.9]}, according to the new ranking method

Table 2.5 Superior-inferior table for six pairs of alternatives and four criteria

	A_1/A_2	A_1/A_3	A_1/A_4	A_2/A_3	A_2/A_4	A_3/A_4
C_1	$_{1/2}I_1$	$_{1/3}S_1$	$_{1/4}S_1$	$_{2/3}S_1$	$_{2/4}S_1$	$_{3/4}S_1$
C_2	$_{1/2}I_2$	$_{1/3}I_2$	$_{1/4}I_2$	$_{2/3}I_2$	$_{2/4}I_2$	$_{3/4}S_2$
C_3	$_{1/2}S_3$	$_{1/3}S_3$	$_{1/4}S_3$	$_{2/3}S_3$	$_{2/4}S_3$	$_{3/4}S_3$
C_4	$_{1/2}S_4$	$_{1/3}I_4$	$_{1/4}S_4$	$_{2/3}I_4$	$_{2/4}S_4$	$_{3/4}S_4$

Note: *'S' denotes "superior to", 'I' denotes "inferior to", "E" denotes "equal to"*

Table 2.6 Hesitant fuzzy heterogeneous decision matrix

	C_1	C_2
A_1	$\tilde{H}$ {[0.4, 0.55], [0.4, 0.55], [0.7, 0.9]}	$H\{0.6, 0.6, 0.8\}$
A_2	$\tilde{H}$ {[0.5, 0.6], [0.5, 0.6], [0.8, 0.9]}	$H\{0.5, 0.8, 0.9\}$
A_3	$\tilde{H}$ {[0.3, 0.4], [0.5, 0.65], [0.8, 0.9]}	$H\{0.6, 0.7, 0.95\}$
A_4	$\tilde{H}$ {[0.4, 0.5], [0.4, 0.5], [0.6, 0.7]}	$H\{0.7, 0.7, 0.9\}$
	C_3	C_4
A_1	$\tilde{H}$ {[0.35, 0.4], [0.5, 0.65], [0.8, 0.95]}	$H\{0.4, 0.5, 0.9\}$
A_2	$\tilde{H}$ {[0.4, 0.5], [0.4, 0.5], [0.65, 0.9]}	$H\{0.4, 0.6, 0.7\}$
A_3	$\tilde{H}$ {[0.3, 0.45], [0.3, 0.45], [0.6, 0.8]}	$H\{0.5, 0.5, 0.8\}$
A_4	$\tilde{H}$ {[0.2, 0.4], [0.2, 0.4], [0.6, 0.7]}	$H\{0.4, 0.4, 0.5\}$

of IVHFEs, it is easy to obtain $Ir_{0.1}(\tilde{h}_{11}) = 1.2369 < Ir_{0.1}(\tilde{h}_{21}) = 1.3706$. Similar logic is used to determine the remaining entries in Table 2.5.

Considering that the numbers of values in different HFEs and IVHFEs in Table 2.4 are different, in order to accurately calculate their distances, we should extend the shorter one until both of them have the same length. According to the regulations mentioned in Definitions 1.3 and 1.10, we assume that the experts are pessimistic in the above decision making problem, and normalize the decision data in Table 2.4 by adding the minimal values as listed in Table 2.6 (Zhang and Xu 2014a).

Then, based on the analysis results in Table 2.5, using Eqs. (2.13) and (2.14) we can calculate the gains and losses of each alternative over the others concerning each criterion, listed in Tables 2.7, 2.8, 2.9 and 2.10 (Zhang and Xu 2014a).

Furthermore, by aggregating the gains and losses of the alternative A_ξ over the alternative A_ζ concerning each criterion C_j using Eq. (2.15), we can obtain the prospect value of each alternative over the others, which are listed in Table 2.11 (Zhang and Xu 2014a).

Thirdly, based on the data in Table 2.11, the overall prospect values of the alternatives are obtained by Eq. (2.16), namely, $\mathcal{Q}^h(A_1) = 0.9390$, $\mathcal{Q}^h(A_2) = 0.9233$, $\mathcal{Q}^h(A_3) = 1.0$, and $\mathcal{Q}^h(A_4) = 0.0$. Finally, according to the overall values, the ranking of the four domestic airlines is determined, i.e., $A_3 \succ A_1 \succ A_2 \succ A_4$. Apparently, A_3 (Mandarin) is the most desirable domestic airline.

Table 2.7 Gains and losses of each alternative over the others regarding the criterion C_1

	A_1	A_2	A_3	A_4
A_1	0	−0.7136	0.1237	0.1198
A_2	0.1070	0	0.1326	0.1456
A_3	−0.8249	−0.8842	0	0.1493
A_4	−0.7989	−0.9710	−0.9953	0

Table 2.8 Gains and losses of each alternative over the others regarding the criterion C_2

	A_1	A_2	A_3	A_4
A_1	0	−0.7521	−0.6452	−0.6325
A_2	0.1881	0	−0.5886	−0.7186
A_3	0.1613	0.1471	0	0.1270
A_4	0.1581	0.1797	−0.5081	0

Table 2.9 Gains and losses of each alternative over the others regarding the criterion C_3

	A_1	A_2	A_3	A_4
A_1	0	0.1944	0.2343	0.2744
A_2	−0.5555	0	0.1663	0.2322
A_3	−0.6693	−0.4753	0	0.1635
A_4	−0.7839	−0.6636	−0.4761	0

Table 2.10 Gains and losses of each alternative over the others regarding the criterion C_4

	A_1	A_2	A_3	A_4
A_1	0	0.1797	−0.5715	0.2440
A_2	−0.7186	0	−0.6325	0.2021
A_3	0.1429	0.1581	0	0.2188
A_4	−0.9758	−0.8082	−0.8752	0

Table 2.11 The prospect value of each alternative over the others

	A_1	A_2	A_3	A_4
A_1	0	−1.0916	−0.8587	0.0057
A_2	−0.9791	0	−0.9222	−0.1387
A_3	−1.1900	−1.0543	0	0.6586
A_4	−2.4005	−2.2631	−2.8457	0

2.4.3 Sensitivity Analysis of Parameters

The sensitivity analysis is usually performed by modifying the parameter θ (i.e., the attenuation factor of the losses). By increasing or decreasing the values of the parameter θ, we recalculate the ranking orders of alternatives.

By changing θ from 1 to 4, we can obtain the change results of ranking orders of alternatives, listed in Table 2.12 (Zhang and Xu 2014a).

From the sensitivity analysis results presented in Table 2.12, we notice that the ranking orders of alternatives are not sensitive to the value of θ. In other words, in

Table 2.12 Ranking orders of alternatives with different values of θ

Different values of θ	Ranking orders of alternatives
$\theta = 1$	$A_3 \succ A_1 \succ A_2 \succ A_4$
$\theta = 2$	$A_3 \succ A_1 \succ A_2 \succ A_4$
$\theta = 3$	$A_3 \succ A_1 \succ A_2 \succ A_4$
$\theta = 4$	$A_3 \succ A_1 \succ A_2 \succ A_4$

Table 2.13 Ranking orders of alternatives with different values of δ

Different values of δ	Ranking orders of alternatives
$\delta = 0.001$	$A_3 \succ A_1 \succ A_2 \succ A_4$
$\delta = 0.01$	$A_3 \succ A_1 \succ A_2 \succ A_4$
$\delta = 0.1$	$A_3 \succ A_1 \succ A_2 \succ A_4$

spite of the alteration in the value of the parameter θ, the obtained rankings are usually consistent.

In addition, it is worth pointing out that the above sensitivity analysis is based on $\delta = 0.1$. In the following, we do the sensitivity analysis about the parameter δ. We assume $\theta = 1$, the sensitivity analysis is performed by modifying (increasing or decreasing) the value of δ, and recalculating the ranking orders of alternatives with different values of δ. By changing δ from 0.001 to 0.1, we can obtain the change results of the ranking orders of alternatives. These calculation results are listed in Table 2.13 (Zhang and Xu 2014a).

Based on the results of sensitivity analysis presented in Table 2.13, it is easy to see that the rankings of alternatives are not sensitive to the value of the parameter δ. That is to say, in spite of the alteration in the values of δ, the obtained rankings are often consistent.

2.4.4 Comparative Analysis and Discussions

A comparative study was conducted by Zhang and Xu (2014a) to validate the results of the HF-TODIM method with those from another approach. With the analysis on the same decision making problem mentioned in Sect. 2.4.1, the HF-TOPSIS approach proposed by Xu and Zhang (2013) was selected to facilitate the comparative analysis. However, it is worth pointing out that the HF-TOPSIS approach is just suitable to deal with the MCDM problems with HFEs but fail to handle the above problem where the criteria values are measured in two different formats, i.e., HFEs and IVHFEs. Therefore, Zhang and Xu (2014a) presented a modified HF-TOPSIS approach to tackle appropriately the MCDM problems in which the criteria values take the form of HFEs or IVHFEs and applied it to the decision making problem mentioned in Sect. 2.4.1.

The modified HF-TOPSIS method starts with the determination of the hesitant fuzzy heterogeneous PIS A^+ and the hesitant fuzzy heterogeneous NIS A^-, which are defined as follows:

(1) if $C_j \in \mathbf{C}^{HF}$, then $x_{ij} = h_{ij}$ $(i \in \{1, 2, \ldots, m\}, j \in \{1, 2, \ldots, j_1\})$. Thus, we have

$$A^{HF+} = \left(H\{\max_{i=1}^m \gamma_{ij}^1, \max_{i=1}^m \gamma_{ij}^2, \ldots, \max_{i=1}^m \gamma_{ij}^{\#h_{ij}}\} | C_j \in \mathbf{C}^{HF}\right) \quad (2.17)$$

$$A^{HF-} = \left(H\{\min_{i=1}^m \gamma_{ij}^1, \min_{i=1}^m \gamma_{ij}^2, \ldots, \min_{i=1}^m \gamma_{ij}^{\#h_{ij}}\} | C_j \in \mathbf{C}^{HF}\right) \quad (2.18)$$

(2) if $C_j \in \mathbf{C}^{IVHF}$, then $x_{ij} = \tilde{h}_{ij}$ $(i \in \{1, 2, \ldots, m\}, j \in \{j_1 + 1, j_1 + 2, \ldots, n\})$, and we have

$$A^{IVHF+} = \left(\tilde{H}\{\max_{i=1}^m \tilde{\gamma}_{ij}^1, \max_{i=1}^m \tilde{\gamma}_{ij}^2, \ldots, \max_{i=1}^m \tilde{\gamma}_{ij}^{\#\tilde{h}_{ij}}\} | C_j \in \mathbf{C}^{IVHF}\right) \quad (2.19)$$

$$A^{IVHF-} = \left(\tilde{H}\{\min_{i=1}^m \tilde{\gamma}_{ij}^1, \min_{i=1}^m \tilde{\gamma}_{ij}^2, \ldots, \min_{i=1}^m \tilde{\gamma}_{ij}^{\#\tilde{h}_{ij}}\} | C_j \in \mathbf{C}^{IVHF}\right) \quad (2.20)$$

Obviously, $A^+ = A^{HF+} \cup A^{IVHF+}$ and $A^- = A^{HF-} \cup A^{IVHF-}$.

Thus, the corresponding hesitant fuzzy heterogeneous PIS and NIS in the above decision making problem can be obtained by Eqs. (2.17)–(2.20) and are listed in Table 2.14 (Zhang and Xu 2014a).

In order to measure the distance between the alternative A_i and the hesitant fuzzy heterogeneous PIS A^+ as well as the hesitant fuzzy heterogeneous NIS A^-, respectively, we adopt the hesitant fuzzy Euclidean distance proposed by Xu and Xia (2011b) and the interval-valued hesitant fuzzy Euclidean distance proposed by Chen et al. (2013b). Thus, for the criterion C_j, the distances between the alternative A_i and the hesitant fuzzy heterogeneous PIS A^+ as well as the hesitant fuzzy heterogeneous NIS A^-, respectively, are calculated by the following equations (Zhang and Xu 2014a):

Table 2.14 The hesitant fuzzy heterogeneous PIS and NIS

	A^+	A^-
C_1	$\tilde{H}$ {[0.5, 0.6], [0.5, 0.65], [0.8, 0.9]}	$\tilde{H}$ {[0.3, 0.4], [0.4, 0.5], [0.6, 0.7]}
C_2	H {0.7, 0.8, 0.95}	H {0.5, 0.6, 0.8}
C_3	$\tilde{H}$ {[0.4, 0.5], [0.5, 0.65], [0.8, 0.95]}	$\tilde{H}$ {[0.2, 0.4], [0.2, 0.4], [0.6, 0.7]}
C_4	H {0.5, 0.6, 0.9}	H {0.4, 0.4, 0.5}

(1) if $C_j \in \boldsymbol{C}^{HF}$, then we obtain

$$d_{ij}^{+} = d\left(h_{ij}, h_j^{+}\right) = \sqrt{\sum\nolimits_{\lambda=1}^{\#h_{ij}} (\gamma_{ij}^{\lambda} - (\lambda_j^{\lambda})^{+})^2 \Big/ \#h_{ij}} \tag{2.21}$$

and

$$d_{ij}^{-} = d\left(h_{ij}, h_j^{-}\right) = \sqrt{\sum\nolimits_{\lambda=1}^{\#h_{ij}} (\gamma_{ij}^{\lambda} - (\lambda_j^{\lambda})^{-})^2 \Big/ \#h_{ij}} \tag{2.22}$$

(2) if $C_j \in \boldsymbol{C}^{IVHF}$, then we obtain

$$d_{ij}^{+} = d\left(\tilde{h}_{ij}, \tilde{h}_j^{+}\right) = \sqrt{\sum\nolimits_{\lambda=1}^{\#\tilde{h}_{ij}} \left(\begin{array}{c} (\tilde{\gamma}_{ij}^{\lambda L} - (\tilde{\gamma}_j^{\lambda L})^{+})^2 \\ + (\tilde{\gamma}_{ij}^{\lambda U} - (\tilde{\gamma}_j^{\lambda U})^{+})^2 \end{array} \right) \Big/ (2\#\tilde{h}_{ij})} \tag{2.23}$$

and

$$d_{ij}^{-} = d\left(\tilde{h}_{ij}, \tilde{h}_j^{-}\right) = \sqrt{\sum\nolimits_{\lambda=1}^{\#\tilde{h}_{ij}} \left(\begin{array}{c} (\tilde{\gamma}_{ij}^{\lambda L} - (\tilde{\gamma}_j^{\lambda L})^{-})^2 \\ + (\tilde{\gamma}_{ij}^{\lambda U} - (\tilde{\gamma}_j^{\lambda U})^{-})^2 \end{array} \right) \Big/ (2\#\tilde{h}_{ij})} \tag{2.24}$$

For all criteria, the weighted distances between the alternative A_i $(i \in \{1, 2, \ldots, m\})$ and the hesitant fuzzy heterogeneous PIS A^+ as well as the hesitant fuzzy heterogeneous NIS A^-, respectively, are derived from the following equations:

$$\begin{aligned} d^{hh}(A_i, A^{+}) = & \sum_{j=1}^{j_1} w_j \sqrt{\sum\nolimits_{\lambda=1}^{\#h_{ij}} (\gamma_{ij}^{\lambda} - (\lambda_j^{\lambda})^{+})^2 \Big/ \#h_{ij}} \\ & + \sum_{j=j_1+1}^{n} w_j \sqrt{\sum\nolimits_{\lambda=1}^{\#\tilde{h}_{ij}} \left(\begin{array}{c} (\tilde{\gamma}_{ij}^{\lambda L} - (\tilde{\gamma}_j^{\lambda L})^{+})^2 \\ + (\tilde{\gamma}_{ij}^{\lambda U} - (\tilde{\gamma}_j^{\lambda U})^{+})^2 \end{array} \right) \Big/ (2\#\tilde{h}_{ij})} \end{aligned} \tag{2.25}$$

and

$$\begin{aligned} d^{hh}(A_i, A^{-}) = & \sum_{j=1}^{j_1} w_j \sqrt{\sum\nolimits_{\lambda=1}^{\#h_{ij}} (\gamma_{ij}^{\lambda} - (\lambda_j^{\lambda})^{-})^2 \Big/ \#h_{ij}} \\ & + \sum_{j=j_1+1}^{n} w_j \sqrt{\sum\nolimits_{\lambda=1}^{\#\tilde{h}_{ij}} \left(\begin{array}{c} (\tilde{\gamma}_{ij}^{\lambda L} - (\tilde{\gamma}_j^{\lambda L})^{-})^2 \\ + (\tilde{\gamma}_{ij}^{\lambda U} - (\tilde{\gamma}_j^{\lambda U})^{-})^2 \end{array} \right) \Big/ (2\#\tilde{h}_{ij})} \end{aligned} \tag{2.26}$$

Table 2.15 Closeness indices and the ranking of alternatives

	$d^{hh}(A_i, A^+)$	$d^{hh}(A_i, A^-)$	$CI^{hh}(A_i)$	Ranking
$\mathbf{A_1}$	**0.0953**	**0.1235**	**0.5645**	**1**
A_2	0.0978	0.1179	0.5467	2
A_3	0.1439	0.0975	0.4039	4
A_4	0.1392	0.1436	0.5076	3

The relative closeness index of the alternative A_i to the hesitant fuzzy heterogeneous PIS A^+ is defined as the following formula:

$$CI^{hh}(A_i) = \frac{d^{hh}(A_i, A^-)}{d^{hh}(A_i, A^+) + d^{hh}(A_i, A^-)} \tag{2.27}$$

Using Eqs. (2.25)–(2.27), the distances $d^{hh}(A_i, A^+)$ and $d^{hh}(A_i, A^-)$, and the relative closeness index $CI^{hh}(A_i)$ in the above problem can be obtained, respectively. The results are presented in Table 2.15 (Zhang and Xu 2014a), together with the corresponding rankings on the basis of $CI^{hh}(A_i)$.

It is easy to see that the optimal order for these four major domestic airlines is $A_1 \succ A_2 \succ A_4 \succ A_3$, and thus the UNI Air ($A_1$) is the most desirable domestic airline.

To provide a better view of the comparison results, we put the results of the ranking of alternatives obtained by the HF-TODIM and the modified HF-TOPSIS methods into Fig. 2.1 (Zhang and Xu 2014a).

From Fig. 2.1, it is easily observed that the ranking orders of alternatives obtained by these two techniques are remarkable different. Using the HF-TODIM technique, the best suitable alternative in the above decision making problem is A_3, while using the modified HF-TOPSIS technique the best alternative is A_1. The main reason is that the HF-TODIM technique, which can take into account the decision maker's behavior in the decision making process, can yield more persuasive results which is more in line with the decision maker's actual experience, while the modified HF-TOPSIS technique, which is based on the strict assumption regarding

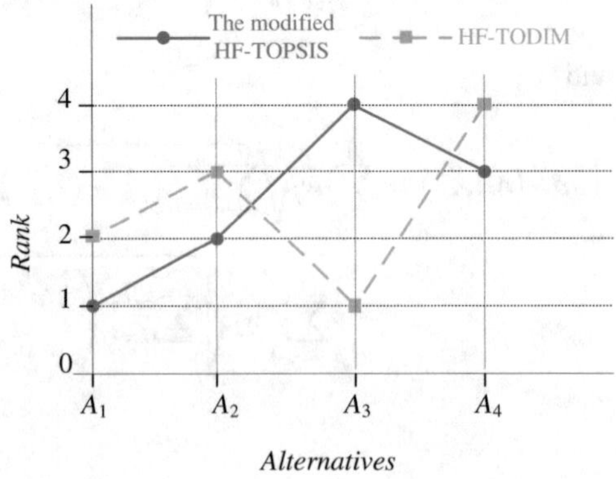

Fig. 2.1 The pictorial representation of the HF-TOPSIS and HF-TODIM rankings

complete rationality of the decision maker, fails to take the decision maker's psychological behavior into account. Obviously, for the practical decision making problems in which the decision maker's psychological behavior should be taken fully into account, compared with the modified HF-TOPSIS approach, the HF-TODIM method can obtain a better final decision result since it effectively captures the decision maker's psychological behavior.

2.5 Extension of the Developed Approach for Handling the MCGDM Problems with HTrFNs

2.5.1 The Concept of HTrFN

In this section, the concept of TrFN is first reviewed. Based on the concepts and operational laws of TrFNs and HFEs, Zhang et al. (2016) presented a new concept of the HTrFN which is good enough to represent the vagueness of the HFLTS.

Definition 2.9 (Zadeh 1975). A fuzzy number $\tilde{\alpha} = Tr(a, b, c, d)$ is said to be a TrFN if its membership function is given as follows:

$$\mu_{\tilde{\alpha}}(t) = \begin{cases} (t-a)/(b-a), & (a \le t < b) \\ 1, & (b \le t \le c) \\ (d-t)/(d-c), & (c < t \le d) \\ 0, & otherwise \end{cases} \tag{2.28}$$

where the closed interval $[b, c]$, a and d are the mode, low and upper limits of $\tilde{\alpha}$, respectively.

Remark 2.3 (Zhang et al. 2016). It is noted that a TrFN $\tilde{\alpha} = Tr(a, b, c, d)$ is reduced to a TFN if $b = c$. A TrFN $\tilde{\alpha} = Tr(a, b, c, d)$ is reduced to a real number if $a = b = c = d$. A TrFN $\tilde{\alpha} = Tr(a, b, c, d)$ is the normalized TrFN if $a \ge 0$ and $d \le 1$. Thus, the TrFN $\tilde{1} = Tr(1, 1, 1, 1)$ is the maximal normalized TrFN which is called the positive ideal TrFN, while the TrFN $\tilde{0} = Tr(0, 0, 0, 0)$ is the minimal normalized TrFN which is called the negative ideal TrFN. Usually, the TrFN is well enough to capture the vagueness of linguistic terms, and the relationships between linguistic term set with seven-point rating scale and the TrFNs are shown in Table 2.16.

Definition 2.10 (Zhang et al. 2016). Let T be a fixed set, a HTrFS $\mathscr{H}$ on T is defined as:

$$\mathscr{H} = \{ < t, h_{\mathscr{H}}(t) > | t \in T \} \tag{2.29}$$

where $h_{\mathscr{H}}(t)$ is a set of different normalized TrFNs, representing the possible membership degrees of the element $t \in T$ to $\mathscr{H}$.

Table 2.16 Linguistic terms and the corresponding TrFNs

Ratings	Abbreviation	TrFNs
l_0: Very poor	VP	Tr(0.0, 0.0, 0.1, 0.2)
l_1: Poor	P	Tr(0.1, 0.2, 0.2, 0.3)
l_2: Medium poor	MP	Tr(0.2, 0.3, 0.4, 0.5)
l_3: Fair	F	Tr(0.4, 0.5, 0.5, 0.6)
l_4: Medium good	MG	Tr(0.5, 0.6, 0.7, 0.8)
l_5: Good	G	Tr(0.7, 0.8, 0.8, 0.9)
l_6: Very good	VG	Tr(0.8, 0.9, 1.0, 1.0)

For convenience, $h_{\mathscr{H}}(t)$ is called a HTrFN denoted by $\hbar = \{\tilde{\alpha}^1, \tilde{\alpha}^2, \ldots, \tilde{\alpha}^{\#\hbar}\}$ where the $\tilde{\alpha}^\lambda = Tr(a^\lambda, b^\lambda, c^\lambda, d^\lambda)(\lambda = 1, 2, \ldots, \#\hbar)$ is a normalized TrFN and $\#\hbar$ is the number of all TrFNs in $\hbar$. If $\#\hbar = 1$, then the HTrFN $\hbar$ is reduced a TrFN. If $b^\lambda = c^\lambda(\lambda = 1, 2, \ldots, \#\hbar)$, the HTrFN $\hbar$ is reduced to a hesitant triangular fuzzy number (Zhao et al. 2014).

Example 2.5 (Zhang and Liu 2016). Let $T = \{t_1, t_2, t_3\}$, and let

$$\begin{aligned}
\hbar_{\mathscr{H}}(t_1) &= \{Tr(0.2, 0.3, 0.4, 0.5), Tr(0.3, 0.4, 0.5, 0.6), Tr(0.35, 0.4, 0.45, 0.5)\},\\
\hbar_{\mathscr{H}}(t_2) &= \{Tr(0.1, 0.2, 0.3, 0.5), Tr(0.3, 0.4, 0.4, 0.6)\},\\
\hbar_{\mathscr{H}}(t_3) &= \{Tr(0.2, 0.4, 0.5, 0.6), Tr(0.1, 0.3, 0.4, 0.6)\}
\end{aligned}$$

be three HTrFNs of t_i $(i = 1, 2, 3)$ to a set $h_{\mathscr{H}}(t)$. Thus, $\mathscr{H}$ can be called an HTrFS which is denoted as:

$$\mathscr{H} = \left\{ \begin{array}{l} \langle t_1, \{Tr(0.2, 0.3, 0.4, 0.5), Tr(0.3, 0.4, 0.5, 0.6), Tr(0.35, 0.4, 0.45, 0.5)\}\rangle, \\ \langle t_2, \{Tr(0.1, 0.2, 0.3, 0.5), Tr(0.3, 0.4, 0.4, 0.6)\}\rangle, \\ \langle t_3, \{Tr(0.2, 0.4, 0.5, 0.6), Tr(0.1, 0.3, 0.4, 0.6)\}\rangle \end{array} \right\}.$$

According to the definition of HTrFNs, it is easy to note that HTrFNs are suitable to capture and represent the uncertainty and vagueness of the HFLTS (please see Chap. 6 for the concept of HFLTS).

Example 2.6 (Zhang et al. 2016). Let $L = \{l_0, l_1, l_2, l_3, l_4, l_5, l_6\}$ be a linguistic term set, the linguistic terms and the corresponding TrFNs are shown in Table 2.16. Given two HFLTSs $hl_1 = HL\{l_0, l_1, l_2\}$ and $hl_2 = HL\{l_1, l_2, l_3\}$, their semantics can be captured by the following two HTrFNs (see Fig. 2.2):

$$\hbar_1 = \left\{ \begin{array}{l} Tr(0.0, 0.0, 0.1, 0.2), Tr(0.1, 0.2, 0.2, 0.3), \\ Tr(0.2, 0.3, 0.4, 0.5) \end{array} \right\} \Leftrightarrow hl_1;$$

$$\hbar_2 = \left\{ \begin{array}{l} Tr(0.1, 0.2, 0.2, 0.3), Tr(0.2, 0.3, 0.4, 0.5), \\ Tr(0.4, 0.5, 0.5, 0.6) \end{array} \right\} \Leftrightarrow hl_2;$$

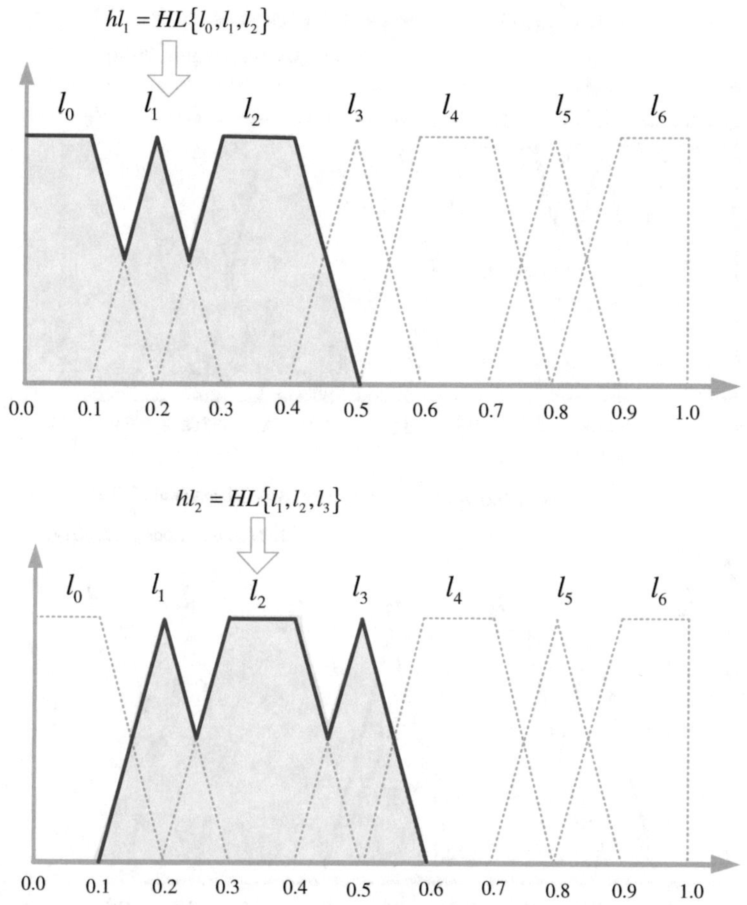

Fig. 2.2 The HFLTSs and the corresponding HTrFNs

As far as we know, Liu and Rodríguez (2014) suggested the use of TrFNs to caputre the vagueness of HFLTSs, and the TrFNs envelopes of hl_1 and hl_2 are obtained as below:

$$env_F(hl_1) = Tr(0, 0, 0.239, 0.5), \quad env_F(hl_2) = Tr(0.1, 0.28, 0.42, 0.6).$$

To provide a better view of the comparison results, we put the results obtained by Zhang et al. (2016), Liu and Rodríguez (2014) in Fig. 2.3.

It can be easily seen from Fig. 2.3 that the HTrFNs take the semantic of each linguistic term of the HFLTSs into account. While in Liu and Rodríguez (2014), the TrFNs were used to represent the semantic of HFLTS, which is relative complex because the TrFNs are obtained by aggregating the fuzzy membership functions of the linguistic terms of the HFLTS using the OWA aggregation operator.

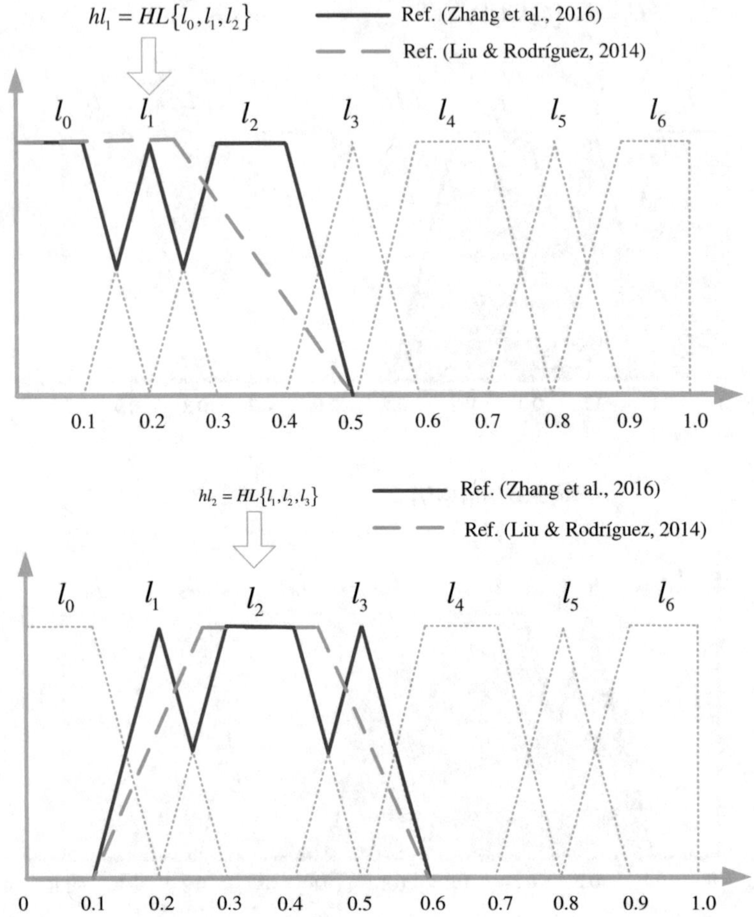

Fig. 2.3 The results obtained by two different methods

Inspired by the operations on HFEs and TrFNs, the basic operations of HTrFNs are introduced as follows:

Definition 2.11 (Zhang et al. 2016). Let $\hbar$, $\hbar_1$ and $\hbar_2$ be three HTrFNs, the basic operations of HTrFNs are defined as:

(1) $\theta\hbar = \cup_{\tilde{\alpha}\in\hbar}\left\{Tr\begin{pmatrix}1-(1-a)^\theta, 1-(1-b)^\theta, \\ 1-(1-c)^\theta, 1-(1-d)^\theta\end{pmatrix}\right\}\ (\theta > 0);$

(2) $\hbar^\theta = \cup_{\tilde{\alpha}\in\hbar}\left\{Tr\left(a^\theta, b^\theta, c^\theta, d^\theta\right)\right\}\ (\theta > 0);$

(3) $\hbar_1 \oplus \hbar_2 = \cup_{\tilde{\alpha}_1\in\hbar_1, \tilde{\alpha}_2\in\hbar_2}\left\{Tr\begin{bmatrix}a_1 + a_2 - a_1 \times a_2,\ b_1 + b_2 - b_1 \times b_2, \\ c_1 + c_2 - c_1 \times c_2,\ d_1 + d_2 - d_1 \times d_2\end{bmatrix}\right\};$

(4) $\hbar_1 \otimes \hbar_2 = \cup_{\tilde{\alpha}_1\in\hbar_1, \tilde{\alpha}_2\in\hbar_2}\{Tr(a_1 \times a_2,\ b_1 \times b_2,\ c_1 \times c_2,\ d_1 \times d_2)\}.$

Proposition 2.4 (Zhang et al. 2016). Let h, h_1 and h_2 be three HTrFNs, then

(1) $h_1 \oplus h_2 = h_2 \oplus h_1$;
(2) $h_1 \otimes h_2 = h_2 \otimes h_1$;
(3) $\theta(h_1 \oplus h_2) = \theta h_1 \oplus \theta h_2 \ (\theta > 0)$;
(4) $(h_1 \otimes h_2)^{\theta} = h_1^{\theta} \otimes h_2^{\theta} \ (\theta > 0)$;
(5) $(\theta_1 + \theta_2)h = \theta_1 h \oplus \theta_2 h \ (\theta_1, \theta_2 > 0)$;
(6) $h^{(\theta_1 + \theta_2)} = h^{\theta_1} \otimes h^{\theta_2} \ (\theta_1, \theta_2 > 0)$.

According to Definition 2.11, it is not hard to obtain the conclusions in Proposition 2.4 (Proof is omitted).

It is worthwhile to point out that the number of TrFNs in different HTrFNs may be different. In such cases, we should extend the shorter one until both of them have the same length when we compare them. To extend the shorter one, the best way is to add some TrFNs in it. Inspired by the similar approaches in the references (Liu and Wang 2007; Xu and Xia 2011a; Xu and Zhang 2013), we extend the shorter one by adding the TrFN in it which mainly depends on the DMs' risk preferences. The optimists anticipate the desirable outcomes and may add the maximum TrFN, while the pessimists expect the unfavorable outcomes and may add the minimum TrFN. Here we employ the sign distance method proposed by Abbasbandy and Asady (2006) to compare the magnitude of TrFNs and further to identify the maximum TrFN or minimum TrFN.

Definition 2.12 (Zhang and Liu 2016). Let $h_j = \{\tilde{\alpha}_j^1, \tilde{\alpha}_j^2, \ldots, \tilde{\alpha}_j^{\#h_j}\}$ $(j = 1, 2)$ be two HTrFNs, and $\#h_1 = \#h_2 = \#h$, then a nature quasi-ordering on HTrFNs is defined as follows.

$$h_1 \le h_2 \text{ if and only if } \tilde{\alpha}_1^{\lambda} \le \tilde{\alpha}_2^{\lambda} \quad (\lambda = 1, 2, \ldots, \#h).$$

It is easily observed from Definition 2.12 that the HTrFN $h^+ = \{Tr(1,1,1,1), \ldots, Tr(1,1,1,1)\}$ is the biggest HTrFN and the HTrFN $h^- = \{Tr(0,0,0,0), \ldots, Tr(0,0,0,0)\}$ is the smallest HTrFN, respectively. We also call h^+ the positive ideal HTrFN and h^- the negative ideal HTrFN, respectively.

Zhang et al. (2016) developed the hesitant trapezoidal Hamming distance for HTrFNs as follows:

Definition 2.13 (Zhang et al. 2016). Given two HTrFNs $h_j = \{\tilde{\alpha}_j^1, \tilde{\alpha}_j^2, \ldots, \tilde{\alpha}_j^{\#h_j}\}$ $(j = 1, 2)$ with $\#h_1 = \#h_2 = \#h$, the hesitant trapezoidal Hamming distance between them is defined as follows:

$$d(h_1, h_2) = \frac{1}{6\#h} \sum_{\lambda=1}^{\#h} \left(\begin{array}{c} |a_1^{\lambda} - a_2^{\lambda}| + 2|b_1^{\lambda} - b_2^{\lambda}| \\ + 2|c_1^{\lambda} - c_2^{\lambda}| + |d_1^{\lambda} - d_2^{\lambda}| \end{array} \right) \tag{2.30}$$

Example 2.7 (Zhang and Liu 2016). For two HTrFNs

$$\begin{aligned} \hbar_1 &= \{Tr(0.1,0.2,0.3,0.5), Tr(0.3,0.4,0.4,0.6)\}, \\ \hbar_2 &= \{Tr(0.1,0.3,0.4,0.6), Tr(0.2,0.4,0.5,0.6)\}, \end{aligned}$$

the following result based on Definition 2.13 is obtained:

$$\begin{aligned} d(\hbar_1, \hbar_2) &= \frac{1}{6 \times 2}\left(\begin{array}{l} |0.1-0.1| + 2|0.2-0.3| + 2|0.3-0.4| + |0.5-0.6| + \\ |0.3-0.2| + 2|0.4-0.4| + 2|0.4-0.5| + |0.6-0.6| \end{array}\right) \\ &= 0.0667. \end{aligned}$$

Then, the distance between the HTrFN $\hbar = \{\tilde{\alpha}^1, \tilde{\alpha}^2, \ldots, \tilde{\alpha}^{\#\hbar}\}$ and the positive ideal HTrFN $\hbar^+$ can be calculated as follows:

$$\begin{aligned} d(\hbar, \hbar^+) &= \frac{1}{6\#\hbar}\sum_{\lambda=1}^{\#\hbar}(1 - a^\lambda + 2(1-b^\lambda) + 2(1-c^\lambda) + 1 - d^\lambda) \\ &= \frac{1}{6\#\hbar}\sum_{\lambda=1}^{\#\hbar}(6 - a^\lambda - 2b^\lambda - 2c^\lambda - d^\lambda) \end{aligned} \tag{2.31}$$

and the distance between the HTrFN $\hbar$ and the negative ideal HTrFN $\hbar^-$ can be computed as:

$$d(\hbar, \hbar^-) = \frac{1}{6\#\hbar}\sum_{\lambda=1}^{\#\hbar}(a^\lambda + 2b^\lambda + 2c^\lambda + d^\lambda) \tag{2.32}$$

In general, the smaller the distance $d(\hbar, \hbar^+)$ is, the bigger the HTrFN $\hbar$ is; and the larger the distance $d(\hbar, \hbar^-)$ is, the bigger the HTrFN $\hbar$ is. Motivated by the idea of TOPSIS method (Hwang and Yoon 1981), the closeness index for the HTrFN is developed as follows.

Definition 2.14 (Zhang and Liu 2016). Let $\hbar = \{\tilde{\alpha}^1, \tilde{\alpha}^2, \ldots, \tilde{\alpha}^{\#\hbar}\}$ be a HTrFN, $\hbar^+$ be the positive ideal HTrFN and $\hbar^-$ be the negative ideal HTrFN; then the closeness index of $\hbar$ can be defined as:

$$\begin{aligned} \varphi(\hbar) &= \frac{d(\hbar, \hbar^-)}{d(\hbar, \hbar^-) + d(\hbar, \hbar^+)} = \frac{\sum_{\lambda=1}^{\#\hbar}(a^\lambda + 2b^\lambda + 2c^\lambda + d^\lambda)}{\sum_{\lambda=1}^{\#\hbar}(6 - a^\lambda - 2b^\lambda - 2c^\lambda - d^\lambda) + \sum_{\lambda=1}^{\#\hbar}(a^\lambda + 2b^\lambda + 2c^\lambda + d^\lambda)} \\ &= \frac{1}{6\#\hbar}\sum_{\lambda=1}^{\#\hbar}(a^\lambda + 2b^\lambda + 2c^\lambda + d^\lambda) \end{aligned} \tag{2.33}$$

Obviously, if $\hbar = \hbar^-$, then $\varphi(\hbar) = 0$; while if $\hbar = \hbar^+$, then $\varphi(\hbar) = 1$.

Based on the closeness indices of HTrFNs, a comparison law for HTrFNs is introduced.

Definition 2.15 (Zhang and Liu 2016). Given two HTrFNs $\hbar_j = \{\tilde{\alpha}_j^1, \tilde{\alpha}_j^2, \ldots, \tilde{\alpha}_j^{\#\hbar_j}\}$ $(j = 1, 2)$, $\varphi(\hbar_1)$ and $\varphi(\hbar_2)$ be the closeness indices of $\hbar_1$ and $\hbar_2$, respectively, then

(1) if $\varphi(\hbar_1) < \varphi(\hbar_2)$, then $\hbar_1 \prec \hbar_2$;
(2) if $\varphi(\hbar_1) > \varphi(\hbar_2)$, then $\hbar_1 \succ \hbar_2$;
(3) if $\varphi(\hbar_1) = \varphi(\hbar_2)$, then $\hbar_1 \sim \hbar_2$.

Example 2.8 (Zhang and Liu 2016). For two HTrFNs

$$\hbar_1 = \{Tr(0.1, 0.2, 0.3, 0.5), Tr(0.3, 0.4, 0.4, 0.6)\},$$
$$\hbar_2 = \{Tr(0.1, 0.3, 0.4, 0.6), Tr(0.2, 0.4, 0.5, 0.6)\},$$

the following result based on Definition 2.14 is obtained:

$$\varphi(\hbar_1) = \frac{1}{12}(0.1 + 0.4 + 0.6 + 0.5 + 0.3 + 0.8 + 0.8 + 0.6) = 0.2563,$$
$$\varphi(\hbar_2) = \frac{1}{12}(0.1 + 0.6 + 0.8 + 0.6 + 0.2 + 0.8 + 1.0 + 0.6) = 0.3912.$$

According to Definition 2.15, it is observed that $\varphi(h_1) < \varphi(\hbar_2)$, i.e., $\hbar_1 \prec \hbar_2$.

2.5.2 Hesitant Trapezoidal Fuzzy TODIM Decision Analysis Method

Consider a decision environment based on HTrFNs for the MCGDM problems in which the criteria values of alternatives take the form of comparative linguistic expressions (please see Chap. 6 for the concept of comparative linguistic expressions). Let $\boldsymbol{A} = \{A_1, A_2, \ldots, A_m\}$ be a discrete set of m $(m \geq 2)$ feasible alternatives, $\boldsymbol{C} = \{C_1, C_2, \ldots, C_n\}$ be a finite set of criteria, and $\boldsymbol{E} = \{e_1, e_2, \ldots, e_g\}$ be a group of experts. The criteria values of the alternative $A_i \in \boldsymbol{A}$ with respect to the criterion $C_j \in \boldsymbol{C}$ provided by the expert $e_k \in \boldsymbol{E}$ can be represented by comparative linguistic expressions ll_{ij}^k. Usually, the semantics of ll_{ij}^k can be captured by a HTrFN $\hbar_{ij}^k = \{\tilde{\alpha}_{ij}^{k(1)}, \tilde{\alpha}_{ij}^{k(2)}, \ldots, \tilde{\alpha}_{ij}^{k(\#\hbar_{ij}^k)}\}$. Thus, the MCGDM problem is expressed in the matrix format $R^k = (\hbar_{ij}^k)_{m \times n}$ $(e_k \in \boldsymbol{E})$. We also denote the weighting vector of criteria by $\boldsymbol{w} = (w_1, w_2, \ldots, w_n)^T$, where w_j is the weight of the criterion C_j, satisfying the normalization condition: $\sum_{j=1}^n w_j = 1$ and $w_j \geq 0$. Meanwhile, we denote the weighting vector of experts by $\boldsymbol{\varpi} = (\varpi_1, \varpi_2, \ldots, \varpi_g)^T$, where ϖ_k is the weight of the expert e_k, satisfying the normalization condition: $\sum_{k=1}^g \varpi_k = 1$ and $\varpi_k \geq 0$. In this chapter, the weights of criteria are completely known in advance and the weights of experts are completely unknown or partially known. Let Γ be a set of the known weight information of experts, and $\Gamma = \Gamma_1 \cup \Gamma_2 \cup \Gamma_3 \cup \Gamma_4 \cup \Gamma_5$ can be constructed by the following forms (Zhang and Xu 2014b, 2015):

(1) a weak ranking: $\Gamma_1 = \{\varpi_i \geq \varpi_j\}$;
(2) a strict ranking: $\Gamma_2 = \{\varpi_i - \varpi_j \geq \kappa_{ij}\}\ (\kappa_{ij} > 0)$;
(3) a ranking of differences: $\Gamma_3 = \{\varpi_i - \varpi_j \geq \varpi_\xi - \varpi_\zeta\}\ (i \neq j \neq \xi \neq \zeta)$;
(4) a ranking with multiples: $\Gamma_4 = \{\varpi_i \geq \kappa_{ij}\varpi_j\}\ (0 \leq \kappa_{ij} \leq 1)$;
(5) an interval form: $\Gamma_5 = \{\kappa_i \leq \varpi_i \leq \kappa_i + \varepsilon_i\}\ (0 \leq \kappa_i \leq \kappa_i + \varepsilon_i \leq 1)$.

To deal effectively with the above MCGDM problem, Zhang and Liu (2016) developed a hesitant trapezoidal fuzzy TODIM method. The focus of the proposed method is to measure the dominance degree of each alternative over the others by constructing the prospect value function based on prospect theory. For this purpose, we need to identify the reference criterion and calculate the relative weight of each criterion to the reference criterion. According to the idea of the classical TODIM, the criterion with the highest weight is usually regarded as the reference criterion C_R, namely,

$$C_R = \{C_j : \max_{j=1}^{n} w_j\} \tag{2.34}$$

Then, the relative weight w_{jR} of the criterion $C_j \in \boldsymbol{C}$ to the reference criterion C_R can be obtained by the following equation:

$$w_{jR} = w_j / w_R, \quad j \in \{1, 2, \ldots, n\} \tag{2.35}$$

where w_R is the weight of the reference criterion C_R.

Furthermore, by employing the closeness index-based ranking method of HTrFNs we compare with the magnitude of the criteria values of alternatives with respect to each criterion. Then, for the expert $e_k \in \boldsymbol{E}$, the dominance value of the alternative $A_\xi \in \boldsymbol{A}$ over the alternative $A_\zeta \in \boldsymbol{A}$ concerning the criterion $C_j \in \boldsymbol{C}$ can be calculated by using the following expression (Zhang and Liu 2016):

$$\mathscr{D}_j^k(A_\xi, A_\zeta) = \begin{cases} \sqrt{\dfrac{w_{jR}(\varphi(\hbar_{\xi j}^k) - \varphi(\hbar_{\zeta j}^k))}{\sum_{j=1}^{n} w_{jR}}}, & \text{if } \varphi(\hbar_{\xi j}^k) - \varphi(\hbar_{\zeta j}^k) > 0 \\ 0, & \text{if } \varphi(\hbar_{\xi j}^k) - \varphi(\hbar_{\zeta j}^k) = 0 \\ -\dfrac{1}{\theta}\sqrt{\dfrac{(\sum_{j=1}^{n} w_{jR})(\varphi(\hbar_{\zeta j}^k) - \varphi(\hbar_{\xi j}^k))}{w_{jR}}}, & \text{if } \varphi(\hbar_{\xi j}^k) - \varphi(\hbar_{\zeta j}^k) < 0 \end{cases} \tag{2.36}$$

where $\varphi(\hbar_{\xi j}^k)$ and $\varphi(\hbar_{\zeta j}^k)$ are respectively the closeness indices of the criteria values $\hbar_{\xi j}^k$ and $\hbar_{\zeta j}^k$, and the parameter $\theta \in [1, 10]$ represents the attenuation factor of the losses.

From Eq. (2.36), it is easily observed that:

(1) if $\varphi(\hbar_{\xi j}^k) - \varphi(\hbar_{\zeta j}^k) > 0$, then $\mathscr{D}_j^k(A_\xi, A_\zeta)$ represents a gain;
(2) if $\varphi(\hbar_{\xi j}^k) - \varphi(\hbar_{\zeta j}^k) = 0$, then $\mathscr{D}_j^k(A_\xi, A_\zeta)$ represents a nil;
(3) if $\varphi(\hbar_{\xi j}^k) - \varphi(\hbar_{\zeta j}^k) < 0$, then $\mathscr{D}_j^k(A_\xi, A_\zeta)$ represents a loss.

For the expert $e_k \in \boldsymbol{E}$, the collective dominance value of the alternative $A_\xi \in \boldsymbol{A}$ over the alternative $A_\zeta \in \boldsymbol{A}$ can be obtained as follows (Zhang and Liu 2016):

$$\mathcal{H}^k(A_\xi, A_\zeta) = \sum_{j=1}^{n} \mathcal{D}_j^k(A_\xi, A_\zeta) \tag{2.37}$$

The overall dominance value of the alternative $A_\xi \in \boldsymbol{A}$ for the expert $e_k \in \boldsymbol{E}$ can be obtained by the following equation and listed in Table 2.17 (Zhang and Liu 2016):

$$\mathcal{Q}^k(A_\xi) = \frac{\sum_{\zeta=1}^{m} \mathcal{H}^k(A_\xi, A_\zeta) - \min_{\xi=1}^{m}\left\{\sum_{\zeta=1}^{m} \mathcal{H}^k(A_\xi, A_\zeta)\right\}}{\max_{\xi=1}^{m}\left\{\sum_{\zeta=1}^{m} \mathcal{H}^k(A_\xi, A_\zeta)\right\} - \min_{\xi=1}^{m}\left\{\sum_{\zeta=1}^{m} \mathcal{H}^k(A_\xi, A_\zeta)\right\}} \tag{2.38}$$

After obtaining the overall dominance value $\mathcal{Q}^k(A_\xi)$, we need to determine the overall dominance value for the group which is represented by $\mathcal{Q}^*(A_\xi)$ ($\xi \in \{1, 2, \ldots, m\}$). Based on the decision data in Table 2.17, we further establish a nonlinear programming model to calculate $\mathcal{Q}^*(A_\xi)$ ($\xi \in \{1, 2, \ldots, m\}$) for the group as follows (Zhang and Liu 2016):

$$\begin{cases} \min \quad Z = \sum\limits_{\xi=1}^{m} \sum\limits_{k=1}^{g} \varpi_k \left| \mathcal{Q}^k(A_\xi) - \mathcal{Q}^*(A_\xi) \right| \\ s.t. \quad \sum\limits_{k=1}^{g} \varpi_k = 1, \ \varpi_k \geq 0, \ k \in \{1, 2, \ldots, g\} \end{cases} \tag{MOD-2.1}$$

To solve the model (MOD-2.1), let

$$\eta_i^k = \frac{1}{2}\left(\left|\mathcal{Q}^k(A_\xi) - \mathcal{Q}^*(A_\xi)\right| + \left(\mathcal{Q}^k(A_\xi) - \mathcal{Q}^*(A_\xi)\right)\right) \tag{2.39}$$

and

$$\rho_i^k = \frac{1}{2}\left(\left|\mathcal{Q}^k(A_\xi) - \mathcal{Q}^*(A_\xi)\right| - \left(\mathcal{Q}^k(A_\xi) - \mathcal{Q}^*(A_\xi)\right)\right) \tag{2.40}$$

Then, the optimal model (MOD-2.1) is transformed into the following optimal model (Zhang and Liu 2016):

Table 2.17 The overall dominance values of alternatives for each expert

Alternatives	Experts			
	e_1	e_2	…	e_g
A_1	$\mathcal{Q}^1(A_1)$	$\mathcal{Q}^2(A_1)$	…	$\mathcal{Q}^g(A_1)$
A_2	$\mathcal{Q}^1(A_2)$	$\mathcal{Q}^2(A_2)$	…	$\mathcal{Q}^g(A_2)$
…	…	…	…	…
A_m	$\mathcal{Q}^1(A_m)$	$\mathcal{Q}^2(A_m)$	…	$\mathcal{Q}^g(A_m)$

$$\begin{cases} \min \quad Z = \sum_{i=1}^{m} \sum_{k=1}^{g} \varpi_k(\eta_i^k + \rho_i^k) \\ s.t. \\ \quad \mathcal{Q}^k(A_\xi) - \mathcal{Q}^*(A_\xi) - \eta_i^k + \rho_i^k = 0 \quad \begin{pmatrix} i \in \{1,2,\ldots,m\}, \\ k \in \{1,2,\ldots,g\} \end{pmatrix} \\ \quad \eta_i^k \geq 0, \ \rho_i^k \geq 0, \ \eta_i^k \rho_i^k = 0 \quad \begin{pmatrix} i \in \{1,2,\ldots,m\}, \\ k \in \{1,2,\ldots,g\} \end{pmatrix} \\ \quad \sum_{k=1}^{g} \varpi_k = 1, \ \varpi_k \geq 0, \quad k \in \{1,2,\ldots,g\} \end{cases} \tag{MOD-2.2}$$

It is observed that the model (MOD-2.2) is a linear programming model and can be easily executed by using the MATLAB 7.4.0 or LINGO 11.0. By solving this model, we get the $\mathcal{Q}^*(A_\xi)$ $(\xi \in \{1,2,\ldots,m\})$ and $\boldsymbol{\varpi} = (\varpi_1, \varpi_2, \ldots, \varpi_g)^T$.

In addition, there are real-world situations that the weights of experts are not completely unknown but partially known. For these cases, based on the set of the known weight information of experts Γ, we construct the following optimization model to get $\mathcal{Q}^*(A_\xi)$ $(\xi \in \{1,2,\ldots,m\})$ and $\boldsymbol{\varpi} = (\varpi_1, \varpi_2, \ldots, \varpi_g)^T$ (Zhang and Liu 2016):

$$\begin{cases} \min \quad Z = \sum_{i=1}^{m} \sum_{k=1}^{g} \varpi_k(\eta_i^k + \rho_i^k) \\ s.t. \\ \quad \mathcal{Q}^k(A_\xi) - \mathcal{Q}^*(A_\xi) - \eta_i^k + \rho_i^k = 0 \quad \begin{pmatrix} i \in \{1,2,\ldots,m\}, \\ k \in \{1,2,\ldots,g\} \end{pmatrix} \\ \quad \eta_i^k \geq 0, \ \rho_i^k \geq 0, \ \eta_i^k \rho_i^k = 0 \quad \begin{pmatrix} i \in \{1,2,\ldots,m\}, \\ k \in \{1,2,\ldots,g\} \end{pmatrix} \\ \quad (\varpi_1, \varpi_2, \ldots, \varpi_g) \in \Gamma \end{cases} \tag{MOD-2.3}$$

where Γ is a set of constraint conditions that the expert weights ϖ_k $(k \in \{1,2,\ldots,g\})$ should satisfy according to the requirements in the real-world situations.

Obviously, the greater the value of $\mathcal{Q}^*(A_\xi)$ $(\xi \in \{1,2,\ldots,m\})$ is, the better the alternative A_ξ will be. Therefore, we can determine the ranking order of alternatives according to the increasing orders of $\mathcal{Q}^*(A_\xi)$ $(\xi \in \{1,2,\ldots,m\})$, and select the best alternative from the alternative set $\boldsymbol{A}$. According to the above analysis, the steps of the proposed method are summarized as follows (Algorithm 2.3):

Step 1. Identify the criteria values of alternatives on criteria under each expert and the weights of criteria, respectively.

Step 2. For each expert, we employ Eq. (2.36) to calculate the dominance value of the alternative $A_\xi \in \boldsymbol{A}$ over the alternative $A_\zeta \in \boldsymbol{A}$ concerning the criterion $C_j \in \boldsymbol{C}$.

Step 3. For each expert, we use Eq. (2.37) to compute the collective dominance value of the alternative $A_\xi \in \boldsymbol{A}$ over the alternative $A_\zeta \in \boldsymbol{A}$.

Step 4. We utilize Eq. (2.38) to determine the overall dominance value of the alternative $A_\xi \in \boldsymbol{A}$ for the expert $e_k \in \boldsymbol{E}$.

Step 5. If the weights of experts are completely unknown, according to the model (MOD-2.2) we construct a linear programming model to determine the overall dominance value of the alternative $A_\xi \in \boldsymbol{A}$ for the group; if the weights of experts are partially known, based on the model (MOD-2.3) we construct an optimal model to determine the overall dominance value of the alternative $A_\xi \in \boldsymbol{A}$ for the group.

Step 6. Rank the alternatives by comparing the magnitudes of the overall dominance value of the alternative $A_\xi \in \boldsymbol{A}$ for the group.

2.5.3 Case Illustration

To demonstrate the decision process and the applicability of the proposed approach, Zhang and Liu (2016) modified the real-life MCDM problem introduced in Sect. 2.4, and assumed that the CAA invites a committee including three experts (e_1, e_2, e_3) to investigate four major Taiwan airlines according to their four criteria. The weight vector of the criteria is $\boldsymbol{w} = (0.2, 0.25, 0.35, 0.2)^{\mathrm{T}}$. The weight vector of the experts is given as follows:

$$\Gamma = \left\{ \begin{array}{l} \varpi_3 \geq \varpi_1, 0.15 \leq \varpi_2 - \varpi_1 \leq 0.25, \varpi_1 + \varpi_3 \geq \varpi_2, 0.2 \leq \varpi_2 \leq 0.35, \\ \varpi_1 + \varpi_2 + \varpi_3 = 1, \varpi_1 \geq 0, \varpi_2 \geq 0, \varpi_3 \geq 0 \end{array} \right\}$$

All experts employ the linguistic terms or comparison linguistic expressions to provide the assessment values of alternatives with respect to each criterion as shown in Table 2.18.

The top-left cell "*Between MG and G*" in Table 2.18 indicates that the degree to which the alternative A_1 (UNI Air) satisfies the criterion C_1 (Booking and ticketing service) is *between Medium Good and Good*. The others in Table 2.18 have the similar meanings. Then, comparative linguistic expressions are transformed into HTrFNs which are listed in Table 2.19.

In what follows, we employ the proposed hesitant trapezoidal fuzzy TODIM method to solve the above group decision making problem. According to the decision steps of the proposed method, we first take the criterion C_3 as the reference criterion and thus the weight of the reference criterion is $w_R = 0.35$. Based on the closeness index-based ranking method of HTrFNs, we compare with the

Table 2.18 The linguistic criteria values of alternatives for each expert

Experts	Alternatives	Criteria			
		C_1	C_2	C_3	C_4
e_1	A_1	Between MG and G	MG	Between MG and G	At most MP
	A_2	MG	Between MP and F	MP	Between F and MG
	A_3	Between MP and F	Between P and MP	At least MG	MG
	A_4	MG	F	G	Between P and MP
e_2	A_1	Between F and MG	MP	P	G
	A_2	Between G and VG	At least MG	F	F
	A_3	G	MG	MP	MP
	A_4	MP	G	F	Between P and MP
e_3	A_1	At least MG	F	At most MP	MG
	A_2	F	MG	At least G	P
	A_3	Between F and G	Between MP and F	Between P and MP	At least G
	A_4	MG	G	Between F and G	G

magnitudes of the criteria values and obtain the superior-inferior table under the expert e_1 as in Table 2.20 (Zhang and Liu 2016).

The top-left cell "$_{1/2}S_1$" in Table 2.20 indicates that for the expert e_1 and under the criterion C_1 the alternative A_1 is superior to the alternative A_2 because of $\varphi(h_{11}^1) = 0.725 > \varphi(h_{21}^1) = 0.65$. Similar logic is used to determine the remaining entries in Table 2.20. Without loss of generality, we take the value of the parameter θ as 3 and the dominance value of the alternative $A_\xi \in \boldsymbol{A}$ over the alternative $A_\zeta \in \boldsymbol{A}$ under the criterion $C_j \in \boldsymbol{C}$ can be calculated by using Eq. (2.36). The calculated results are listed in Tables 2.21, 2.22, 2.23 and 2.24 (Zhang and Liu 2016).

Furthermore, by aggregating the gains and losses of the alternative $A_\xi \in \boldsymbol{A}$ over the alternative $A_\zeta \in \boldsymbol{A}$ under the criterion $C_j \in \boldsymbol{C}$ using Eq. (2.37), we can obtain the weighted dominance value of each alternative over the others, listed in Table 2.25.

Using Eq. (2.38), the overall dominance values of alternatives for the expert e_1 are obtained as follows:

$$\mathcal{Q}^1(A_1) = 0.6150, \quad \mathcal{Q}^1(A_2) = 1.0, \quad \mathcal{Q}^1(A_3) = 0.0, \quad \mathcal{Q}^1(A_4) = 0.5892.$$

Analogously, we can also calculate the overall dominance values of alternatives for the expert e_2 as:

Table 2.19 The hesitant criteria values of alternatives

Experts	Alternatives	Criteria	
		C_1	C_2
e_1	A_1	{Tr(0.5, 0.6, 0.7, 0.8), Tr(0.7, 0.8, 0.8, 0.9)}	Tr(0.5, 0.6, 0.7, 0.8)
	A_2	Tr(0.5, 0.6, 0.7, 0.8)	{Tr(0.2, 0.3, 0.4, 0.5), Tr(0.4, 0.5, 0.5, 0.6)}
	A_3	{Tr(0.2, 0.3, 0.4, 0.5), Tr(0.4, 0.5, 0.5, 0.6)}	{Tr(0.1, 0.2, 0.2, 0.3), Tr(0.2, 0.3, 0.4, 0.5)}
	A_4	Tr(0.5, 0.6, 0.7, 0.8)	Tr(0.4, 0.5, 0.5, 0.6)
e_2	A_1	{Tr(0.4, 0.5, 0.5, 0.6), Tr(0.5, 0.6, 0.7, 0.8)}	Tr(0.2, 0.3, 0.4, 0.5)
	A_2	{Tr(0.7, 0.8, 0.8, 0.9), Tr(0.8, 0.9, 1.0, 1.0)}	{Tr(0.5, 0.6, 0.7, 0.8), Tr(0.7, 0.8, 0.8, 0.9), Tr(0.8, 0.9, 1.0, 1.0)}
	A_3	Tr(0.7, 0.8, 0.8, 0.9)	Tr(0.5, 0.6, 0.7, 0.8)
	A_4	Tr(0.2, 0.3, 0.4, 0.5)	Tr(0.7, 0.8, 0.8, 0.9)
e_3	A_1	{Tr(0.5, 0.6, 0.7, 0.8), Tr(0.7, 0.8, 0.8, 0.9), Tr(0.8, 0.9, 1.0, 1.0)}	Tr(0.4, 0.5, 0.5, 0.6)
	A_2	Tr(0.4, 0.5, 0.5, 0.6)	Tr(0.5, 0.6, 0.7, 0.8)
	A_3	{Tr(0.4, 0.5, 0.5, 0.6), Tr(0.5, 0.6, 0.7, 0.8), Tr(0.7, 0.8, 0.8, 0.9)}	{Tr(0.2, 0.3, 0.4, 0.5), Tr(0.4, 0.5, 0.5, 0.6)}
	A_4	Tr(0.5, 0.6, 0.7, 0.8)	Tr(0.7, 0.8, 0.8, 0.9)
		C_3	C_4
e_1	A_1	{Tr(0.5, 0.6, 0.7, 0.8), Tr(0.7, 0.8, 0.8, 0.9)}	{Tr(0.0, 0.0, 0.1, 0.2), Tr(0.1, 0.2, 0.2, 0.3), Tr(0.2, 0.3, 0.4, 0.5)}
	A_2	Tr(0.7, 0.8, 0.8, 0.9)	{Tr(0.4, 0.5, 0.5, 0.6), Tr(0.5, 0.6, 0.7, 0.8)}
	A_3	{Tr(0.5, 0.6, 0.7, 0.8), Tr(0.7, 0.8, 0.8, 0.9), Tr(0.8, 0.9, 1.0, 1.0)}	Tr(0.5, 0.6, 0.7, 0.8)
	A_4	Tr(0.7, 0.8, 0.8, 0.9)	{Tr(0.1, 0.2, 0.2, 0.3), Tr(0.2, 0.3, 0.4, 0.5)}
e_2	A_1	Tr(0.1, 0.2, 0.2, 0.3)	Tr(0.7, 0.8, 0.8, 0.9)
	A_2	Tr(0.4, 0.5, 0.5, 0.6)	Tr(0.4, 0.5, 0.5, 0.6)
	A_3	Tr(0.2, 0.3, 0.4, 0.5)	Tr(0.2, 0.3, 0.4, 0.5)
	A_4	Tr(0.4, 0.5, 0.5, 0.6)	{Tr(0.4, 0.5, 0.5, 0.6), Tr(0.5, 0.6, 0.7, 0.8)}
e_3	A_1	{Tr(0.0, 0.0, 0.1, 0.2), Tr(0.1, 0.2, 0.2, 0.3), Tr(0.2, 0.3, 0.4, 0.5)}	Tr(0.5, 0.6, 0.7, 0.8)
	A_2	{Tr(0.7, 0.8, 0.8, 0.9), Tr(0.8, 0.9, 1.0, 1.0)}	Tr(0.1, 0.2, 0.2, 0.3)
	A_3	{Tr(0.4, 0.5, 0.5, 0.6), Tr(0.5, 0.6, 0.7, 0.8)}	{Tr(0.7, 0.8, 0.8, 0.9), Tr(0.8, 0.9, 1.0, 1.0)}
	A_4	{Tr(0.4, 0.5, 0.5, 0.6), Tr(0.5, 0.6, 0.7, 0.8), Tr(0.7, 0.8, 0.8, 0.9)}	Tr(0.7, 0.8, 0.8, 0.9)

Table 2.20 A superior-inferior table over alternatives with criteria for the expert e_1

	A_1/A_2	A_1/A_3	A_1/A_4	A_2/A_3	A_2/A_4	A_3/A_4
C_1	${}_{1/2}S_1$	${}_{1/3}S_1$	${}_{1/4}S_1$	${}_{2/3}S_1$	${}_{2/4}E_1$	${}_{3/4}I_1$
C_2	${}_{1/2}S_2$	${}_{1/3}S_2$	${}_{1/4}S_2$	${}_{2/3}S_2$	${}_{2/4}I_2$	${}_{3/4}I_2$
C_3	${}_{1/2}I_3$	${}_{1/3}I_3$	${}_{1/4}I_3$	${}_{2/3}S_3$	${}_{2/4}E_3$	${}_{3/4}I_3$
C_4	${}_{1/2}I_4$	${}_{1/3}I_4$	${}_{1/4}I_4$	${}_{2/3}I_4$	${}_{2/4}S_4$	${}_{3/4}S_4$

Table 2.21 Gains and losses of alternatives over the others for the criterion C_1 and the expert e_1

	A_1	A_2	A_3	A_4
A_1	0	0.1225	0.2449	0.1225
A_2	−0.2041	0	0.2121	0
A_3	−0.4082	−0.3536	0	−0.3536
A_4	−0.2041	0	0.2121	0

Table 2.22 Gains and losses of alternatives over the others for the criterion C_3 and the expert e_1

	A_1	A_2	A_3	A_4
A_1	0	−0.1543	−0.1484	−0.1543
A_2	0.1620	0	0.0443	0
A_3	0.1559	−0.0422	0	−0.0422
A_4	0.1620	0	0.0443	0

Table 2.23 Gains and losses of alternatives over the others for the criterion C_2 and the expert e_1

	A_1	A_2	A_3	A_4
A_1	0	0.2372	0.3062	0.1936
A_2	−0.3162	0	0.1936	−0.1826
A_3	−0.4082	−0.2582	0	−0.3162
A_4	−0.2582	0.1369	0.2372	0

Table 2.24 Gains and losses of alternatives over the others for the criterion C_4 and the expert e_1

	A_1	A_2	A_3	A_4
A_1	0	−0.4530	−0.4969	−0.1964
A_2	0.2718	0	−0.2041	0.2449
A_3	0.2981	0.1225	0	0.2739
A_4	0.1178	−0.4082	−0.4564	0

Table 2.25 Weighted dominance values of alternatives over others for the expert e_1

	A_1	A_2	A_3	A_4
A_1	0	−0.2476	−0.0942	−0.0346
A_2	−0.0865	0	0.2459	0.0623
A_3	−0.3624	−0.5315	0	−0.4381
A_4	−0.1825	−0.2713	0.0372	0

$$\mathcal{Q}^2(A_1) = 0.0,\ \ \mathcal{Q}^2(A_2) = 1.0,\ \ \mathcal{Q}^2(A_3) = 0.0218,\ \ \mathcal{Q}^2(A_4) = 0.3991,$$

and the overall dominance values of alternatives for the expert e_3 as follows:

$$\mathcal{Q}^3(A_1) = 0.0210,\ \ \mathcal{Q}^3(A_2) = 0.0,\ \ \mathcal{Q}^3(A_3) = 0.3669,\ \ \mathcal{Q}^3(A_4) = 1.0.$$

According to the model (MOD-2.3), we construct the following optimal model (Zhang and Liu 2016):

$$\begin{cases} \min \quad Z = \sum_{i=1}^{4}\sum_{k=1}^{3} \varpi_k(\eta_i^k + \rho_i^k) \\ s.t. \\ \quad 0.615 - \mathcal{Q}^*(A_1) - \eta_1^1 + \rho_1^1 = 0; \quad 0.0 - \mathcal{Q}^*(A_1) - \eta_1^2 + \rho_1^2 = 0; \\ \quad 0.021 - \mathcal{Q}^*(A_1) - \eta_1^3 + \rho_1^3 = 0; \quad 1.0 - \mathcal{Q}^*(A_2) - \eta_2^1 + \rho_2^1 = 0; \\ \quad 1.0 - \mathcal{Q}^*(A_2) - \eta_2^2 + \rho_2^2 = 0; \quad 0.0 - \mathcal{Q}^*(A_2) - \eta_2^3 + \rho_2^3 = 0; \\ \quad 0.0 - \mathcal{Q}^*(A_3) - \eta_3^1 + \rho_3^1 = 0; \quad 0.0218 - \mathcal{Q}^*(A_3) - \eta_3^2 + \rho_3^2 = 0; \\ \quad 0.3669 - \mathcal{Q}^*(A_3) - \eta_3^3 + \rho_3^3 = 0; \quad 0.5892 - \mathcal{Q}^*(A_4) - \eta_4^1 + \rho_4^1 = 0; \\ \quad 0.3991 - \mathcal{Q}^*(A_4) - \eta_4^2 + \rho_4^2 = 0; \quad 1.0 - \mathcal{Q}^*(A_4) - \eta_4^3 + \rho_4^3 = 0; \\ \quad \varpi_3 \geq \varpi_1, 0.15 \leq \varpi_2 - \varpi_1 \leq 0.25,\ \varpi_1 + \varpi_3 \geq \varpi_2, 0.2 \leq \varpi_2 \leq 0.35 \\ \quad \sum_{k=1}^{3} \varpi_k = 1,\ \varpi_k \geq 0,\ \ k \in \{1,2,3\} \\ \quad \eta_i^k \geq 0,\ \rho_i^k \geq 0,\ \eta_i^k \rho_i^k = 0;\ \ i \in \{1,2,3,4\}, k \in \{1,2,3\} \end{cases}$$

By solving the above model, the weights of experts and the overall dominance values of the alternatives for the group can be obtained as follows (Zhang and Liu 2016):

$$\varpi_1 = 0.20,\ \ \varpi_2 = 0.35,\ \ \varpi_3 = 0.45$$
$$\mathcal{Q}^*(A_1) = 0.0210,\ \mathcal{Q}^*(A_2) = 1.0000,$$
$$\mathcal{Q}^*(A_3) = 0.0218,\ \mathcal{Q}^*(A_4) = 0.5892.$$

Apparently, the ranking of alternatives is obtained as $A_2 \succ A_4 \succ A_3 \succ A_1$, and the best alternative is A_2.

2.6 Conclusions

The classical TODIM method is a helpful tool for solving the classical MCDM problems in case of considering the decision maker's psychological behavior, but it cannot be used to directly handle the MCDM problems with fuzzy information. This chapter introduced a HF-TODIM method developed by Zhang and Xu (2014a) for solving the MCDM problems with hesitant fuzzy information in case of considering the decision maker's psychological behavior. The main advantages of the HF-TODIM approach are that (1) it can handle the MCDM problems in which the ratings of alternatives with respect to each criterion are represented by HFEs or IVHFEs and (2) it can take the decision maker's psychological behavior into account. On the other hand, this chapter also introduced a hesitant trapezoidal fuzzy TODIM approach developed by Zhang and Liu (2016) to handle the MCGDM problems in which the decision data are expressed as comparative linguistic expressions based on HTrFNs. Both the HF-TODIM method and the hesitant trapezoidal fuzzy TODIM approach can be further extended to deal with the MCDM or MCGDM problems with interdependent criteria, and can also be expected to be applicable to other similar decision making problems, such as performance evaluation, supply chain management, risk investment, etc.

References

Abbasbandy, S., & Asady, B. (2006). Ranking of fuzzy numbers by sign distance. *Information Sciences, 176*, 2405–2416.

Chen, T. Y., Chang, C. H., & Lu, J. F. R. (2013a). The extended QUALIFLEX method for multiple criteria decision analysis based on interval type-2 fuzzy sets and applications to medical decision making. *European Journal of Operational Research, 226*, 615–625.

Chen, N., Xu, Z. S., & Xia, M. M. (2013b). Interval-valued hesitant preference relations and their applications to group decision making. *Knowledge-Based Systems, 37*, 528–540.

Fan, Z. P., Zhang, X., Chen, F.-D., & Liu, Y. (2013). Extended TODIM method for hybrid multiple attribute decision making problems. *Knowledge-Based Systems, 42*, 40–48.

Farhadinia, B. (2013). A novel method of ranking hesitant fuzzy values for multiple attribute decision-making problems. *International Journal of Intelligent Systems, 28*, 752–767.

Gomes, L. F. A. M., & Lima, M. M. P. P. (1991). TODIM: Basics and application to multicriteria ranking of projects with environmental impacts. *Foundations of Computing and Decision Sciences, 16*, 113–127.

Gomes, L. F. A. M., & Lima, M. M. P. P. (1992). From modeling individual preferences to multicriteria ranking of discrete alternatives: a look at prospect theory and the additive difference model. *Foundations of Computing and Decision Sciences, 17*, 171–184.

Gomes, L. F. A. M., & Rangel, D. (2009). An application of the TODIM method to the multicriteria rental evaluation of residential properties. *European Journal of Operational Research, 193*, 204–211.

Gomes, L. F. A. M., Rangel, L. A. D., & Maranhão, F. J. C. (2009). Multicriteria analysis of natural gas destination in Brazil: An application of the TODIM method. *Mathematical and Computer Modelling, 50*, 92–100.

Gomes, L. F. A. M., Machado, M. A. S., & Rangel, L. A. D. (2013). Behavioral multi-criteria decision analysis: the TODIM method with criteria interactions. *Annals of Operations Research, 211*(1), 531–548.

Hwang, C. L., & Yoon, K. (1981). *Multiple attribute decision making methods and applications*. Berlin: Springer.

Kahneman, D., & Tversky, A. (1979). Prospect theory: An analysis of decision under risk. *Econometrica, 47*, 263–291.

Krohling, R. A., & de Souza, T. (2012). Combining prospect theory and fuzzy numbers to multi-criteria decision making. *Expert Systems with Applications, 39*, 11487–11493.

Liao, H. C., & Xu, Z. S. (2013). A VIKOR-based method for hesitant fuzzy multi-criteria decision making. *Fuzzy Optimization and Decision Making, 12*, 373–392.

Liou, J. J., Tsai, C. Y., Lin, R. H., & Tzeng, G. H. (2011). A modified VIKOR multiple-criteria decision method for improving domestic airlines service quality. *Journal of Air Transport Management, 17*, 57–61.

Liu, H. B., & Rodríguez, R. M. (2014). A fuzzy envelope for hesitant fuzzy linguistic term set and its application to multicriteria decision making. *Information Sciences, 258*, 220–238.

Liu, H. W., & Wang, G. J. (2007). Multi-criteria decision-making methods based on intuitionistic fuzzy sets. *European Journal of Operational Research, 179*, 220–233.

Lourenzutti, R., & Krohling, R. A. (2013). A study of TODIM in a intuitionistic fuzzy and random environment. *Expert Systems with Applications, 40*, 6459–6468.

Passos, A. C., Teixeira, M. G., Garcia, K. C., Cardoso, A. M., & Gomes, L. F. A. M. (2014). Using the TODIM-FSE method as a decision-making support methodology for oil spill response. *Computers and Operations Research, 42*, 40–48.

Xia, M. M., & Xu, Z. S. (2011). Hesitant fuzzy information aggregation in decision making. *International Journal of Approximate Reasoning, 52*, 395–407.

Xu, Z. S., & Xia, M. M. (2011a). Distance and similarity measures for hesitant fuzzy sets. *Information Sciences, 181*, 2128–2138.

Xu, Z. S., & Xia, M. M. (2011b). On distance and correlation measures of hesitant fuzzy information. *International Journal of Intelligent Systems, 26*, 410–425.

Xu, Z. S., & Zhang, X. L. (2013). Hesitant fuzzy multi-attribute decision making based on TOPSIS with incomplete weight information. *Knowledge-Based Systems, 52*, 53–64.

Zadeh, L. A. (1975). The concept of a linguistic variable and its application to approximate reasoning. Information Sciences, Part I, II, III (8, 9), 199–249, 301–357, 143–180.

Zhang, Z. M. (2013). Hesitant fuzzy power aggregation operators and their application to multiple attribute group decision making. *Information Sciences, 234*, 150–181.

Zhang, X. L., & Liu, M. F. (2016). Hesitant trapezoidal fuzzy TODIM approach with a closeness index-based ranking method for qualitative group decision making. Technique Report.

Zhang, X. L., & Xu, Z. S. (2014a). The TODIM analysis approach based on novel measured functions under hesitant fuzzy environment. *Knowledge-Based Systems, 61*, 48–58.

Zhang, X. L., & Xu, Z. S. (2014b). Deriving experts' weights based on consistency maximization in intuitionistic fuzzy group decision making. *Journal of Intelligent and Fuzzy Systems, 27*, 221–233.

Zhang, X. L., & Xu, Z. S. (2015). Soft computing based on maximizing consensus and fuzzy TOPSIS approach to interval-valued intuitionistic fuzzy group decision making. *Applied Soft Computing, 26*, 42–56.

Zhang, X. L., Xu, Z. S., & Liu, M. F. (2016). Hesitant trapezoidal fuzzy QUALIFLEX method and its application in the evaluation of green supply chain initiatives. Sustainability, accepted.

Zhao, X., Lin, R., & Wei, G. (2014). Hesitant triangular fuzzy information aggregation based on Einstein operations and their application to multiple attribute decision making. *Expert Systems with Applications, 41*(4), 1086–1094.

Zhu, B., Xu, Z. S., & Xia, M. M. (2012). Hesitant fuzzy geometric Bonferroni means. *Information Sciences, 205*, 72–85.

Chapter 3
Hesitant Fuzzy Multiple Criteria Decision Analysis Based on QUALIFLEX

Abstract The QUALIFLEX is a very useful outranking method to deal simultaneously with the cardinal and ordinal information in decision making process. The purpose of this chapter is to develop the hesitant fuzzy QUALIFLEX with a signed distance-based comparison method for handling MCDM problems in which both the assessments of alternatives on criteria and the weights of criteria are expressed by HFEs. In this chapter, we propose the novel concept of hesitancy index for the HFE to measure the degree of hesitancy of the decision-maker or the decision organization. By taking their hesitancy indices into account, we present a signed distance-based ranking method to compare the magnitude of HFEs. By adopting such a comparison approach, the hesitant fuzzy QUALIFLEX technique is developed. Compared with the hesitant fuzzy ELECTRE method, to handle the MCDM problems where the number of criteria markedly exceeds the number of alternatives the developed technique does not require the complicated computation procedures but still yields a reasonable and credible solution. Finally, the developed technique is extended to manage the heterogeneous information including real numbers, interval numbers, TFNs, IFNs, and HFEs.

The classical QUALIFLEX method initially developed by Paelinck (1976, 1977, 1978) is one of outranking methods to solve the MCDM problems with crisp numbers, especially suitable to deal with the decision making problems where the number of criteria markedly exceeds the number of alternatives. This classical QUALIFLEX method is based on the pair-wise comparisons of alternatives with respect to each criterion under all possible permutations of the alternatives and identifies the optimal permutation that maximizes the value of concordance/discordance index (Martel and Matarazzo 2005). The most characteristic of the classical QUALIFLEX method is the correct treatment of cardinal and ordinal information (Rebai et al. 2006). Recently, the classical QUALIFLEX method has been extended into various fuzzy environments because the crisp data are usually inadequate or insufficient to model the real-life complex decision making problems, for example, Chen et al. (2013) proposed an extended QUALIFLEX method for handling the MCDM problems with interval type-2 trapezoidal fuzzy numbers (IT2TrFNs). Chen (2014) developed an

X. Zhang and Z. Xu, *Hesitant Fuzzy Methods for Multiple Criteria Decision Analysis*, Studies in Fuzziness and Soft Computing 345,
DOI 10.1007/978-3-319-42001-1_3

interval-valued intuitionistic fuzzy QUALIFLEX method to address the MCDM problems with IVIFNs. Zhang (2016) developed a closeness index-based Pythagorean fuzzy QUALIFLEX method to address the hierarchical MCDM problems within Pythagorean fuzzy environment based on Pythagorean fuzzy numbers (PFNs) and interval-valued Pythagorean fuzzy numbers (IVPFNs).

In many complex MCDM problems, both the assessments of alternatives on criteria and the weights of criteria are usually expressed by HFEs when the decision maker has hesitation in providing the assessments. For this kind of problems, Zhang and Xu (2015) employed the main structure of the classical QUALIFLEX method to develop a hesitant fuzzy QUALIFLEX with a signed distance-based comparison method. In this approach, we propose a concept of hesitancy index for the HFE to measure the degree of hesitancy of the decision maker or the decision organization. By taking their hesitancy indices into account, we present a signed distance-based method to compare the magnitude of HFEs. Using the signed distance-based comparison approach, we define the concordance/discordance index, the weighted concordance/discordance index and the comprehensive concordance/discordance index. By investigating all possible permutations of alternatives with respect to the level of concordance/discordance of the complete preference order, the optimal ranking orders of alternatives can be obtained. An application study of the proposed method on green supplier selection is conducted, which indicates that the proposed method does not require the complicated computation procedures but still yields a reasonable and credible solution. At length, the hesitant fuzzy QUALIFLEX with a signed distance-based comparison method is extended to manage the heterogeneous information including real numbers, interval numbers, TFNs, IFNs, and HFEs.

3.1 The Concept of a Hesitancy Index for HFE

To take into account the degrees of deviation among different possible values in HFEs, Zhang and Xu (2015) introduced a concept of a hesitancy index for a HFE. By taking their hesitancy indices into account, Zhang and Xu (2015) further presented a signed distance-based comparison method and two improved hesitant fuzzy distance measures for HFEs.

3.1.1 The Hesitancy Index

Given a HFE, there exists a degree of deviation among different possible values that compound the HFE. It is easy to note that the deviation degrees of different HFEs are usually different and they can be interpreted as the decision maker's (or decision organization's) degree of hesitancy. The deviation degree of a HFE is called the hesitancy index of this HFE (Zhang and Xu 2015). The notion of the hesitancy index of a HFE is presented as follows:

Definition 3.1 (Zhang and Xu 2015). Given a HFE $h = H\{\gamma^{\lambda}|\lambda = 1, 2, \ldots, \#h\}$, the hesitancy index of h is defined as:

$$\hbar(h) = \begin{cases} \left(\sum_{\xi > \zeta = 1}^{\#h} \left| \gamma^{\xi} - \gamma^{\zeta} \right| \right) \Big/ \binom{\#h}{2}, & \text{if } \#h > 1 \\ 0, & \text{if } \#h = 1 \end{cases} \tag{3.1}$$

where $\gamma^{\xi} \in [0, 1]$ and $\gamma^{\zeta} \in [0, 1]$ are the ξth and ζth smallest values in h, respectively, and $\binom{\#h}{2} = \frac{1}{2}\#h \times (\#h - 1)$, which will be used thereafter.

Proposition 3.1 (Zhang and Xu 2015). *Let h be a HFE and h^c be the complement of h, then the hesitancy indices of the HFEs h and h^c, denoted by $\hbar(h)$ and $\hbar(h^c)$ respectively, satisfy the following properties:*

(1) $0 \leq \hbar(h) \leq 1$;
(2) $\hbar(h) = \hbar(h^c)$.

Proof

(1) It is straightforward.
(2) Because

$$\begin{aligned} \hbar(h^c) &= \begin{cases} \left(\sum_{\xi > \zeta = 1}^{\#h} \left| (1 - \gamma^{\zeta}) - (1 - \gamma^{\xi}) \right| \right) \Big/ \binom{\#h}{2}, & \text{if } \#h > 1 \\ 0, & \text{if } \#h = 1 \end{cases} \\ &= \begin{cases} \left(\sum_{\xi > \zeta = 1}^{\#h} \left| \gamma^{\xi} - \gamma^{\zeta} \right| \right) \Big/ \binom{\#h}{2}, & \text{if } \#h > 1 \\ 0, & \text{if } \#h = 1 \end{cases} \\ &= \hbar(h). \end{aligned}$$

This completes the proof (Zhang and Xu 2015). □

As we know that the hesitancy index $\hbar(h)$ is a measure of the pair-wise deviations among possible values in the HFE h. In practical decision making problems, the hesitancy index can be interpreted as the degree of hesitation of an organization or individual. For example, a decision organization including several experts is authorized to estimate the degree to which an alternative should satisfy a criterion. The organization may hesitate among 0.3, 0.5 and 0.6, and provides the assessment represented by a HFE as $h = H\{0.3, 0.5, 0.6\}$. Using Eq. (3.1), we can calculate the hesitancy index of h as:

$$\hbar(h) = \frac{2}{3 \times 2} \times (|0.3 - 0.5| + |0.3 - 0.6| + |0.5 - 0.6|) = 0.2$$

which indicates that the degree of hesitation of the organization is 0.2. Generally, the larger the range among these different values in a HFE is, the greater the hesitancy index of a HFE is.

Considering the impact of hesitancy indices of HFEs, Zhang and Xu (2015) defined a partial order of HFEs:

Definition 3.2 (Zhang and Xu 2015). For any two HFEs $h_i = H\{\gamma_i^\lambda | \lambda = 1, 2, \ldots, \#h_i\} (i = 1, 2)$ with $\#h = \#h_i$, the partial order of HFEs denoted by $\leq_2$ is defined as:

$$h_1 \leq_2 h_2 \text{ if and only if } \gamma_1^\lambda \leq \gamma_2^\lambda (\lambda = 1, 2, \ldots, \#h) \text{ and } \hbar(h_1) \geq \hbar(h_2)$$

It is easily observed from Definition 3.2 that the HFE $\tilde{1} = \{\gamma^\lambda = 1 | \lambda = 1, 2, \ldots, \#\tilde{1}\}$ is the maximal HFE, which is also called an ideal HFE.

3.1.2 *The Improved Distance Measures for HFEs*

In the process of handling hesitant fuzzy information, such as the comparison of HFEs and the distance measure between HFEs, etc., the hesitancy index plays an important role and makes a crucial impact on the final process results. Thus, Zhang and Xu (2015) put forward two improved distance measures for HFEs by taking their hesitancy indices into account.

Definition 3.3 (Zhang and Xu 2015). For two HFEs $h_i = H\{\gamma_i^\lambda | \lambda = 1, 2, \ldots, \#h_i\} (i = 1, 2)$ with $\#h = \#h_i$, the improved hesitant fuzzy Hamming distance between them can be defined as:

$$d_{NH}(h_1, h_2) = \frac{1}{2}\left(\frac{1}{\#h}\sum_{\lambda=1}^{\#h} |\gamma_1^\lambda - \gamma_2^\lambda| + |\hbar(h_1) - \hbar(h_2)|\right) \tag{3.2}$$

and the improved hesitant fuzzy Euclidean distance as:

$$d_{NE}(h_1, h_2) = \sqrt{\frac{1}{2}\left(\frac{1}{\#h}\sum_{\lambda=1}^{\#h} \left(\gamma_1^\lambda - \gamma_2^\lambda\right)^2 + \left(\hbar(h_1) - \hbar(h_2)\right)^2\right)} \tag{3.3}$$

where

$$\hbar(h_i) = \begin{cases} \left(\sum_{\xi > \zeta = 1}^{\#h_i} \left|\gamma_i^\xi - \gamma_i^\zeta\right|\right) \Big/ \binom{\#h_i}{2}, & \text{if } \#h_i > 1 \\ 0, & \text{if } \#h_i = 1 \end{cases}, \quad (i = 1, 2) \tag{3.4}$$

Theorem 3.1 (Zhang and Xu 2015). *For three HFEs $h_i = H\{\gamma_i^\lambda | \lambda = 1, 2, \ldots, \#h_i\}$ $(i = 1, 2, 3)$ with $\#h = \#h_i$, d_{NH} and d_{NE} are the improved hesitant fuzzy Hamming and Euclidean distance measures, respectively, then they possess the following properties:*

(1) $0 \le d_{NH}(h_1, h_2), d_{NE}(h_1, h_2) \le 1$;
(2) $d_{NH}(h_1, h_2) = d_{NE}(h_1, h_2) = 0$, *if and only if* $h_1 = h_2$;
(3) $d_{NH}(h_1, h_2) = d_{NH}(h_2, h_1)$ *and* $d_{NE}(h_1, h_2) = d_{NE}(h_2, h_1)$;
(4) *if* $h_1 \le_2 h_2 \le_2 h_3$, then $d_{NH}(h_1, h_2) \le d_{NH}(h_1, h_3), d_{NH}(h_2, h_3) \le d_{NH}(h_1, h_3), d_{NE}(h_1, h_2) \le d_{NE}(h_1, h_3)$ *and* $d_{NE}(h_2, h_3) \le d_{NE}(h_1, h_3)$.

Proof

(1) It is straightforward;
(2) If $d_{NH}(h_1, h_2) = d_{NE}(h_1, h_2) = 0$, that is to say, $\frac{1}{\#h}\sum_{\lambda=1}^{\#h} |\gamma_1^\lambda - \gamma_2^\lambda| = 0$ and $|\hbar(h_1) - \hbar(h_2)| = 0$, then $\gamma_1^\lambda - \gamma_2^\lambda = 0 (\lambda = 1, 2, \ldots, \#h)$, thus $h_1 = h_2$. While if $h_1 = h_2$, i.e., $\gamma_1^\lambda - \gamma_2^\lambda = 0$ $(\lambda = 1, 2, \ldots, \#h)$, it can be easily derived that $d_{NH}(h_1, h_2) = d_{NE}(h_1, h_2) = 0$. Thus, we can conclude that $d_{NH}(h_1, h_2) = d_{NE}(h_1, h_2) = 0$ if and only if $h_1 = h_2$.
(3) Since

$$\begin{aligned} d_{NH}(h_2, h_1) &= \frac{1}{2}\left(\frac{1}{\#h}\sum_{\lambda=1}^{\#h} |\gamma_2^\lambda - \gamma_1^\lambda| + |\hbar(h_2) - \hbar(h_1)|\right) \\ &= \frac{1}{2}\left(\frac{1}{\#h}\sum_{\lambda=1}^{\#h} |\gamma_1^\lambda - \gamma_2^\lambda| + |\hbar(h_1) - \hbar(h_2)|\right) \\ &= d_{NH}(h_1, h_2) \end{aligned}$$

and

$$\begin{aligned} d_{NE}(h_2, h_1) &= \sqrt{\frac{1}{2}\left(\frac{1}{\#h}\sum_{\lambda=1}^{\#h} \left(\gamma_2^\lambda - \gamma_1^\lambda\right)^2 + (\hbar(h_2) - \hbar(h_1))^2\right)} \\ &= \sqrt{\frac{1}{2}\left(\frac{1}{\#h}\sum_{\lambda=1}^{\#h} \left(\gamma_1^\lambda - \gamma_2^\lambda\right)^2 + (\hbar(h_1) - \hbar(h_2))^2\right)} \\ &= d_{NE}(h_1, h_2). \end{aligned}$$

Thus, $d_{NH}(h_1, h_2) = d_{NH}(h_2, h_1)$ and $d_{NE}(h_1, h_2) = d_{NE}(h_2, h_1)$.
(4) If $h_1 \le_2 h_2 \le_2 h_3$, according to Definition 3.2, then we can obtain $\gamma_1^\lambda \le \gamma_2^\lambda \le \gamma_3^\lambda (\lambda = 1, 2, \ldots, \#h)$ and $\hbar(h_1) \ge \hbar(h_2) \ge \hbar(h_3)$, thus

$$\begin{aligned} &|\gamma_1^\lambda - \gamma_2^\lambda| \le |\gamma_1^\lambda - \gamma_3^\lambda|, \quad |\gamma_2^\lambda - \gamma_3^\lambda| \le |\gamma_1^\lambda - \gamma_3^\lambda|, \\ &|\hbar(h_1) - \hbar(h_2)| \le |\hbar(h_1) - \hbar(h_3)|, \quad |\hbar(h_2) - \hbar(h_3)| \le |\hbar(h_1) - \hbar(h_3)|. \end{aligned}$$

Furthermore, we can obtain

$$\frac{1}{2}\left(\frac{1}{\#h}\sum_{\lambda=1}^{\#h}\left|\gamma_1^{\lambda}-\gamma_2^{\lambda}\right|+\left|\hbar(h_1)-\hbar(h_2)\right|\right)$$
$$\leq\frac{1}{2}\left(\frac{1}{\#h}\sum_{\lambda=1}^{\#h}\left|\gamma_1^{\lambda}-\gamma_3^{\lambda}\right|+\left|\hbar(h_1)-\hbar(h_3)\right|\right),$$
$$\frac{1}{2}\left(\frac{1}{\#h}\sum_{\lambda=1}^{\#h}\left|\gamma_2^{\lambda}-\gamma_3^{\lambda}\right|+\left|\hbar(h_2)-\hbar(h_3)\right|\right)$$
$$\leq\frac{1}{2}\left(\frac{1}{\#h}\sum_{\lambda=1}^{\#h}\left|\gamma_1^{\lambda}-\gamma_3^{\lambda}\right|+\left|\hbar(h_1)-\hbar(h_3)\right|\right),$$
$$\sqrt{\frac{1}{2}\left(\frac{1}{\#h}\sum_{\lambda=1}^{\#h}\left(\gamma_1^{\lambda}-\gamma_2^{\lambda}\right)^2+\left(\hbar(h_1)-\hbar(h_2)\right)^2\right)}$$
$$\leq\sqrt{\frac{1}{2}\left(\frac{1}{\#h}\sum_{\lambda=1}^{\#h}\left(\gamma_1^{\lambda}-\gamma_3^{\lambda}\right)^2+\left(\hbar(h_1)-\hbar(h_3)\right)^2\right)},$$
$$\sqrt{\frac{1}{2}\left(\frac{1}{\#h}\sum_{\lambda=1}^{\#h}\left(\gamma_2^{\lambda}-\gamma_3^{\lambda}\right)^2+\left(\hbar(h_2)-\hbar(h_3)\right)^2\right)}$$
$$\leq\sqrt{\frac{1}{2}\left(\frac{1}{\#h}\sum_{\lambda=1}^{\#h}\left(\gamma_1^{\lambda}-\gamma_3^{\lambda}\right)^2+\left(\hbar(h_1)-\hbar(h_3)\right)^2\right)}.$$

Namely, $d_{NH}(h_1,h_2)\leq d_{NH}(h_1,h_3)$, $d_{NH}(h_2,h_3)\leq d_{NH}(h_1,h_3)$, $d_{NE}(h_1,h_2)\leq d_{NE}(h_1,h_3)$ and $d_{NE}(h_2,h_3)\leq d_{NE}(h_1,h_3)$.

This completes the proof (Zhang and Xu 2015). □

Zhang and Xu (2015) gave a numerical example to demonstrate the advantage of the improved hesitant fuzzy Hamming and Euclidean distance measures compared with the corresponding distance measures introduced in Eqs. (1.1) and (1.2).

Example 3.1 (Zhang and Xu 2015). Given three HFEs:

$$h_1=H\{0.3,0.5\},h_2=H\{0.2,0.3\}\quad\text{and}\quad h_3=H\{0.2,0.7\},$$

we need to calculate the distances between h_1 than h_2 as well as between h_1 and h_3, respectively.

Using the hesitant fuzzy distance measures presented in Eqs. (1.1)–(1.2) and in Eqs. (3.2)–(3.3) respectively, it is easy to calculate the corresponding distances between h_1 than h_2 as well as between h_1 and h_3. The derived results are listed in Table 3.1 (Zhang and Xu 2015).

Table 3.1 The distances of HFEs obtained by different distance measures

The different distance measures	The distances of HFEs
Hesitant fuzzy Hamming distance (Eq. 1.1)	$d_H(h_1, h_2) = d_H(h_1, h_3) = 0.15$
Hesitant fuzzy Euclidean distance (Eq. 1.2)	$d_E(h_1, h_2) = d_E(h_1, h_3) = 0.1581$
Improved hesitant fuzzy Hamming distance (Eq. 3.2)	$d_{NH}(h_1, h_2) = 0.125 < d_{NH}(h_1, h_3) = 0.225$
Improved hesitant fuzzy Euclidean distance (Eq. 3.3)	$d_{NE}(h_1, h_2) = 0.1323 < d_{NE}(h_1, h_3) = 0.24$

It is easily seen from Table 3.1 that we cannot carry out the comparison in such a case by using the hesitant fuzzy distance measures presented in Eqs. (1.1) and (1.2). This will get the decision maker into trouble in practical application. While using the hesitant fuzzy distance measures presented in Eqs. (3.2) and (3.3), we can clearly know that the HFE h_1 is farther from h_3 than h_2, which are in accordance with people's intuition. Therefore, the proposed distance measures are more reasonable than the distance measures presented in Definition 1.5.

3.1.3 *The Comparison Methods Based on Hesitancy Indices for HFEs*

By simultaneously considering the scores and hesitancy indices of HFEs, Zhang and Xu (2015) presented a new method to compare the magnitude of HFEs.

Definition 3.4 (Zhang and Xu 2015). Let h_1 and h_2 be two HFEs, $s(h_1)$ and $s(h_2)$ be the scores of h_1 and h_2, respectively, $\hbar(h_1)$ and $\hbar(h_2)$ be hesitancy indices of h_1 and h_2, respectively. Then, we have

(1) if $s(h_1) < s(h_2)$, then $h_1 \prec_s h_2$;

(2) if $s(h_1) = s(h_2)$ then $\begin{cases} \hbar(h_1) = \hbar(h_2) \Rightarrow h_1 \sim_{s\hbar} h_2 \\ \hbar(h_1) > \hbar(h_2) \Rightarrow h_1 \prec_{s\hbar} h_2 \end{cases}$;

(3) if $s(h_1) > s(h_2)$, then $h_1 \succ_s h_2$.

Example 3.2 (Zhang and Xu 2015). Given three HFEs:

$$h_1 = H\{0.3, 0.5\},\ h_2 = H\{0.4\},\ h_3 = H\{0.2, 0.4, 0.6\},$$

and we need to rank these three HFEs.

According to the scores and hesitancy indices of HFEs, we can obtain:

$$s(h_1) = s(h_2) = s(h_3) = 0.4,\ \hbar(h_1) = 0.2,\ \hbar(h_2) = 0,\ \hbar(h_3) = 0.2667.$$

By using the proposed comparison approach defined in Definition 3.4, we get $h_2 \succ_{s\hbar} h_1 \succ_{s\hbar} h_3$.

It is noted that the above comparison approach of HFEs (Definition 3.4) is the algorithmic ranking approach which may make the decision making process more time-consuming. To avoid this shortcoming, Zhang and Xu (2015) further developed a signed distance-based comparison approaches for HFEs which is not only valuable for modeling decision making problems, but is also directly applicable to deal with ordering HFEs in the practical decision making process.

Motivated by the concept of signed distance (Chiang 2001), Zhang and Xu (2015) introduced a signed distance of HFEs.

Definition 3.5 (Zhang and Xu 2015). Let $h = H\{\gamma^\lambda | \lambda = 1, 2, \ldots, \#h\}$ be a HFE and $\tilde{1}$ be an ideal HFE, then the signed distance from h to $\tilde{1}$ can be defined as follows:

$$d_S(h, \tilde{1}) = \begin{cases} \frac{1}{2}\left(\frac{\sum_{\lambda=1}^{\#h}(1-\gamma^\lambda)}{\#h} + \frac{\sum_{\xi>\zeta=1}^{\#h}|\gamma^\xi - \gamma^\zeta|}{\binom{\#h}{2}}\right), & \#h > 1 \\ \frac{1-\gamma}{2}, & \#h = 1 \end{cases} \tag{3.5}$$

Proposition 3.2 (Zhang and Xu 2015). *Let h, h_1 and h_2 be three HFEs, and $\tilde{1}$ be an ideal HFE, then the signed distance $d_S(h, \tilde{1})$ satisfies the following properties:*

(1) $0 \le d_S(h, \tilde{1}) < 1$;
(2) *h is located at $\tilde{1}$ if and only if $d_S(h, \tilde{1}) = 0$;*
(3) *h_1 is farther from $\tilde{1}$ than h_2 if and only if $d_S(h_1, \tilde{1}) > d_S(h_2, \tilde{1})$.*

Proof

(1) It is straightforward;
(2) For the necessity, if h is located at $\tilde{1}$, then $\gamma^\lambda = 1 (\lambda = 1, 2, \ldots, \#h)$, thus the signed distance $d_S(h, \tilde{1}) = 0$. For the sufficiency, if $d_S(h, \tilde{1}) = 0$, it implies that $\frac{1}{\#h}\sum_{\lambda=1}^{\#h}(1-\gamma^\lambda) = 0$ and $\frac{1}{\binom{\#h}{2}}\sum_{\lambda>\delta=1}^{\#h}|\gamma^\lambda - \gamma^\delta| = 0$, obviously, we have $\gamma^\lambda = 1 (\lambda = 1, 2, \ldots, \#h)$. Therefore, h is located at $\tilde{1}$.
(3) For the necessity, because the signed distance between h_1 and $\tilde{1}$ is $d_S(h_1, \tilde{1})$, and similarly, the signed distance between h_2 and $\tilde{1}$ is $d_S(h_2, \tilde{1})$, thus if h_2 is farther from $\tilde{1}$ than h_2, then $d_S(h_1, \tilde{1}) > d_S(h_2, \tilde{1})$. For the sufficiency, because $0 \le d_S(h_1, \tilde{1}) \le 1$ and $0 \le d_S(h_2, \tilde{1}) \le 1$, thus if $d_S(h_1, \tilde{1}) > d_S(h_2, \tilde{1})$, then h_1 is farther from $\tilde{1}$ than h_2.

This completes the proof (Zhang and Xu 2015). □

It is easily seen that for any two HFEs $h_i (i = 1, 2)$, the signed distances $d_S(h_1, \tilde{1})$ and $d_S(h_2, \tilde{1})$ are real numbers, and they satisfy linear ordering, namely, one of the following three conditions must hold: $d_S(h_1, \tilde{1}) > d_S(h_2, \tilde{1})$, $d_S(h_1, \tilde{1}) = d_S(h_2, \tilde{1})$ or $d_S(h_1, \tilde{1}) < d_S(h_2, \tilde{1})$. Therefore, a signed distance-based comparison approach for HFEs is presented as follows.

Table 3.2 The comparison results of orders of HFEs

The comparison approach	Ranking orders of HFEs
Definition 3.4	$h_2 \succ_{sh} h_1 \succ_{sh} h_3$
Definition 3.6	$h_2 \succ_{d} h_1 \succ_{d} h_3$

Definition 3.6 (Zhang and Xu 2015). Let $h_i(i = 1, 2)$ be two HFEs and $\tilde{1}$ be an ideal HFE, $d_S(h_1, \tilde{1})$ and $d_S(h_2, \tilde{1})$ be the signed distances, and then the ranking of HFEs can be defined as:

(1) If $d_S(h_1, \tilde{1}) > d_S(h_2, \tilde{1})$, then h_1 is inferior to h_2, denoted by $h_1 \prec_d h_2$;
(2) If $d_S(h_1, \tilde{1}) < d_S(h_2, \tilde{1})$, then h_1 is superior to h_2, denoted by $h_1 \succ_d h_2$;
(3) If $d_S(h_1, \tilde{1}) = d_S(h_2, \tilde{1})$, then h_1 is indifferent to h_2, denoted by $h_1 \sim_d h_2$.

Example 3.3 (Zhang and Xu 2015). Consider three HFEs $h_i(i = 1, 2, 3)$ given in Example 3.2, and let $\tilde{1}$ be an ideal HFE. We need to rank these three HFEs.

Using Eq. (3.5), we can obtain:

$$d_S(h_1, \tilde{1}) = 0.4, \quad d_S(h_2, \tilde{1}) = 0.3 \quad \text{and} \quad d_S(h_3, \tilde{1}) = 0.4335.$$

According to the comparison approach given in Definition 3.6, we get $h_2 \succ_d h_1 \succ_d h_3$.

In order to provide a synthetic view of the comparison results, we put all the results of the ranking of alternatives with different approaches into Table 3.2.

As shown in Table 3.2, the ranking of HFEs obtained by Definition 3.6 is the same as the result obtained by Definition 3.4. However, the use of the approach developed in Definition 3.4 makes the decision making process more time-consuming. Therefore, we can very confidently say that the ranking approach proposed in Definition 3.6 is much superior to the approach in Definition 3.4 in the practical decision making process.

3.2 Hesitant Fuzzy QUALIFLEX Approach with a Signed Distance-Based Ranking Method

The MCDM is to identify the desirable compromise solution from the set of all feasible alternatives which are assessed based on a set of conflicting criteria. We here consider a MCDM problem under the hesitant fuzzy environment. Let $\boldsymbol{A} = \{A_1, A_2, \ldots, A_m\}$ $(m \geq 2)$ be a discrete set of m feasible alternatives, $\boldsymbol{C} = \{C_1, C_2, \ldots, C_n\}$ be a finite set of criteria. Each alternative is assessed on each criterion and the assessment value of the alternative $A_i \in \boldsymbol{A}$ with respect to the criterion $C_j \in \boldsymbol{C}$ is expressed by a HFE $h_{ij} = H\left\{\gamma_{h_{ij}}^1, \gamma_{h_{ij}}^2, \ldots, \gamma_{h_{ij}}^{\#h_{ij}}\right\}$. Therefore, the MCDM problem with hesitant fuzzy information can be concisely expressed in the matrix format as

$\Re = \left(h_{ij}\right)_{m\times n}$. In the real-life decision making process, the weights of criteria should be taken into account. Here we denote the criteria weight vector by $\boldsymbol{W} = (W_1, W_2, \ldots, W_n)^T$, where W_j is the relative weight of the criterion C_j. Similarly, the HFEs can be used to express the importance weight for various decision criteria during the evaluation process. That is to say, each $W_j (j = 1, 2, \ldots, n)$ can also be expressed as a HFE denoted by $W_j = H\left\{\gamma_{W_j}^1, \gamma_{W_j}^2, \ldots, \gamma_{W_j}^{\#W_j}\right\}$.

To solve the MCDM problems in which both the assessments of alternatives with respect to criteria and the weights of criteria are expressed by HFEs, Zhang and Xu (2015) proposed a hesitant fuzzy QUALIFLEX approach with the signed distance-based comparison method. We call it the HF-QUALIFLEX approach. The HF-QUALIFLEX method starts with the computation of the concordance/discordance index based on successive permutations of all possible rankings of alternatives. Considering that the decision data takes the form of HFEs, Zhang and Xu (2015) utilized the signed distance-based comparison approach of HFEs proposed in Sect. 3.1.3 to identify the corresponding concordance/discordance index.

Given the set of alternatives $\boldsymbol{A} = \{A_1, A_2, \ldots, A_m\}$, there exist $m!$ permutations of the ranking of the alternatives. Let $\mathscr{L}_\rho$ denote the ρth permutation as:

$$\mathscr{L}_\rho = (\ldots, A_\xi, \ldots, A_\zeta, \ldots), \quad \rho = 1, 2, \ldots, m! \tag{3.6}$$

where $A_\xi, A_\zeta \in \boldsymbol{A}$ and the alternative A_ξ is ranked higher than or equal to A_ζ.

The concordance/discordance index $\mathscr{D}_j^\rho(A_\xi, A_\zeta)$ for each pair of alternatives (A_ξ, A_ζ), A_ξ, $A_\zeta \in \boldsymbol{A}$, at the level of preorder according to the jth criterion and the ranking corresponding to the ρth permutation, can be defined as follows (Zhang and Xu 2015):

$$\mathscr{D}_j^\rho(A_\xi, A_\zeta) = d_S\left(h_{\zeta j}, \tilde{1}\right) - d_S\left(h_{\xi j}, \tilde{1}\right) \tag{3.7}$$

Based on the signed distance-based comparison method of HFEs, it is easily concluded from Eq. (3.7) that (Zhang and Xu 2015):

(1) if $\mathscr{D}_j^\rho(A_\xi, A_\zeta) > 0$, that is, $d_S(h_{\xi j}, \tilde{1}) < d_S(h_{\zeta j}, \tilde{1})$, then A_ξ ranks over A_ζ under the jth criterion, thus there is concordance between the signed distance-based ranking orders and the preorders of A_ξ and A_ζ under the ρth permutation.
(2) if $\mathscr{D}_j^\rho(A_\xi, A_\zeta) = 0$, that is, $d_S(h_{\xi j}, \tilde{1}) = d_S(h_{\zeta j}, \tilde{1})$, then both A_ξ and A_ζ have the same rank in the signed distance-based ranking and in the ρth permutation, thus there is ex aequo.
(3) if $\mathscr{D}_j^\rho(A_\xi, A_\zeta) < 0$, that is, $d_S(h_{\xi j}, \tilde{1}) > d_S(h_{\zeta j}, \tilde{1})$, then A_ζ ranks over A_ξ, thus there is discordance between the signed distance-based ranking orders and the preorders of A_ξ and A_ζ under the ρth permutation.

For convenience of understanding, the concordance/discordance index $\mathscr{D}_j^\rho(A_\xi, A_\zeta)$ can also be rewritten as follows (Zhang and Xu 2015):

$$\mathscr{D}_j^\rho(A_\xi, A_\zeta) = \begin{cases} d_S(h_{\zeta j}, \tilde{1}) - d_S(h_{\xi j}, \tilde{1}) > 0 \Leftrightarrow \textit{there is concordance} \\ d_S(h_{\zeta j}, \tilde{1}) - d_S(h_{\xi j}, \tilde{1}) = 0 \Leftrightarrow \textit{there is ex aequo} \\ d_S(h_{\zeta j}, \tilde{1}) - d_S(h_{\xi j}, \tilde{1}) < 0 \Leftrightarrow \textit{there is discordance} \end{cases} \tag{3.8}$$

Due to the weights of criteria being expressed as HFEs, we also utilize the signed distance of HFEs to calculate the weighted concordance/discordance index $\mathscr{D}_j^\rho(A_\xi, A_\zeta)$ for each pair of alternatives (A_ξ, A_ζ), at the level of preorder with respect to the n criteria and the ranking corresponding to the ρ permutation as (Zhang and Xu 2015):

$$\begin{aligned} \mathscr{D}^\rho(A_\xi, A_\zeta) &= \sum_{j=1}^{n} \mathscr{D}_j^\rho(A_\xi, A_\zeta) \;*\; \left(1 - d_S(W_j, \tilde{1})\right) \\ &= \sum_{j=1}^{n} \left(1 - d_S(W_j, \tilde{1})\right) \;*\; \left(d_S(h_{\zeta j}, \tilde{1}) - d_S(h_{\xi j}, \tilde{1})\right) \end{aligned} \tag{3.9}$$

At length, the comprehensive concordance/discordance index $\mathscr{D}^\rho$ for the ρth permutation can be defined as follows (Zhang and Xu 2015):

$$\mathscr{D}^\rho = \sum_{A_\xi, A_\zeta \in A} \sum_{j=1}^{n} \left(1 - d_S(W_j, \tilde{1})\right) * \left(d_S(h_{\zeta j}, \tilde{1}) - d_S(h_{\xi j}, \tilde{1})\right) \tag{3.10}$$

According to the signed distance-based comparison method of HFEs, it is easily seen that the bigger the comprehensive concordance/discordance index is, the better the final ranking result of alternatives is. Therefore, the optimal ranking order of alternatives can be determined by comparing the values $\mathscr{D}^\rho$ of each permutation $\mathscr{L}_\rho$, which is the permutation with the maximal comprehensive concordance/discordance index $\mathscr{D}^\rho$, namely,

$$\mathscr{L}^* = \left\{ \mathscr{D}^* : \max_{\rho=1}^{m!} \mathscr{D}^\rho \right\} \tag{3.11}$$

Based on the above analysis, the algorithm (Algorithm 3.1) of the HF-QUALIFLEX method for solving the hesitant fuzzy MCDM problems can be summarized as (Zhang and Xu 2015):

Step 1. Formulate the hesitant fuzzy MCDM problem and identify the hesitant fuzzy decision matrix as well as hesitant fuzzy weighted values of criteria.

Step 2. List all of the possible $m!$ permutation of the m alternatives that should be tested in the next steps. Let $\mathscr{L}_\rho$ denote the ρth permutation using Eq. (3.6).

Step 3. Calculate the concordance/discordance index $\mathscr{D}_j^\rho(A_\xi, A_\zeta)$ using Eq. (3.7).

Step 4. Compute the weighted concordance/discordance indices $\mathscr{D}^{\rho}(A_{\xi}, A_{\zeta})$ using Eq. (3.9).

Step 5. Calculate the comprehensive concordance/discordance index $\mathscr{D}^{\rho}$ for the permutation $\mathscr{L}_{\rho}$ by using Eq. (3.10).

Step 6. Determine the optimal ranking order of all alternatives, which is the permutation with the maximal comprehensive concordance/discordance index by using Eq. (3.11).

3.3 A Case Study for Green Supplier Selection

Zhang and Xu (2015) gave a green supplier selection problem in an automobile manufacturing company to demonstrate the applicability and the implementation process of the proposed method.

3.3.1 Decision Context

Owing to the increasing public awareness of the need to protect the environment and regulatory pressures coming from governments and NGOs, businesses are progressively promoting practices to help ease further negative impacts on the environment (Zhu and Sarkis 2004). Automobile manufacturing companies have begun to implement green practices at all stages of the manufacturing process to achieve profit and market share objectives by lowering their environmental impacts and increasing their ecological efficiency. Selecting green suppliers according to their environmental criteria will improve the company's environmental performance and by extension, their competitive advantages because many automobile parts are outsourced to suppliers (Shen et al. 2013). Thus, we here consider a real-life MCDM problem that is how an automobile manufacturing company to select a suitable green supplier from several potential suppliers. Assume that there are three possible green suppliers $\{A_1, A_2, A_3\}$ to be selected. According to the discussion of the experts, the major nine criteria including pollution production, resource consumption, eco-design, green image, environmental management system, commitment of GSCM from managers, use of environmentally friendly technology, use of environmentally friendly materials and staff environmental training, are determined to evaluate these three possible suppliers, which are listed in Table 3.3 (Shen et al. 2013). A decision organization including an operations manager (e_1), a purchasing manager (e_2) and an environmental manager (e_3) is invited to assess the performance of these three potential suppliers under each criterion and to provide the weights of criteria.

Table 3.3 Criteria for selecting and evaluating green suppliers

Criteria	Name	Definition
C_1	Pollution production	Average volume of air emission pollutant, waste water, solid wastes and harmful materials releases per day during measurement period
C_2	Resource consumption	Resource consumption in terms of raw material, energy and water during the measurement period
C_3	Eco-design	Design of products for reduced consumption of material/energy, design of products for reuse, recycle, recovery of material, design of products
C_4	Green image	The ratio of green customers to total customers
C_5	Environmental management system	Environmental certifications such as ISO 14000, environmental policies, planning of environmental objectives, checking and control of environmental activities
C_6	Commitment of GSCM from managers	Senior and mid-level managers commitment and support to improve green supply chain management practices and environmental performance
C_7	Use of environmentally friendly technology	The application of the environmental science to conserve the natural environment and resources, and to curb the negative impacts of human involvement
C_8	Use of environmentally friendly materials	The level of green recyclable material used in packaging and manufacturing of goods
C_9	Staff environmental training	Staff training on environmental targets

The original assessments of suppliers on each criterion and the weight information of criteria provided by the three experts are listed in Table 3.4, respectively. For an alternative under a criterion, although all of the experts provide their evaluation values, some of these values may be repeated. It is noted that a value repeated more times does not indicate that it has more importance than other values repeated less times. Because the value repeated one time may be provided by an expert who is an expert at this area, while the value repeated twice may be provided by two experts who are not familiar with this area. In such cases, the value repeated one time may be more important than the one repeated twice. To get a more reasonable result, it is better that the experts give their evaluations anonymously. We only collect all of the possible values for an alternative under a criterion, and each value provided only means that it is a possible value, but its importance is unknown. Thus, the times that the values repeated are unimportant, and it is reasonable to allow these values repeated many times appear only once. The HFE is just a tool to deal with such cases, and all possible evaluations for an alternative under each criterion can be considered as a HFE (Xu and Xia 2011). The collective opinions of the original assessments of suppliers with respect to criteria and the weights of criteria provided by the decision organization are taken as HFEs, listed in Table 3.4 (Zhang and Xu 2015), respectively.

Table 3.4 Ratings of the green suppliers by experts under various criteria

Criteria	Supplier A_1				Supplier A_2			
	e_1	e_2	e_3	Collective opinion	e_1	e_2	e_3	Collective opinion
C_1	0.3	0.6	0.7	H{0.3, 0.6, 0.7}	0.3	0.3	0.5	H{0.3, 0.5}
C_2	0.5	0.7	0.7	H{0.5, 0.7}	0.4	0.4	0.4	H{0.4}
C_3	0.4	0.6	0.6	H{0.4, 0.6}	0.5	0.6	0.7	H{0.5, 0.6, 0.7}
C_4	0.5	0.6	0.7	H{0.5, 0.6, 0.7}	0.2	0.4	0.3	H{0.2, 0.3, 0.4}
C_5	0.5	0.5	0.5	H{0.5}	0.5	0.3	0.3	H{0.3, 0.5}
C_6	0.7	0.6	0.7	H{0.6, 0.7}	0.6	0.6	0.6	H{0.6}
C_7	0.4	0.6	0.9	H{0.4, 0.6, 0.9}	0.8	0.6	0.8	H{0.6, 0.8}
C_8	0.8	0.6	0.8	H{0.6, 0.8}	0.9	0.9	0.7	H{0.7, 0.9}
C_9	0.7	0.8	0.9	H{0.7, 0.8, 0.9}	0.7	0.6	0.8	H{0.6, 0.7, 0.8}
Criteria	Supplier A_3				Weights of criteria			
	e_1	e_2	e_3	Collective opinion	e_1	e_2	e_3	Collective opinion
C_1	0.4	0.6	0.5	H{0.4, 0.5, 0.6}	0.6	0.7	0.6	H{0.6, 0.7}
C_2	0.6	0.6	0.4	H{0.4, 0.6}	0.5	0.6	0.6	H{0.5, 0.6}
C_3	0.5	0.5	0.5	H{0.5}	0.5	0.5	0.5	H{0.5}
C_4	0.4	0.6	0.3	H{0.3, 0.4, 0.6}	0.5	0.4	0.3	H{0.3, 0.4, 0.5}
C_5	0.4	0.4	0.4	H{0.4}	0.5	0.2	0.2	H{0.2, 0.5}
C_6	0.5	0.6	0.8	H{0.5, 0.6, 0.8}	0.4	0.4	0.4	H{0.4}
C_7	0.5	0.8	0.7	H{0.5, 0.7, 0.8}	0.3	0.4	0.6	H{0.3, 0.4, 0.6}
C_8	0.7	0.7	0.7	H{0.7}	0.8	0.8	0.7	H{0.7, 0.8}
C_9	0.9	0.9	0.6	H{0.6, 0.9}	0.2	0.4	0.4	H{0.2, 0.4}

3.3.2 Illustration of the Proposed Method

In the following, the HF-QUALIFLEX approach proposed Zhang and Xu (2015) is employed to help the managers of the company select a suitable green supplier from three potential suppliers.

In Steps 1–2, there are 6(=3!) permutations of the rankings for all alternatives that must be tested:

$$\mathscr{L}_1 = (A_1, A_2, A_3), \mathscr{L}_2 = (A_1, A_3, A_2), \mathscr{L}_3 = (A_2, A_1, A_3),$$
$$\mathscr{L}_4 = (A_2, A_3, A_1), \mathscr{L}_5 = (A_3, A_1, A_2), \mathscr{L}_6 = (A_3, A_2, A_1).$$

In Step 3, for each pair of alternatives (A_ξ, A_ζ) in the permutation $\mathscr{L}_\rho(\rho \in \{1, 2, \ldots, 6\})$ with respect to each criterion C_j, the concordance/discordance index $\mathscr{D}_j^\rho(A_\xi, A_\zeta)$ can be calculated by employing Eq. (3.7), and the results are presented in Table 3.5 (Zhang and Xu 2015).

Table 3.5 The results of the concordance/discordance index

$\mathscr{L}_1$	$\mathscr{D}_j^1(A_1,A_2)$	$\mathscr{D}_j^1(A_1,A_3)$	$\mathscr{D}_j^1(A_2,A_3)$	$\mathscr{L}_2$	$\mathscr{D}_j^2(A_1,A_3)$	$\mathscr{D}_j^2(A_1,A_2)$	$\mathscr{D}_j^2(A_3,A_2)$
C_1	0.067	−0.1	−0.167	C_1	−0.1	0.067	0.167
C_2	0	0.1	0.1	C_2	0.1	0	−0.1
C_3	−0.167	−0.2	−0.33	C_3	−0.2	−0.167	0.33
C_4	0.3	0.234	−0.066	C_4	0.234	0.3	0.066
C_5	0.3	−0.1	−0.4	C_5	−0.1	0.3	0.4
C_6	−0.35	−0.183	−0.167	C_6	−0.183	−0.35	0.167
C_7	−0.167	−0.134	0.033	C_7	−0.134	−0.167	−0.033
C_8	−0.1	−0.2	−0.1	C_8	−0.2	−0.1	0.1
C_9	0.1	0.117	0.017	C_9	0.117	0.1	−0.017
$\mathscr{L}_3$	$\mathscr{D}_j^3(A_2,A_1)$	$\mathscr{D}_j^3(A_2,A_3)$	$\mathscr{D}_j^3(A_1,A_3)$	$\mathscr{L}_3$	$\mathscr{D}_j^4(A_2,A_3)$	$\mathscr{D}_j^4(A_2,A_1)$	$\mathscr{D}_j^4(A_3,A_1)$
C_1	−0.067	−0.167	−0.1	C_1	−0.167	−0.067	0.1
C_2	0	0.1	0.1	C_2	0.1	0	-0.1
C_3	0.167	−0.33	−0.2	C_3	−0.33	0.167	0.2
C_4	−0.3	−0.066	0.234	C_4	−0.066	−0.3	−0.234
C_5	−0.3	−0.4	−0.1	C_5	−0.4	−0.3	0.1
C_6	0.35	−0.167	−0.183	C_6	−0.167	0.35	0.183
C_7	0.167	0.033	−0.134	C_7	0.033	0.167	0.134
C_8	0.1	−0.1	−0.2	C_8	−0.1	0.1	0.2
C_9	−0.1	0.017	0.117	C_9	0.017	−0.1	0.117
$\mathscr{L}_5$	$\mathscr{D}_j^5(A_3,A_1)$	$\mathscr{D}_j^5(A_3,A_2)$	$\mathscr{D}_j^5(A_1,A_2)$	$\mathscr{L}_6$	$\mathscr{D}_j^6(A_3,A_2)$	$\mathscr{D}_j^6(A_3,A_1)$	$\mathscr{D}_j^6(A_2,A_1)$
C_1	0.1	0.167	0.067	C_1	0.167	0.1	−0.067
C_2	−0.1	−0.1	0	C_2	−0.1	−0.1	0
C_3	0.2	0.33	−0.167	C_3	0.33	0.2	0.167
C_4	−0.234	0.066	0.3	C_4	0.066	−0.234	−0.3
C_5	0.1	0.4	0.3	C_5	0.4	0.1	−0.3
C_6	0.183	0.167	−0.35	C_6	0.167	0.183	0.35
C_7	0.134	−0.033	−0.167	C_7	−0.033	0.134	0.167
C_8	0.2	0.1	−0.1	C_8	0.1	0.2	0.1
C_9	0.1	−0.017	0.1	C_9	−0.017	0.1	−0.1

In Step 4, we utilize Eq. (3.9) to calculate the weighted concordance/discordance index $\mathscr{D}^{\rho}(A_{\xi},A_{\zeta})$, which are listed in Table 3.6 (Zhang and Xu 2015).

In Step 5, we compute the comprehensive concordance/discordance index $\mathscr{D}^{\rho}(\rho \in \{1,2,3,4,5,6\})$ by using Eq. (3.10) as follows:

$$\mathscr{D}^1 = -0.6339, \mathscr{D}^2 = 0.1087, \mathscr{D}^3 = -0.4977,$$
$$\mathscr{D}^4 = -0.0312, \mathscr{D}^5 = 0.5573, \mathscr{D}^6 = 0.6936.$$

Table 3.6 The results of the weighted concordance/discordance index

$\mathscr{L}_1$	$\mathscr{D}^1(A_1,A_2)$	$\mathscr{D}^1(A_1,A_3)$	$\mathscr{D}^1(A_2,A_3)$	$\mathscr{L}_2$	$\mathscr{D}^2(A_1,A_3)$	$\mathscr{D}^2(A_1,A_2)$	$\mathscr{D}^2(A_3,A_2)$
	−0.0681	−0.1945	−0.3713		−0.1945	−0.0681	0.3713
$\mathscr{L}_3$	$\mathscr{D}^3(A_2,A_1)$	$\mathscr{D}^3(A_2,A_3)$	$\mathscr{D}^3(A_1,A_3)$	$\mathscr{L}_4$	$\mathscr{D}^4(A_2,A_3)$	$\mathscr{D}^4(A_2,A_1)$	$\mathscr{D}^4(A_3,A_1)$
	0.0681	−0.3713	−0.1945		−0.3582	0.0681	0.2588
$\mathscr{L}_5$	$\mathscr{D}^5(A_3,A_1)$	$\mathscr{D}^5(A_3,A_2)$	$\mathscr{D}^5(A_1,A_2)$	$\mathscr{L}_6$	$\mathscr{D}^6(A_3,A_2)$	$\mathscr{D}^6(A_3,A_1)$	$\mathscr{D}^6(A_2,A_1)$
	0.2541	0.3713	−0.0681		0.3713	0.2541	0.0681

In Step 6, according to the derived comprehensive concordance/discordance index ϕ^{ρ}, it is easily seen that $\mathscr{L}^* = \mathscr{L}_6$, namely, the ranking order of the three potential suppliers is: $A_3 \succ A_2 \succ A_1$. Therefore, the best suitable green supplier for the automobile manufacturing company is the supplier A_3.

3.3.3 *Comparative Analysis and Discussions*

To validate the results of the HF-QUALIFLEX method, Zhang and Xu (2015) made a comparative study with the hesitant fuzzy ELECTRE (HF-ELECTRE) method developed by Wang et al. (2014). This HF-ELECTRE method is mainly used to deal with the MCDM problems in which the criteria values are expressed by HFEs and weights of criteria are expressed by crisp numbers, which is the closest to the proposed HF-QUALIFLEX method. But this HF-ELECTRE method fails to handle the MCDM problems in which both the criteria values and weights of criteria are expressed as HFEs. To this end, Zhang and Xu (2015) modified the HF-ELECTRE method to facilitate the comparative analysis conducted on the same selection problem of green suppliers.

In the modified HF-ELECTRE method, we also utilize the signed distance-based comparison approach of HFEs to identify the concordance set and the discordance set. For each pair of alternatives (A_ξ, A_ζ), the concordance set $\mathbb{C}_{\xi\zeta}$ is composed of all criteria where A_ξ outranks A_ζ (A_ξ is at least as good as A_ζ) by comparing their signed distances, can be formulated as:

$$\mathbb{C}_{\xi\zeta} = \left\{ j | d_S(h_{\xi j}, \tilde{1}) \leq d_S(h_{\zeta j}, \tilde{1}) \right\} \tag{3.12}$$

and on the contrary, the discordance set $\mathbb{N}_{\xi\zeta}$ is formulated as:

$$\mathbb{N}_{\xi\zeta} = \left\{ j | d_S(h_{\xi j}, \tilde{1}) > d_S(h_{\zeta j}, \tilde{1}) \right\} \tag{3.13}$$

In the green supplier selection problem, the concordance and discordance sets can be obtained by using Eqs. (3.12) and (3.13), respectively (Zhang and Xu 2015):

$$\mathbb{C}_{12} = \{1,2,3,4,5,9\}, \mathbb{C}_{13} = \{2,4,9\}, \mathbb{C}_{21} = \{2,3,6,7,8\},$$
$$\mathbb{C}_{23} = \{2,6,9\}, \mathbb{C}_{31} = \{1,3,5,6,7,8\}, \mathbb{C}_{32} = \{1,3,4,5,8\},$$
$$\mathbb{N}_{12} = \{3,6,7,8\}, \mathbb{N}_{13} = \{1,3,5,6,7,8\}, \mathbb{N}_{21} = \{1,4,5,9\},$$
$$\mathbb{N}_{23} = \{1,3,4,5,8\}, \mathbb{N}_{31} = \{2,4,9\}, \mathbb{N}_{32} = \{2,6,7,9\}.$$

The relative value of the concordance set is measured by means of the concordance index, which is equal to the sum of the weights associated with those criteria that are contained in the concordance sets (Figueira et al. 2005). Owing to the fact that the weights of criteria are also expressed as HFEs, the concordance index $cc_{\xi\zeta}$ of the pair of (A_ξ, A_ζ) is defined on the basis of the proposed signed distances as below:

$$cc_{\xi\zeta} = \sum_{j \in \mathbb{C}_{\xi\zeta}} \left(1 - d_S(W_j, \tilde{1})\right) \tag{3.14}$$

The concordance index reflects the relative importance of A_ξ with respect to A_ζ. Obviously, $0 \leq cc_{\xi\zeta} \leq 1$. A large value of $cc_{\xi\zeta}$ indicates that A_ξ is superior to A_ζ. Then the concordance threshold value can be defined as the average of all concordance indices, namely,

$$\bar{cc} = \sum_{\xi=1,\xi\neq\zeta}^{m} \sum_{\zeta=1,\zeta\neq\xi}^{m} \frac{cc_{\xi\zeta}}{m(m-1)} \tag{3.15}$$

On the other hand, the discordance index $dc_{\xi\zeta}$ which is a reflection of the relative difference of A_ξ with respect to A_ζ in terms of discordance criteria can be defined as:

$$dc_{\xi\zeta} = \frac{\max_{j \in \mathbb{N}_{\xi\zeta}} \left(1 - d_S(W_j, \tilde{1})\right) * d_E(h_{\xi j}, h_{\zeta j})}{\max_{j=1}^{n} \left(1 - d_S(W_j, \tilde{1})\right) * d_E(h_{\xi j}, h_{\zeta j})} \tag{3.16}$$

and the discordance threshold value $\bar{dc}$, which is the average of all discordance indices, is defined as:

$$\bar{dc} = \sum_{\xi=1,\xi\neq\zeta}^{m} \sum_{\zeta=1,\zeta\neq\xi}^{m} \frac{dc_{\xi\zeta}}{m(m-1)} \tag{3.17}$$

By comparing the concordance index $cc_{\xi\zeta}$ and the concordance threshold $\bar{cc}$ as well as the discordance index $dc_{\xi\zeta}$ and the discordance threshold $\bar{dc}$, the concordance dominance matrix $\mathbb{Q}$ and the discordance dominance matrix $\mathbb{R}$ can be constructed, respectively, as follows (Zhang and Xu 2015):

$$\mathbb{Q} = \begin{bmatrix} - & \cdots & ccd_{1\zeta} & \cdots & ccd_{1(m-1)} & ccd_{1m} \\ \vdots & \ddots & \vdots & \ddots & \vdots & \vdots \\ ccd_{\xi 1} & \cdots & ccd_{\xi\zeta} & \cdots & ccd_{\xi(m-1)} & ccd_{\xi m} \\ \vdots & \ddots & \vdots & \ddots & \vdots & \vdots \\ ccd_{m1} & \cdots & ccd_{m\zeta} & \cdots & ccd_{m(m-1)} & - \end{bmatrix} \tag{3.18}$$

and

$$\mathbb{R} = \begin{bmatrix} - & \cdots & dcd_{1\zeta} & \cdots & dcd_{1(m-1)} & dcd_{1m} \\ \vdots & \ddots & \vdots & \ddots & \vdots & \vdots \\ dcd_{\xi 1} & \cdots & dcd_{\xi\zeta} & \cdots & dcd_{\xi(m-1)} & dcd_{\xi m} \\ \vdots & \ddots & \vdots & \ddots & \vdots & \vdots \\ dcd_{m1} & \cdots & dcd_{m\zeta} & \cdots & dcd_{m(m-1)} & - \end{bmatrix} \tag{3.19}$$

whose elements satisfy

$$\begin{cases} ccd_{\xi\zeta} = 1, & if\ cc_{\xi\zeta} \geq \bar{c}c \\ ccd_{\xi\zeta} = 0, & if\ cc_{\xi\zeta} < \bar{c}c \end{cases} \quad \begin{cases} dcd_{\xi\zeta} = 1, & if\ dc_{\xi\zeta} \leq \bar{d}c \\ dcd_{\xi\zeta} = 0, & if\ dc_{\xi\zeta} > \bar{d}c \end{cases}, \tag{3.20}$$

Finally, the aggregation dominance matrix $\mathbb{S}$ is constructed based on the elements of the matrix $\mathbb{Q}$ and the matrix $\mathbb{R}$ through the following formula:

$$\mathbb{S} = \mathbb{Q} \otimes \mathbb{R} \tag{3.21}$$

where each element $ad_{\xi\zeta}$ of $\mathbb{S}$ is obtained by

$$ad_{\xi\zeta} = ccd_{\xi\zeta} \times dcd_{\xi\zeta} \tag{3.22}$$

In the green supplier selection problem, the concordance indices and the concordance threshold value are obtained by using Eqs. (3.14) and (3.15), respectively (Zhang and Xu 2015):

$$cc_{12} = 3.2085, \quad cc_{13} = 1.9085, \quad cc_{21} = 3.7165, \quad cc_{23} = 2.6915,$$
$$cc_{31} = 4.2915, \quad cc_{32} = 3.5085, \quad \bar{c}c = 3.2208.$$

and the discordance indices and the discordance threshold value are obtained by using Eqs. (3.16) and (3.17), respectively, as (Zhang and Xu 2015):

$$dc_{12} = 0.6521, \quad dc_{13} = 0.7521, \quad dc_{21} = 1.0, \quad dc_{23} = 1.0,$$
$$dc_{31} = 1.0, \quad dc_{32} = 0.8790, \quad \bar{dc} = 0.8656.$$

Combining with Eqs. (3.18)–(3.22), we further obtain the concordance dominance matrix $\mathbb{Q}$, the discordance dominance matrix $\mathbb{R}$ and the aggregation dominance matrix $\mathbb{S}$ as follows (Zhang and Xu 2015):

$$\mathbb{Q} = \begin{bmatrix} - & 0 & 0 \\ 1 & - & 0 \\ 1 & 1 & - \end{bmatrix}, \quad \mathbb{R} = \begin{bmatrix} - & 1 & 1 \\ 0 & - & 0 \\ 0 & 1 & - \end{bmatrix}, \quad \mathbb{S} = \begin{bmatrix} - & 0 & 0 \\ 0 & - & 0 \\ 0 & 1 & - \end{bmatrix}.$$

According to the aggregation dominance matrix $\mathbb{S}$, it is easy to obtain the following outranking relationship: $A_3 \succ A_2$. But the preference relations between A_1 and A_3 as well as between A_1 and A_2 cannot be discerned. In other words, for this green supplier selection problem we cannot utilize the modified HF-ELECTRE approach to determine the ranking orders of three potential suppliers.

It is not hard to see that compared with the modified HF-ELECTRE approach, the HF-QUALIFLEX approach can yield the distinct ranking results of the alternatives, i.e., $A_3 \succ A_2 \succ A_1$, which is more reasonable. Moreover, we also notice that the computation process of the modified HF-ELECTRE method is more complex and cumbersome than the HF-QUALIFLEX approach. That is to say, for solving the green supplier selection problem described in Sect. 3.3.1 the HF-QUALIFLEX method can not only obtain a reasonable and credible solution but also have the relative simple computation procedures.

The ELECTRE method often consists of two steps: (1) the construction of one or several outranking relations and (2) the derivation of a recommendation based on the outranking relations (Hatami-Marbini and Tavana 2011); while the QUALIFLEX method is based on the pairwise criterion comparison of alternatives, but no outranking relation is constructed (Figueira et al. 2005). Generally speaking, the preferred alternatives obtained by the proposed HF-QUALIFLEX method and the modified HF-ELECTRE method are in agreement for a given problem with a suitable number of alternatives and criteria. But in the HF-QUALIFLEX approach, the number of permutations increases tremendously with the number of alternatives, which results in the tedious computations required for handling the MCDM problem with a sufficiently large number of alternatives. This is also a limitation of this method. Fortunately, Ancot (2013) formulated this problem as a particular case of Quadratic Assignment Problem and took the MICROQUALIFLEX software to deal with such tedious computation requirement. Nevertheless, to avoid the troublesome computations, Chen et al. (2013) suggested that the QUALIFLEX method should be inapplicable to the decision making problems with large numbers of alternatives, but preferring to be used to solve the problems where the number of criteria markedly exceeds the number of alternatives. While the modified

Table 3.7 Problem requirements of different MCDM methods

MCDM methods	Problems with different characteristics			
	Performance ratings	Criteria weights	Problem with a few alternatives and a large number of criteria	Problem with a large number of alternatives and a few criteria
The HF-QUALIFLEX (Zhang and Xu 2015)	HFEs	HFEs	√	×
IVIF-QUALIFLEX (Chen 2014)	IVIFNs	Crisp numbers	√	×
IT2TrF-QUALIFLEX (Chen et al. 2013)	IT2TrFNs	IT2TrFNs	√	×
The modified HF-ELECTRE	HFEs	HFEs	×	√
The HF-ELECTRE (Wang et al. 2014)	HFEs	Crisp numbers	×	√

Note "√" denotes "suitable" meaning the simple computation process and the desirable decision results; and "×" denotes "unsuitable" meaning the complex computation process and the undesirable decision results

HF-ELECTRE approach is the preferred method for the MCDM problems with a large set of alternatives and a few criteria (Hatami-Marbini and Tavana 2011). As the aforementioned comparison analysis, for the green supplier selection problem with few alternatives and a large number of criteria described in Sect. 3.3.1, if we employ the modified HF-ELECTRE method to deal with it, we fail to obtain the distinct ranking results of the alternatives and its computation process is also very complex and cumbersome; while using the HF- QUALIFLEX approach we can get the desirable final decision results.

In addition, Chen et al. (2013) and Chen (2014) extended the QUALIFLEX method to accommodate the IT2TrF context (i.e., the IT2TrF-QUALIFLEX method) and the IVIF context (i.e., the IVIF-QUALIFLEX method), respectively. These two methods mainly focus on the IT2TrFNs and IVIFNs and fail to deal with the HFEs. In Table 3.7, we present a side-by-side comparison analysis among the HF-QUALIFLEX method (Zhang and Xu 2015), the IVIF-QUALIFLEX method (Chen 2014), the IT2TrF-QUALIFLEX method (Chen et al. 2013), the modified HF-ELECTRE method and the HF-ELECTRE method (Wang, et al. 2014).

As shown in Table 3.7 (Zhang and Xu 2015), we know that:

(1) Both the HF-QUALIFLEX method and the modified HF-ELECTRE use HFEs to express the performance ratings of alternatives on each criterion and the weights of criteria, while the IT2TrF-QUALIFLEX uses IT2TrFNs to express the performance ratings and the criteria weights, the IVIF-QUALIFLEX method and the HF-ELECTRE, respectively, use IVIFNs and HFEs to express the performance ratings and crisp numbers to express the criteria weights.

(2) The HF-QUALIFLEX method, the IVIF-QUALIFLEX method and the IT2TrF-QUALIFLEX method are more suitable to solve the MCDM problems where the number of criteria markedly exceeds the number of alternatives; while both the modified HF-ELECTRE method and the HF-ELECTRE method are appropriate to be used to handle the MCDM problems with a large set of alternatives and a few criteria.

3.4 Extension of the Proposed Method for Heterogeneous Information

So far, we have presented a hesitant fuzzy QUALIFLEX approach with a signed distance-based ranking method for managing HFEs. In what follows, we further extend the proposed method to manage heterogeneous information which includes five distinct forms of information, i.e., real numbers, interval numbers, TFNs, IFNs, and HFEs.

The criteria set $\boldsymbol{C}$ in the aforementioned MCDM problem is divided into the five subsets $\boldsymbol{C}_\eta(\eta = 1,2,3,4,5)$ in which the criteria values are expressed as real numbers, interval numbers, TFNs, IFNs, and HFEs, respectively. We stipulate that

- if $C_j \in \boldsymbol{C}_1$, then $x_{ij} = a_{ij}$ is expressed as a real number;
- if $C_j \in \boldsymbol{C}_2$, then $x_{ij} = [\underline{a}_{ij}, \bar{a}_{ij}]$ is expressed as an interval number;
- if $C_j \in \boldsymbol{C}_3$, then $x_{ij} = T(a_{ij}, b_{ij}, c_{ij})$ is expressed as a TFN;
- if $C_j \in \boldsymbol{C}_4$, then $x_{ij} = I(\mu_{ij}, \nu_{ij})$ is expressed as an IFN;
- if $C_j \in \boldsymbol{C}_5$, then $x_{ij} = H\left\{(\gamma_{ij})^1, (\gamma_{ij})^2, \ldots, (\gamma_{ij})^{\#x_{ij}}\right\}$ is expressed as a HFE.

For convenience, all criteria values are assumed to be normalized. If not, we can employ the normalized method developed by Zhang et al. (2015) to normalize the heterogeneous criteria values. In these contexts, we first need to define the ranking method for heterogeneous information. Motivated by the concept of signed distance (Chiang 2001), we introduce the concepts of the signed distances for interval numbers, TFNs, and IFNs, respectively.

Definition 3.7 Given a normalized heterogeneous decision matrix $\Re = (x_{ij})_{m\times n}$, if $C_j \in \boldsymbol{C}_2$, then $x_{ij} = [\underline{a}_{ij}, \bar{a}_{ij}]$ is a normalized interval number,[1] its signed distance is defined as follows:

$$d_S(x_{ij}, \tilde{1}_{\tilde{a}}) = \sqrt{\frac{1}{2}\left((1 - \underline{a}_{ij})^2 + (1 - \bar{a}_{ij})^2\right)} \tag{3.23}$$

[1]The interval number $[\underline{a}, \bar{a}]$ is called a normalized interval number if $0 \le \underline{a} \le \bar{a} \le 1$.

where $\tilde{1}_{\tilde{\alpha}} = [1, 1]$ is an ideal normalized interval number.

Definition 3.8 Given a normalized heterogeneous decision matrix $\Re = (x_{ij})_{m \times n}$, if $C_j \in C_3$, then $x_{ij} = T(a_{ij}, b_{ij}, c_{ij})$ is a normalized TFN,[2] the signed distance of the TFN is defined as follows:

$$d_S\left(x_{ij}, \tilde{1}_{\tilde{\beta}}\right) = \sqrt{\frac{1}{3}\left((1 - a_{ij})^2 + (1 - b_{ij})^2 + (1 - c_{ij})^2\right)} \tag{3.24}$$

where $\tilde{1}_{\tilde{\beta}} = T(1, 1, 1)$ is an ideal normalized TFN.

Definition 3.9 Given a normalized heterogeneous decision matrix $\Re = (x_{ij})_{m \times n}$, if $C_j \in C_4$, then $x_{ij} = I(\mu_{ij}, \nu_{ij})$ is an IFN, its signed distance is defined as follows:

$$d_S(x_{ij}, \tilde{1}_{\tilde{\chi}}) = \sqrt{\frac{1}{2}\left((1 - u_{ij})^2 + (v_{ij})^2 + (1 - u_{ij} - v_{ij})^2\right)} \tag{3.25}$$

where $\tilde{1}_{\tilde{\chi}} = I(1, 0)$ is an ideal IFN.

It is noted that the signed distance of the HFE is introduced in Definition 3.5, and meanwhile let $d_S(x_{ij}, \tilde{1}_j) = -x_{ij}$ if $C_j \in C_1$. Based on the concepts of the signed distance $d_S(x_{ij}, \tilde{1}_j)$, we next present a new ranking method for heterogeneous information.

Definition 3.10 Given a normalized heterogeneous decision matrix $\Re = (x_{ij})_{m \times n}$, for two criteria values $x_{\xi j}$ and $x_{\zeta j}$, it is easily observed that:

(1) If $d_S(x_{\xi j}, \tilde{1}_j) < d_S(x_{\zeta j}, \tilde{1}_j)$, then $x_{\xi j} \succ x_{\zeta j}$;
(2) If $d_S(x_{\xi j}, \tilde{1}_j) = d_S(x_{\zeta j}, \tilde{1}_j)$, then $x_{\xi j} \sim x_{\zeta j}$;
(3) If $d_S(x_{\xi j}, \tilde{1}_j) > d_S(x_{\zeta j}, \tilde{1}_j)$, then $x_{\xi j} \prec x_{\zeta j}$.

Afterwards, we employ the developed signed distance-based ranking method to identify the concordance/discordance index. Analogously, there are $m!$ permutations of the ranking of alternatives, and let $\mathscr{L}_\rho$ denote the ρth permutation. Then, the concordance/discordance index $\mathscr{D}_j^\rho(A_\xi, A_\zeta)$ is defined by the following expression:

[2]The TFN $T(a, b, c)$ is called a normalized TFN if $0 \le a \le b \le c \le 1$.

$$\mathscr{D}_j^{\rho}(A_\xi, A_\zeta) = d_S(x_{\zeta j}, \tilde{1}_j) - d_S(x_{\xi j}, \tilde{1}_j)$$
$$= \begin{cases} a_{\xi j} - a_{\zeta j},\ \mathit{if}\ C_j \in \boldsymbol{C}_1 \\ \left(\begin{array}{l} \sqrt{\frac{1}{2}\left((1-\underline{a}_{\zeta j})^2 + (1-\bar{a}_{\zeta j})^2\right)} \\ -\sqrt{\frac{1}{2}\left((1-\underline{a}_{\xi j})^2 + (1-\bar{a}_{\xi j})^2\right)} \end{array} \right),\ \mathit{if}\ C_j \in \boldsymbol{C}_2 \\ \left(\begin{array}{l} \sqrt{\frac{1}{3}\left((1-a_{\zeta j})^2 + (1-b_{\zeta j})^2 + (1-c_{\zeta j})^2\right)} \\ -\sqrt{\frac{1}{3}\left((1-a_{\xi j})^2 + (1-b_{\xi j})^2 + (1-c_{\xi j})^2\right)} \end{array} \right),\ \mathit{if}\ C_j \in \boldsymbol{C}_3 \\ \left(\begin{array}{l} \sqrt{\frac{1}{2}\left((1-u_{\zeta j})^2 + (v_{\zeta j})^2 + (1-\pi_{\zeta j})^2\right)} \\ -\sqrt{\frac{1}{2}\left((1-u_{\xi j})^2 + (v_{\xi j})^2 + (1-\pi_{\xi j})^2\right)} \end{array} \right),\ \mathit{if}\ C_j \in \boldsymbol{C}_4 \\ d_S(h_{\zeta j}, \tilde{1}) - d_S(h_{\xi j}, \tilde{1}),\ \mathit{if}\ C_j \in \boldsymbol{C}_5 \end{cases} \tag{3.26}$$

Once the concordance/discordance index $\mathscr{D}_j^{\rho}(A_\xi, A_\zeta)$ has been identified, the remaining of the decision making process is similar to the process presented in Sect. 3.2. Namely, we can employ Eq. (3.9) to calculate the weighted concordance/discordance index $\mathscr{D}^{\rho}(A_\xi, A_\zeta)$ and use Eq. (3.10) to obtain the comprehensive concordance/discordance index $\mathscr{D}^{\rho}$. Finally, the optimal ranking order of the alternatives is obtained by comparing the magnitude of $\mathscr{D}^{\rho}$ and the decision is made.

3.5 Conclusions and Future Research Directions

In this chapter, we have presented a HF-QUALIFLEX approach with a signed distance-based comparison method developed by Zhang and Xu (2015) for handling the MCDM problems in which both the assessments of alternatives on criteria and the weights of criteria are expressed by HFEs. We have first presented a concept of hesitancy index for the HFE, and meanwhile by taking the hesitancy index into account, we have further developed two improved distance measures and a signed distance-based comparison approaches for HFEs, respectively. Meanwhile, we have modified the HF-ELECTRE method (Wang et al. 2014) to conduct a comparative study. Compared with the modified HF-ELECTRE method, the HF-QUALIFLEX approach for handling the MCDM problems where the number of criteria markedly exceeds the number of alternatives does not require the complicated computation procedures but still yields a reasonable and credible solution. Finally, we have extended this HF-QUALIFLEX technique to manage the heterogeneous information including real numbers, interval numbers, TFNs, IFNs, and HFEs.

On the other hand, this HF-QUALIFLEX technique also has several limitations which may sever as suggestions for further research, listed as follows:

(1) In some practical decision making problems, various types of relationships may exist among the criteria, while the proposed approach under the hypothesis that all criteria are independent cannot deal with such problems. It would be interesting to integrate the Choquet integral into the HF-QUALIFLEX approach to take the dependency between criteria into account.
(2) In the HF-QUALIFLEX approach, the weights of criteria are assumed to be completely known. However, in the practical decision making process the weight information may be incomplete or incomplete and inconsistence. In further research, it would be interesting to combine the fuzzy AHP approach or goal programming approaches into the HF-QUALIFLEX approach to effectively determine the weights of criteria.
(3) As Cheng et al. (2003) pointed out that each MCDM method reflects different characteristics and assumptions, using a single method may not give satisfactory results. In further studies, it would be interesting to investigate the potential of combining the HF-QUALIFLEX method with other useful MCDM techniques within the environment of HFEs.
(4) It would be interesting to develop a decision support system based on the HF-QUALIFLEX approach to solve the practical MCDM problems within the environment of HFEs and expend their application ranges, such as in the fields of performance evaluation, risk investment, etc.

References

Ancot, J. P. (2013). Micro-QUALIFLEX: *An interactive software package for the determination and analysis of the optimal solution to decision problems*. Berlin: Springer Science & Business Media.

Chen, T. Y. (2014). Interval-valued intuitionistic fuzzy QUALIFLEX method with a likelihood-based comparison approach for multiple criteria decision analysis. *Information Sciences, 261*, 149–169.

Chen, T. Y., Chang, C. H., & Lu, J. F. R. (2013). The extended QUALIFLEX method for multiple criteria decision analysis based on interval type-2 fuzzy sets and applications to medical decision making. *European Journal of Operational Research, 226*, 615–625.

Cheng, S., Chan, C. W., & Huang, G. H. (2003). An integrated multi-criteria decision analysis and inexact mixed integer linear programming approach for solid waste management. *Engineering Applications of Artificial Intelligence, 16*, 543–554.

Chiang, J. (2001). Fuzzy linear programming based on statistical confidence interval and interval-valued fuzzy set. *European Journal of Operational Research, 129*, 65–86.

Figueira, J., Greco, S., & Ehrgott, M. (2005). *Multiple criteria decision analysis: state of the art surveys* (Vol. 78). Berlin: Springer Science & Business Media.

Hatami-Marbini, A., & Tavana, M. (2011). An extension of the Electre I method for group decision-making under a fuzzy environment. *Omega, 39*, 373–386.

Martel, J. M., & Matarazzo, B. (2005). *Other outranking approaches in multiple criteria decision analysis: State of the art surveys* (pp. 197–259). Berlin: Springer.

Paelinck, J. H. P. (1976). Qualitative multiple criteria analysis, environmental protection and multiregional development. *Papers in Regional Science, 36*, 59–76.

Paelinck, J. H. P. (1977). Qualitative multicriteria analysis: an application to airport location. *Environment and Planning A, 9*, 883–895.

Paelinck, J. H. P. (1978). Qualiflex: A flexible multiple-criteria method. *Economics Letters, 1*, 193–197.

Rebai, A., Aouni, B., & Martel, J.-M. (2006). A multi-attribute method for choosing among potential alternatives with ordinal evaluation. *European Journal of Operational Research, 174*, 360–373.

Shen, L., Olfat, L., Govindan, K., Khodaverdi, R., & Diabat, A. (2013). A fuzzy multi criteria approach for evaluating green supplier's performance in green supply chain with linguistic preferences. *Resources, Conservation and Recycling, 74*, 170–179.

Wang, J. Q., Wang, D. D., Zhang, Y. H., & Chen, X. H. (2014). Multi-criteria outranking approach with hesitant fuzzy sets. *OR Spectrum, 36*, 1001–1019.

Xu, Z. S., & Xia, M. M. (2011). Distance and similarity measures for hesitant fuzzy sets. *Information Sciences, 181*, 2128–2138.

Zhang, X. L. (2016). Multicriteria Pythagorean fuzzy decision analysis: a hierarchical QUALIFLEX approach with the closeness index-based ranking methods. *Information Sciences, 330*, 104–124.

Zhang, X. L., & Xu, Z. S. (2015). Hesitant fuzzy QUALIFLEX approach with a signed distance-based comparison method for multiple criteria decision analysis. *Expert Systems with Applications, 42*, 873–884.

Zhu, Q., & Sarkis, J. (2004). Relationships between operational practices and performance among early adopters of green supply chain management practices in Chinese manufacturing enterprises. *Journal of operations management, 22*, 265–289.

Chapter 4
Hesitant Fuzzy Multiple Criteria Decision Analysis Based on LINMAP

Abstract The LINMAP technique is one of the most representative methods for handling the MCDM or MCGDM problems with respect to the preference information over alternatives. This chapter utilizes the main structure of LINMAP to develop the hesitant fuzzy group LINMAP technique with interval programming model and the hesitant fuzzy programming model-based LINMAP technique. The former is mainly used to solve the MCGDM problems in which the ratings of alternatives with respect to criteria are taken as HFEs, and all pair-wise comparison judgments are represented by interval numbers. While the latter is mainly utilized to address the MCDM problems with incomplete weight information in which the ratings of alternatives with each criterion are taken as HFEs and the incomplete judgments on pair-wise comparisons of alternatives with hesitant degrees are also represented by HFEs. The main contributions of this chapter are that (1) the developed techniques not only can take sufficiently into account the experts' hesitancy in expressing their assessment information for criteria values by using HFEs but also can simultaneously consider the uncertainty of preference information over alternatives by using interval numbers or HFEs; (2) the concept of hesitant fuzzy programming model in which both the objective function and the constraints' coefficients take the form of HFEs has been proposed, and an effective technique to solve this sort of model is developed; (3) the bi-objective programming model has been constructed to address the issues of incomplete and inconsistent weights of the criteria.

The LINMAP method initially proposed by Srinivasan and Shocker (1973) proves to be a practical and useful approach for determining the weights of criteria and the PIS. In the classical LINMAP approach, the decision maker is required to not only provide the ratings of alternatives with respect to each criterion but also simultaneously give the incomplete preference relations on pair-wise comparisons of alternatives. The underlying logic of LINMAP is to define the consistency and inconsistency indices based on pair-wise comparisons of alternatives. The consistency index is to measure the degree of consistency between the given preference relations over alternatives by the decision-maker in advance and the derived preference relations by analyzing data in the decision table, while the inconsistency index is to measure the degree of inconsistency between them. Based on the consistency and inconsistency indices, a

X. Zhang and Z. Xu, *Hesitant Fuzzy Methods for Multiple Criteria Decision Analysis*, Studies in Fuzziness and Soft Computing 345,
DOI 10.1007/978-3-319-42001-1_4

linear programming model is constructed to obtain the PIS and weights of criteria. Finally, the best alternative which has the shortest distance to the PIS is identified.

Because the classical LINMAP method can only be used to deal with the MCDM problems with crisp numbers but fail to directly handle the MCDM problems with fuzzy decision information, many researchers have extended the classical LINMAP method for solving the MCDM problems within a variety of different fuzzy environments. For example, considering the uncertainty of criteria values, Li and Yang (2004) proposed the fuzzy LINMAP method for solving the MCDM problems with TFNs. Li et al. (2010) developed an intuitionistic fuzzy LINMAP method for solving the MCGDM problems with IFNs. Wang and Li (2012) also extended the classical LINMAP method for solving the MCGDM problems with interval-valued intuitionistic fuzzy numbers (IVIFNs). Combining with comparison possibilities and hybrid averaging operations, Chen (2013) proposed an interval-valued intuitionistic fuzzy LINMAP method for solving the MCGDM problems with IVIFNs. On the other hand, considering the uncertainty of preference information over alternatives, Sadi-Nezhad and Akhtari (2008) developed a possibility LINMAP method for solving the MCGDM problems with triangular fuzzy truth degree. Li and Wan (2014) proposed a LINMAP-based fuzzy linear programming technique for solving the heterogeneous MCDM problems with trapezoidal fuzzy truth degree. Wan and Li (2013) presented a generalization of the classical LINMAP method for solving the heterogeneous MCDM problems with intuitionistic fuzzy truth degrees.

Considering the fact that the use of hesitant fuzzy assessments makes the experts' judgments more reliable and informative in decision making, Zhang and Xu (2014) extended the classical LINMAP method to put forward a hesitant fuzzy group LINMAP with interval programming model for solving the MCGDM problems in which the ratings of alternatives with respect to each criterion are represented by HFEs and all pair-wise comparison judgments over alternatives take the form of interval numbers. On the other hand, to address some practical MCDM problems in which both the criteria values and the pair-wise comparison judgments over alternatives are expressed in the form of HFEs, Zhang et al. (2015) further developed a hesitant fuzzy programming model-based LINMAP method.

4.1 Some Basic Concepts

In what follows, we introduce some basic concepts and terminologies which will be used in the next sections. Readers familiar with these topics are encouraged to proceed directly to Sect. 4.2.

4.1.1 Description of the Hesitant Fuzzy MCGDM Problems

The MCGDM is the process that a group consisting of $g(g \geq 2)$ experts has to select an optimal alternative from $m(m \geq 2)$ potential alternatives or rank all potential

Table 4.1 Hesitant fuzzy group decision matrix

Experts	Alternatives	Criteria			
		C_1	C_2	...	C_n
e_1	A_1	h_{11}^1	h_{12}^1	...	h_{1n}^1
	A_2	h_{21}^1	h_{22}^1	...	h_{2n}^1
	...	...	...	...	...
	A_m	h_{m1}^1	h_{m2}^1	...	h_{mn}^1
...	A_1	h_{11}^k	h_{12}^k	...	h_{1n}^k
	A_2	h_{21}^k	h_{22}^k	...	h_{2n}^k
	...	...	...	...	...
	A_m	h_{m1}^k	h_{m2}^k	...	h_{mn}^k
e_g	A_1	h_{11}^g	h_{12}^g	...	h_{1n}^g
	A_2	h_{21}^g	h_{22}^g	...	h_{2n}^g
	...	...	...	...	...
	A_m	h_{m1}^g	h_{m2}^g	...	h_{mn}^g

alternatives based on $n(n \geq 2)$ criteria. Let x_{ij}^k be the criterion value of the alternative A_i with respect to the criterion C_j provided by the expert e_k . Then, such a MCGDM problem is expressed by the group decision matrix $\Re^k = (x_{ij}^k)_{m \times n}$ $(k = 1, 2, \ldots, g)$. If the criteria values are represented by HFEs, such a MCGDM problem is called the hesitant fuzzy MCGDM problem which is expressed in the form of the hesitant fuzzy group decision matrix in Table 4.1.

It is noted that the hesitant fuzzy MCGDM problem is reduced to a hesitant fuzzy MCDM problem if the number of experts is equal to 1 (i.e., g = 1). In this chapter, we first explore the hesitant fuzzy MCGDM problem in which the experts in the group are assumed to have the same importance but the weights of criteria are partially known or completely unknown. Next, we study the hesitant fuzzy MCDM problems in which the weights of criteria are partially known or completely unknown and the hesitant fuzzy PIS is completely unknown.

4.1.2 Interval-Valued Preference Relations

One of the most disadvantages in multidimensional analysis of preference is to consider pair-wise comparison between alternatives with the crisp degree of truth (i.e., 0 or 1) (Sadi-Nezhad and Akhtari 2008). In practical decision making problems, the decision makers are not sure enough in all pair-wise comparisons over alternative and usually express their opinions with a degree of insurance. As a result, using interval numbers instead of crisp numbers are more realistic. In other words, the decision maker's preferences can be considered as interval numbers given through pair-wise comparisons between alternatives.

Table 4.2 Linguistic meanings of interval preferences

Linguistic meanings	Interval preferences
Strong preferable	[0.9, 1.0]
Preferable	[0.8, 0.9]
Almost preferable	[0.7, 0.8]
Moderate preferable	[0.6, 0.7]
Scarcely preferable	[0.5, 0.6]

Definition 4.1 (Zhang and Xu 2014). For the expert e_k and each pair of alternatives (A_ξ, A_ζ), if the expert e_k prefers the alternative A_ξ to A_ζ with the interval truth degree $R_k(\xi, \zeta)$, the set of ordered pairs of alternatives (A_ξ, A_ζ) for the expert e_k is defined as:

$$\Omega^k = \left\{(\xi, \zeta)\middle|A_\xi \succeq_{R_k(\xi,\zeta)} A_\zeta,\ (\xi, \zeta \in \{1, 2, \ldots, m\})\right\} \tag{4.1}$$

where the degree of truth $R_k(\xi, \zeta)$ is expressed as an interval number denoted by $R_k(\xi, \zeta) = [\underline{R}^k_{\xi\zeta}, \bar{R}^k_{\xi\zeta}]\ \left(0 \le \underline{R}^k_{\xi\zeta} \le \bar{R}^k_{\xi\zeta} \le 1\right)$.

In the real-life decision making process, the interval preference relationship usually represents a certain kind of meanings. That is to say, there exist some corresponding relations between interval preference information and linguistic meanings. These relations can be described in Table 4.2 (Zhang and Xu 2014). For example, $R_k(\xi, \zeta) = [0.9, 1.0]$ indicates that the expert e_k strongly prefers the alternative A_ξ to A_ζ, while $R_k(\xi, \zeta) = [0.6, 0.7]$ expresses the meanings that the expert e_k moderately prefers the alternative A_ξ to A_ζ.

4.1.3 Hesitant Fuzzy Preference Relation

In the practical decision making problems, the decision makers may have hesitancy for expressing their preference on pair-wise comparisons of alternatives and express their opinions with a degree of hesitation by using HFEs.

Definition 4.2 (Zhang et al. 2015). In a MCDM problem with HFEs, for each pair of alternatives (A_ξ, A_ζ), if the decision maker prefers the alternative A_ξ to A_ζ with the hesitant fuzzy truth degree $\tilde{R}(\xi, \zeta)$, the set of all ordered pairs of alternatives (A_ξ, A_ζ) is defined as:

$$\tilde{\Omega} = \left\{(\xi, \zeta)\middle|A_\xi \succeq_{\tilde{R}(\xi,\zeta)} A_\zeta,\ (\xi, \zeta \in \{1, 2, \ldots, m\})\right\} \tag{4.2}$$

where the degree of truth $\tilde{R}(\xi, \zeta)$ is expressed as a HFE denoted by $\tilde{R}(\xi, \zeta) = H\{\gamma^1, \gamma^2, \ldots, \gamma^{\#\tilde{R}(\xi,\zeta)}\}$.

For example, $\tilde{R}(1,2) = H\{0.5, 0.6, 0.7\}$ means that the decision organization or the individual expert has hesitancy among the values 0.5, 0.6 and 0.7 when providing the degree to which the alternative A_1 is superior to A_2.

Remark 4.1 It is easy to see that the pair-wise preference information provided by the experts is expressed by HFEs, which is simply called hesitant fuzzy preference relation originally developed by Zhu and Xu (2014). The cardinality $|\tilde{\Omega}|$ of $\tilde{\Omega}$, i.e., the number of alternative pairs in $\tilde{\Omega}$, is $\binom{m}{2} = \frac{1}{2}m(m-1)$ when the pair-wise preference information over alternatives in the decision making problem is complete. In this chapter, the pair-wise preference information between alternatives given by the decision maker is allowed to be incomplete (i.e., $|\tilde{\Omega}| < \binom{m}{2}$) and/or intransitive, which cannot be solved by the consistent method developed by Zhu and Xu (2014).

4.1.4 Optimization Problems with Interval-Objective Functions

Let $\tilde{a} = [\underline{a}, \bar{a}] = \{a | \underline{a} \le a \le \bar{a}, a \in R\}$ be an interval number, where $\underline{a}$ and $\bar{a}$ are the lower and upper limits of the interval number $\tilde{a}$ on the real number set R, respectively. If $\underline{a} = \bar{a}$, then $\tilde{a} = [\underline{a}, \bar{a}]$ reduces to a real number $a(a = \underline{a} = \bar{a})$. Given that $m(\tilde{a}) = (\underline{a} + \bar{a})/2$ and $n(\tilde{a}) = (\bar{a} - \underline{a})/2$ are respectively the midpoint and half-width of $\tilde{a}$, thus the interval $\tilde{a}$ can be alternatively represented as $\tilde{a} = (m(\tilde{a}), n(\tilde{a}))$.

For two interval numbers $\tilde{a} = [\underline{a}, \bar{a}]$ and $\tilde{b} = [\underline{b}, \bar{b}]$, the basic operations on them can be described as (Tong 1994):

(1) $\tilde{a} \oplus \tilde{b} = [\underline{a} + \underline{b},\ \bar{a} + \bar{b}]$;
(2) $\theta\tilde{a} = [\theta\underline{a}, \theta\bar{a}]\ (\theta > 0)$;
(3) $\tilde{a} \le \tilde{b}$, if and only if $\underline{a} \le \underline{b}$ and $\bar{a} \le \bar{b}$.

Now, we review the following minimization problem with interval-objective function:

$$\begin{aligned} \min \quad & \tilde{a} \\ s.t. \quad & \tilde{a} \in \Omega_1 \end{aligned} \tag{MOD-4.1}$$

which is equivalent to the bi-objective mathematical programming problem as follows (Ishibuchi and Tanaka 1990):

$$\begin{aligned} \min \quad & \{\bar{a}, m(\tilde{a})\} \\ s.t. \quad & \tilde{a} \in \Omega_1 \end{aligned} \tag{MOD-4.2}$$

where Ω_1 is a set constraint conditions which the variable $\tilde{a}$ should satisfy according to requirements in the real situations.

The maximization problem with interval-objective function is described as:

$$\begin{array}{ll} \max & \{\tilde{a}\} \\ s.t. & \tilde{a} \in \Omega_2 \end{array} \qquad \text{(MOD-4.3)}$$

which can be equivalent to the bi-objective mathematical programming problem as (Ishibuchi and Tanaka 1990):

$$\begin{array}{ll} \max & \{\underline{a}, m(\tilde{a})\} \\ s.t. & \tilde{a} \in \Omega_2 \end{array} \qquad \text{(MOD-4.4)}$$

where Ω_2 is a set constraint conditions which the variable $\tilde{a}$ should satisfy.

4.2 Hesitant Fuzzy LINMAP Group Decision Method with Interval Programming Models

To deal with the MCGDM problem with incomplete weight information in which criteria values are expressed by HFEs and the pair-wise comparison preferences over alternatives are denoted by intervals, Zhang and Xu (2014) developed a hesitant fuzzy LINMAP group decision method with interval programming model. We call it Zhang and Xu (2014)'s method.

4.2.1 The Proposed Approach

The proposed approach starts with the definitions of new consistency and inconsistency indices. Then, an interval programming model is constructed to derive the optimal weight vector and the hesitant fuzzy PIS. Afterwards, the distances between the alternatives and the hesitant fuzzy PIS can be calculated and each expert's ranking of all alternatives is also obtained. At length, the collective ranking of alternatives is obtained using social choice functions such as the Borda's function and Copeland's function (Hwang and Yoon 1981). The details of the proposed approach are elaborated as follows:

Firstly, we denote the hesitant fuzzy PIS by $A^+ = (h_1^+, h_2^+, \ldots, h_n^+)$, where h_j^+ $(j \in \{1,2,\ldots,n\})$ is expressed as a HFE $h_j^+ = H\{(\gamma_j^1)^+, (\gamma_j^2)^+, \ldots, (\gamma_j^{\#h_j^+})^+\}$. Meanwhile, we assume that the numbers of the elements in all HFE data are the same, i.e., $\#h_{ij}^k = \#h_j^+ = \#h$ $(i \in \{1,2,\ldots,m\}; j \in \{1,2,\ldots,n\}; k \in \{1,2,\ldots,g\})$. If not, we should extend the shorter one until all of them have the same length by employing the extension law in Definition 1.3. Then, for the experts $e_k (k \in \{1,2,\ldots,g\})$, using Eq. (1.2) the weighted square of the Euclidean distance between the alternatives $A_i (i \in \{1,2,\ldots,m\})$ and the hesitant fuzzy PIS A^+ can be calculated as:

$$D_i^k = \sum_{j=1}^{n} w_j d\left(h_{ij}^k, h_j^+\right)^2 = \sum_{j=1}^{n} w_j \left(\frac{1}{\#h} \sum_{\lambda=1}^{\#h} \left((\gamma_{ij}^{\lambda})^k - (\gamma_j^{\lambda})^+ \right)^2 \right) \tag{4.3}$$

As mentioned previously, in some practical decision making situations, the experts may not only provide the ratings of alternatives with respect to each criterion but also give the incomplete pair-wised preference judgments over alternatives by a set of ordered pairs $\Omega^k = \{(\xi,\zeta)|A_\xi \succeq_{R_k(\xi,\zeta)} A_\zeta,\ \xi,\zeta \in \{1,2,\ldots,m\}\}$ with interval fuzzy truth degrees $R_k(\xi,\zeta)$ $(\xi,\zeta \in \{1,2,\ldots,m\})$. Thus, the weighted square of the Euclidean distance between each pair alternatives $(A_\xi, A_\zeta) \in \Omega^k$ and the hesitant fuzzy PIS A^+ can be calculated as:

$$\begin{aligned} D_\xi^k &= \sum_{j=1}^{n} w_j d\left(h_{\xi j}^k, h_j^+\right)^2 = \sum_{j=1}^{n} w_j \left(\frac{1}{\#h} \sum_{\lambda=1}^{\#h} \left((\gamma_{\xi j}^{\lambda})^k - (\gamma_j^{\lambda})^+ \right)^2 \right) \\ D_\zeta^k &= \sum_{j=1}^{n} w_j d\left(h_{\zeta j}^k, h_j^+\right)^2 = \sum_{j=1}^{n} w_j \left(\frac{1}{\#h} \sum_{\lambda=1}^{\#h} \left((\gamma_{\zeta j}^{\lambda})^k - (\gamma_j^{\lambda})^+ \right)^2 \right) \end{aligned} \tag{4.4}$$

Let $\rho_{\zeta\xi}^k = D_\zeta^k - D_\xi^k$, it is easy to obtain:

$$\begin{aligned} \rho_{\zeta\xi}^k &= \sum_{j=1}^{n} w_j \left(\frac{1}{\#h} \sum_{\lambda=1}^{\#h} \left((\gamma_{\zeta j}^{\lambda})^k - (\gamma_j^{\lambda})^+ \right)^2 \right) \\ &\quad - \sum_{j=1}^{n} w_j \left(\frac{1}{\#h} \sum_{\lambda=1}^{\#h} \left((\gamma_{\xi j}^{\lambda})^k - (\gamma_j^{\lambda})^+ \right)^2 \right) \\ &= \sum_{j=1}^{n} \sum_{\lambda=1}^{\#h} \frac{w_j}{\#h} \left(\left((\gamma_{\zeta j}^{\lambda})^k \right)^2 - \left((\gamma_{\xi j}^{\lambda})^k \right)^2 \right) \\ &\quad - \sum_{j=1}^{n} \sum_{\lambda=1}^{\#h} \frac{2w_j}{\#h} \left(\gamma_j^{\lambda} \right)^+ \left((\gamma_{\zeta j}^{\lambda})^k - (\gamma_{\xi j}^{\lambda})^k \right) \end{aligned} \tag{4.5}$$

It is easy to see that for each pair alternatives $(A_\xi, A_\zeta) \in \Omega^k$ (Zhang and Xu 2014):

(1) if $D_\zeta^k \geq D_\xi^k$ (i.e., $\rho_{\zeta\xi}^k \geq 0$), then the alternative A_ξ is closer to the hesitant fuzzy PIS A^+ than the alternative A_ζ. Thus, the ranking order of the alternatives A_ξ and A_ζ determined by D_ξ^k and D_ζ^k based on $(\boldsymbol{w}, A^+)$ is consistent with the preference relation $(A_\xi, A_\zeta) \in \Omega^k$ given by the expert $e_k \in \boldsymbol{E}$.

(2) if $D_\zeta^k < D_\xi^k$ (i.e., $\rho_{\zeta\xi}^k < 0$), then the alternative A_ξ is farther to the hesitant fuzzy PIS A^+ than the alternative A_ζ. Thus, the ranking order of the alternatives A_ξ and A_ζ determined by D_ξ^k and D_ζ^k based on $(\boldsymbol{w}, A^+)$ is inconsistent with the preference relation $(A_\xi, A_\zeta) \in \Omega^k$ given by the expert $e_k \in \boldsymbol{E}$.

Based on the above analysis, in the following we define an inconsistency index $(D_\zeta^k - D_\xi^k)^-$ to measure the degree of inconsistency between the ranking order of the alternatives A_ξ and A_ζ in which one ranking order is determined by D_ξ^k and D_ζ^k based on $(\boldsymbol{w}, A^+)$, and the other ranking order obtained by the experts' preference relationship in Ω^k as:

$$(D_\zeta^k - D_\xi^k)^- = \begin{cases} 0, & \text{if } D_\zeta^k \geq D_\xi^k \\ R_k(\xi,\zeta) \times (D_\xi^k - D_\zeta^k), & \text{if } D_\zeta^k < D_\xi^k \end{cases} \tag{4.6}$$

For convenience, we rewrite the inconsistency index as:

$$(D_\zeta^k - D_\xi^k)^- = \max\left\{0, R_k(\xi,\zeta) \times (D_\xi^k - D_\zeta^k)\right\} \tag{4.7}$$

Let

$$\begin{aligned} \tilde{B}^k &= \sum_{(A_\xi,A_\zeta)\in\Omega^k} (D_\zeta^k - D_\xi^k)^- \\ &= \sum_{(A_\xi,A_\zeta)\in\Omega^k} \max\left\{0, R_k(\xi,\zeta) \times (D_\xi^k - D_\zeta^k)\right\} \end{aligned} \tag{4.8}$$

which is called the inconsistency index of the expert $e_k \in \boldsymbol{E}$.

Thus, the collective inconsistency index of all experts is defined as:

$$\begin{aligned} \tilde{B} &= \sum_{k=1}^{g} \tilde{B}^k = \sum_{k=1}^{g} \sum_{(A_\xi,A_\zeta)\in\Omega^k} (D_\zeta^k - D_\xi^k)^- \\ &= \sum_{k=1}^{g} \sum_{(A_\xi,A_\zeta)\in\Omega^k} \max\left\{0, R_k(\xi,\zeta) \times (D_\xi^k - D_\zeta^k)\right\} \end{aligned} \tag{4.9}$$

In a similar way, a consistency index $(D_\zeta^k - D_\xi^k)^+$ is defined as:

$$(D_\zeta^k - D_\xi^k)^+ = \begin{cases} R_k(\xi,\zeta) \times \left(D_\zeta^k - D_\xi^k\right), & \text{if } D_\zeta^k \geq D_\xi^k \\ 0, & \text{if } D_\zeta^k < D_\xi^k \end{cases} \tag{4.10}$$

which is to measure the degree of consistency between the ranking order of the alternatives A_ξ and A_ζ in which one ranking order is determined by D_ξ^k and D_ζ^k based on $(\boldsymbol{w}, A^+)$, and the other ranking order obtained by the expert's preference relationship in Ω^k.

Similarly, the consistency index in Eq. (4.10) is rewritten as:

$$\left(D_\zeta^k - D_\xi^k\right)^+ = \max\left\{0, R_k(\xi,\zeta) \times \left(D_\zeta^k - D_\xi^k\right)\right\} \tag{4.11}$$

Hence, the collective consistency index of all experts is defined as:

$$\begin{aligned}\tilde{G} &= \sum_{k=1}^{g} \tilde{G}^k = \sum_{k=1}^{g} \sum_{(A_\xi, A_\zeta)\in\Omega^k} (D_\zeta^k - D_\xi^k)^+ \\ &= \sum_{k=1}^{g} \sum_{(A_\xi, A_\zeta)\in\Omega^k} \max\left\{0, R_k(\xi,\zeta) \times (D_\zeta^k - D_\xi^k)\right\}\end{aligned} \tag{4.12}$$

Remark 4.2 (Zhang and Xu 2014). It is noticed that both the collective consistency index $\tilde{G}$ and the collective inconsistency index $\tilde{B}$ are interval numbers because the experts' preferences on pair-wise comparisons of alternatives are interval numbers.

In the real-world decision making process, the smaller the collective inconsistency index $\tilde{B}$ is, the better the decision result is. In general, the collective inconsistency index $\tilde{B}$ should be no bigger than the collective consistency index $\tilde{G}$. In this sense, a mathematical programming model is constructed to determine the weighting vector $\boldsymbol{w}$ and the hesitant fuzzy PIS A^+ as follows (Zhang and Xu 2014):

$$\min \left\{\tilde{B}\right\}$$
$$s.t. \begin{cases} \tilde{G} - \tilde{B} \succeq \varepsilon_{\tilde{a}} \\ \sum_{j=1}^{n} w_j = 1 \\ 0 \le (\gamma_j^1)^+ \le (\gamma_j^2)^+ \le \cdots \le (\gamma_j^{\#h})^+ \le 1,\ j = 1,2,\ldots,n \\ 0 \le w_j \le 1, j = 1,2,\ldots,n \end{cases} \tag{MOD-4.5}$$

where $\varepsilon_{\tilde{a}} = [\underline{a}, \bar{a}]$ is an arbitrary positive interval number given by the experts in advance.

Apparently, the model (MOD-4.5) intends to minimize the collective inconsistency index $\tilde{B}$ under the condition in which the collective consistency index $\tilde{G}$ is no smaller than the collective inconsistency index $\tilde{B}$ by $\varepsilon_{\tilde{a}}$.

It can be easily derived from Eqs. (4.6) and (4.10) that

$$\left(D_\zeta^k - D_\xi^k\right)^+ - \left(D_\zeta^k - D_\xi^k\right)^- = R_k(\xi,\zeta) \times \left(D_\zeta^k - D_\xi^k\right) \tag{4.13}$$

Combining with Eqs. (4.5), (4.9), (4.12) and (4.13), we get:

$$\begin{aligned}\tilde{G} - \tilde{B} &= \sum_{k=1}^{g} \sum_{(\xi,\zeta)\in\Omega^k} \left(\left(D_\zeta^k - D_\xi^k\right)^+ - \left(D_\zeta^k - D_\xi^k\right)^-\right) \\ &= \sum_{k=1}^{g} \sum_{(\xi,\zeta)\in\Omega^k} R_k(\xi,\zeta) \times \rho_{\zeta\xi}^k\end{aligned} \tag{4.14}$$

Using Eqs. (4.9) and (4.14), the mathematical programming model (MOD-4.5) is rewritten as (Zhang and Xu 2014):

$$\min \left\{ \sum_{k=1}^{g} \sum_{(\xi,\zeta)\in\Omega^k} \max\left\{0, R_k(\xi,\zeta) \times \left(D_\xi^k - D_\zeta^k\right)\right\} \right\}$$

$$s.t. \begin{cases} \sum_{k=1}^{g} \sum_{(\xi,\zeta)\in\Omega^k} R_k(\xi,\zeta) \times \rho_{\zeta\xi}^k \succeq \varepsilon_{\tilde{a}} \\ \sum_{j=1}^{n} w_j = 1 \\ 0 \le (\gamma_j^1)^+ \le (\gamma_j^2)^+ \le \cdots \le (\gamma_j^{\#h})^+ \le 1,\ j = 1, 2, \ldots, n \\ 0 \le w_j \le 1, j = 1, 2, \ldots, n \end{cases} \quad \text{(MOD-4.6)}$$

For each pair alternatives $(A_\xi, A_\zeta) \in \Omega^k$, let $z_{\xi\zeta}^k = \max\left\{0, D_\xi^k - D_\zeta^k\right\}$, then $z_{\xi\zeta}^k \ge D_\xi^k - D_\zeta^k$ and $z_{\xi\zeta}^k \ge 0$. According to Eq. (4.5), it is easily observed that $\rho_{\zeta\xi}^k + z_{\xi\zeta}^k \ge 0$ and $z_{\xi\zeta}^k \ge 0$. Thus, $\max\left\{0, R_k(\xi,\zeta) \times \left(D_\xi^k - D_\zeta^k\right)\right\}$ can be rewritten as $R_k(\xi,\zeta) \times z_{\xi\zeta}^k$.

Then, the mathematical programming model (MOD-4.6) is converted into the following programming model (MOD-4.7) as (Zhang and Xu 2014):

$$\min \left\{ \sum_{k=1}^{g} \sum_{(\xi,\zeta)\in\Omega^k} R_k(\xi,\zeta) \times z_{\xi\zeta}^k \right\}$$

$$s.t. \begin{cases} \sum_{k=1}^{g} \sum_{(\xi,\zeta)\in\Omega^k} R_k(\xi,\zeta) \times \rho_{\zeta\xi}^k \succeq \varepsilon_{\tilde{a}} \\ \rho_{\zeta\xi}^k + z_{\xi\zeta}^k \ge 0 \quad \left((\xi,\zeta) \in \Omega^k; k = 1, 2, \ldots, g\right) \\ \sum_{j=1}^{n} w_j = 1 \\ z_{\xi\zeta}^k \ge 0 \quad \left((\xi,\zeta) \in \Omega^k; k = 1, 2, \ldots, g\right) \\ 0 \le (\gamma_j^1)^+ \le (\gamma_j^2)^+ \le \cdots \le (\gamma_j^{\#h})^+ \le 1,\ j = 1, 2, \ldots, n \\ 0 \le w_j \le 1,\ j = 1, 2, \ldots, n \end{cases} \quad \text{(MOD-4.7)}$$

Because the objective function and the first constraint condition in the model (MOD-4.7) are interval numbers, the model (MOD-4.7) can be further rewritten as the following model (Zhang and Xu 2014):

$$\min \left\{ \left[\sum_{k=1}^{g} \sum_{(\xi,\zeta)\in\Omega^k} \underline{R}^k_{\xi\zeta} \times z^k_{\xi\zeta}, \sum_{k=1}^{g} \sum_{(\xi,\zeta)\in\Omega^k} \bar{R}^k_{\xi\zeta} \times z^k_{\xi\zeta} \right] \right\}$$

$$s.t. \begin{cases} \left[\sum_{k=1}^{g} \sum_{(\xi,\zeta)\in\Omega^k} \underline{R}^k_{\xi\zeta} \times \rho^k_{\zeta\xi}, \sum_{k=1}^{g} \sum_{(\xi,\zeta)\in\Omega^k} \bar{R}^k_{\xi\zeta} \times \rho^k_{\zeta\xi} \right] \geq [\underline{a}, \bar{a}] \\ \rho^k_{\zeta\xi} + z^k_{\xi\zeta} \geq 0, \ (\xi,\zeta) \in \Omega^k; k = 1,2,\ldots,g \\ \sum_{j=1}^{n} w_j = 1 \\ z^k_{\xi\zeta} \geq 0, \ (\xi,\zeta) \in \Omega^k; k = 1,2,\ldots,g \\ 0 \leq (\gamma_j^1)^+ \leq (\gamma_j^2)^+ \leq \cdots \leq (\gamma_j^{\#h})^+ \leq 1, \ j = 1,2,\ldots,n \\ 0 \leq w_j \leq 1, \ j = 1,2,\ldots,n \end{cases} \quad \text{(MOD-4.8)}$$

Obviously, the model (MOD-4.8) is a standard linear interval programming model which can be solved by the existing methods (Ishibuchi and Tanaka 1990; Lai et al. 2002). Here, the model (MOD-4.8) is solved in the sense of the model (MOD-4.2). Thus, the model (MOD-4.8) is transformed into the bi-objective linear programming model (MOD-4.9) as follows (Zhang and Xu 2014):

$$\min \left\{ \sum_{k=1}^{g} \sum_{(\xi,\zeta)\in\Omega^k} \bar{R}^k_{\xi\zeta} \times z^k_{\xi\zeta}, \sum_{k=1}^{g} \sum_{(\xi,\zeta)\in\Omega^k} ((\underline{R}^k_{\xi\zeta} + \bar{R}^k_{\xi\zeta})/2) \times z^k_{\xi\zeta} \right\}$$

$$s.t. \begin{cases} \sum_{k=1}^{g} \sum_{(\xi,\zeta)\in\Omega^k} \underline{R}^k_{\xi\zeta} \times \rho^k_{\zeta\xi} \geq \underline{a} \\ \sum_{k=1}^{g} \sum_{(\xi,\zeta)\in\Omega^k} \bar{R}^k_{\xi\zeta} \times \rho^k_{\zeta\xi} \geq \bar{a} \\ \rho^k_{\zeta\xi} + z^k_{\xi\zeta} \geq 0, \ (\xi,\zeta) \in \Omega^k; k = 1,2,\ldots,g \\ z^k_{\xi\zeta} \geq 0, \ (\xi,\zeta) \in \Omega^k; k = 1,2,\ldots,g \\ 0 \leq (\gamma_j^1)^+ \leq (\gamma_j^2)^+ \leq \cdots \leq (\gamma_j^{\#h})^+ \leq 1, \ j = 1,2,\ldots,n \\ 0 \leq w_j \leq 1, \ \sum_{j=1}^{n} w_j = 1, j = 1,2,\ldots,n \end{cases} \quad \text{(MOD-4.9)}$$

Using the weight average approach, the model (MOD-4.9) is further transformed into the parametric linear programming model (MOD-4.10) (Zhang and Xu 2014):

$$\min \left\{ \theta \times \sum_{k=1}^{g} \sum_{(\xi,\zeta)\in\Omega^k} \bar{R}_{\xi\zeta}^k \times z_{\xi\zeta}^k + (1-\theta) \times \sum_{k=1}^{g} \sum_{(\xi,\zeta)\in\Omega^k} ((\underline{R}_{\xi\zeta}^k + \bar{R}_{\xi\zeta}^k)/2) \times z_{\xi\zeta}^k \right\}$$

$$s.t. \begin{cases} \sum_{k=1}^{g} \sum_{(\xi,\zeta)\in\Omega^k} \underline{R}_{\xi\zeta}^k \times \rho_{\zeta\xi}^k \geq \underline{a} \\ \sum_{k=1}^{g} \sum_{(\xi,\zeta)\in\Omega^k} \bar{R}_{\xi\zeta}^k \times \rho_{\zeta\xi}^k \geq \bar{a} \\ \rho_{\zeta\xi}^k + z_{\xi\zeta}^k \geq 0,\ (\xi,\zeta) \in \Omega^k;\ k = 1,2,\ldots,g \\ z_{\xi\zeta}^k \geq 0,\ (\xi,\zeta) \in \Omega^k;\ k = 1,2,\ldots,g \\ 0 \leq (\gamma_j^1)^+ \leq (\gamma_j^2)^+ \leq \cdots \leq (\gamma_j^{\#h})^+ \leq 1,\ j = 1,2,\ldots,n \\ 0 \leq w_j \leq 1,\ \sum_{j=1}^{n} w_j = 1, j = 1,2,\ldots,n \end{cases}$$

(MOD-4.10)

where $0 \leq \theta \leq 1$ is a parameter determined by the decision makers in advance.

By solving the model (MOD-4.10), we can determine the weight vector $\boldsymbol{w}$ and the hesitant fuzzy PIS A^+. Then, we calculate the distances $D_i^k(i \in \{1,2,\ldots,m\}$, $k \in \{1,2,\ldots,g\})$ of the alternatives $A_i(i \in \{1,2,\ldots,m\})$ to the hesitant fuzzy PIS A^+ using Eq. (4.2). Afterwards, we can obtain the rankings of alternatives for each expert according to the increasing orders of the distances $D_i^k(i \in \{1,2,\ldots,m\},\ k \in \{1,2,\ldots,g\})$, respectively. Finally, the collective ranking of alternatives is obtained by using social choice functions such as the Borda's function and Copeland's function (Hwang and Yoon 1981).

Based on the above models and analysis, an algorithm (Algorithm 4.1) is presented for solving the MCGDM problems with incomplete weight information in which the criteria values take the form of HFEs and the preferences on pair-wise comparison over alternatives are expressed by interval numbers. The steps of the Algorithm 4.1 are summarized as in Table 4.3 (Zhang and Xu 2014).

4.2.2 *Other Generalizations of the Proposed Model*

It is worth pointing out that the MCGDM problem mentioned in Sect. 4.2.1 is a general and complex decision making problem, in real life we may encounter some special cases of this problem, such as in this MCGDM problem the weights of criteria are completely known or partially known, the hesitant fuzzy PIS is known

Table 4.3 The decision steps of Algorithm 4.1

Step 1:	Form a group of experts and identify all alternatives on criteria
Step 2:	Determine the ratings of all alternatives with respect to each criterion by the expert $e_k (k \in \{1, 2, \ldots, g\})$ using the HFEs, and then construct the hesitant fuzzy decision matrices $\Re^k (k \in \{1, 2, \ldots, g\})$
Step 3:	Express pair-wise comparison judgments over alternatives by a set of ordered pairs $\Omega^k = \{(\xi, \zeta) \mid A_\xi \succeq_{R_k(\xi,\zeta)} A_\zeta,\ \xi, \zeta \in \{1, 2, \ldots, m\}\}$ with the interval truth degrees $R_k(\xi, \zeta)$ for all the expert $e_k (k \in \{1, 2, \ldots, g\})$
Step 4:	Define the collective inconsistency index and the collective consistency index using Eqs. (4.9) and (4.12), respectively
Step 5:	Construct the interval mathematical programming model by using the model (MOD-4.8), and transform such a model into the linear programming model (MOD-4.10)
Step 6:	Solve the linear programming model (MOD-4.10) by using the MATLAB and Lingo soft package with the given parameter $\theta (0 \le \theta \le 1)$ by the decision makers in advance, and then determine the optimal weight vector $\boldsymbol{w}$ and the hesitant fuzzy PIS A^+
Step 7:	Calculate the distances $D_i^k (i \in \{1, 2, \ldots, m\},\ k \in \{1, 2, \ldots, g\})$ of the alternative $A_i \in \boldsymbol{A}$ to the hesitant fuzzy PIS A^+ by using Eq. (4.2)
Step 8:	Rank the alternatives $A_i \in \boldsymbol{A}$ for each expert $e_k \in \boldsymbol{E}$ according to the increasing orders of the distances $D_i^k (i \in \{1, 2, \ldots, m\},\ k \in \{1, 2, \ldots, g\})$
Step 9:	Rank the alternatives for the group using social choice functions such as the Borda's function and Copeland's function (Hwang and Yoon 1981), and then determine the optimal alternative from the set of alternatives $\boldsymbol{A}$

in advance, or there may have just one expert (i.e., the MCDM problem), etc. Can the proposed approach [i.e., the model (MOD-4.10)] be generalized to deal with these special cases of such a MCGDM problem under hesitant fuzzy environment? Zhang and Xu (2014) answered this question:

Case 4.1 The MCGDM problem mentioned above only includes the single decision maker, namely, the MCGDM problem is reduced to the MCDM problem, which is similar to the LINMAP-based approaches (Li and Yang 2004; Wan and Li 2013). In this case, $k = 1$, the model (MOD-4.10) can be converted into the following model (Zhang and Xu 2014):

$$\min \left\{ \theta \times \sum_{(\xi,\zeta)\in\Omega} \bar{R}_{\xi\zeta} \times z_{\xi\zeta} + (1-\theta) \times \sum_{(\xi,\zeta)\in\Omega} \left((\underline{R}_{\xi\zeta} + \bar{R}_{\xi\zeta})/2 \right) \times z_{\xi\zeta} \right\}$$

$$s.t. \begin{cases} \sum_{(\xi,\zeta)\in\Omega} \underline{R}_{\xi\zeta} \times \rho_{\zeta\xi} \geq \underline{a} \\ \sum_{(\xi,\zeta)\in\Omega} \bar{R}_{\xi\zeta} \times \rho_{\zeta\xi} \geq \bar{a} \\ \rho_{\zeta\xi} + z_{\xi\zeta} \geq 0 \quad (\xi,\zeta)\in\Omega \\ z_{\xi\zeta} \geq 0 \quad (\xi,\zeta)\in\Omega \\ 0 \leq (\gamma_j^1)^+ \leq (\gamma_j^2)^+ \leq \cdots \leq (\gamma_j^{\#h})^+ \leq 1,\ j=1,2,\ldots,n \\ 0 \leq w_j \leq 1,\ \sum_{j=1}^{n} w_j = 1, j=1,2,\ldots,n \end{cases}$$

(MOD-4.11)

Case 4.2 The interval preference information reduces to the crisp preference information. Namely, $R_k(\xi,\zeta)$ reduces to the real number 0 or 1, which is similar to the LINMAP-based methods (Li and Yang 2004; Li et al. 2010; Wang and Li 2012; Chen 2013). In this case, the model (MOD-4.10) reduces to the following model (Zhang and Xu 2014):

$$\min \left\{ \sum_{k=1}^{g} \sum_{(\xi,\zeta)\in\Omega^k} z_{\xi\zeta}^k \right\}$$

$$s.t. \begin{cases} \sum_{k=1}^{g} \sum_{(\xi,\zeta)\in\Omega^k} \rho_{\zeta\xi}^k \geq \varepsilon_a \\ \rho_{\zeta\xi}^k + z_{\xi\zeta}^k \geq 0 \quad ((\xi,\zeta)\in\Omega^k; k=1,2,\ldots,g) \\ z_{\xi\zeta}^k \geq 0 \quad ((\xi,\zeta)\in\Omega^k; k=1,2,\ldots,g) \\ 0 \leq (\gamma_j^1)^+ \leq (\gamma_j^2)^+ \leq \cdots \leq (\gamma_j^{\#h})^+ \leq 1,\ j=1,2,\ldots,n \\ 0 \leq w_j \leq 1,\ \sum_{j=1}^{n} w_j = 1, j=1,2,\ldots,n \end{cases} \quad \text{(MOD-4.12)}$$

where ε_a is not an interval number but an arbitrary positive real number given by the decision makers in advance.

Remark 4.3 (Zhang and Xu 2014). As mentioned previously, in the practical decision making problems experts are not sure enough in pair-wise comparisons over alternative and usually express their opinions with a degree of insurance. As a result, the decision maker's pair-wise comparison judgments over alternatives may be represented by interval numbers, TFNs, TrFNs, IFNs or HFEs, etc. It is noticed that the decision making problems with the different forms of the experts' preferences on pair-wise comparisons require different LINMAP-based models to solve them. This chapter has introduced an interval programming model-based hesitant fuzzy LINMAP group decision method developed by Zhang and Xu (2014) to solve the MCGDM problems in which the criteria values are taken as HFEs and all pair-wise comparison judgments over alternatives are represented by interval numbers. While if the decision maker's pair-wise comparison judgments are represented by TFNs, TrFNs, IFNs or HFEs, both the collective consistency index $\tilde{G}$ and the inconsistency index $\tilde{B}$ are TFNs, TrFNs, IFNs or HFEs, respectively. It is noted that this method fails to solve them. Nevertheless, Sadi-Nezhad and Akhtari (2008) developed a possibility LINMAP for solving the MCGDM problems with triangular fuzzy truth degree. Li and Wan (2014) proposed a LINMAP-based fuzzy linear programming technique for solving the heterogeneous MCDM problems with trapezoidal fuzzy truth degree. Wan and Li (2013) developed a heterogeneous LINMAP method to deal with intutionsitic fuzy truch degree. For the case that the decision maker's pair-wise comparison judgments are represented by HFEs, in the next section we will introduce a hesitant fuzzy programming model-based LINMAP method proposed by Zhang et al. (2015) to address it.

Case 4.3 The weights of criteria are completely known in advance. In this case, the variables in the model (MOD-4.10) are only the hesitant fuzzy PIS, and thus the model (MOD-4.10) reduces to the following model (Zhang and Xu 2014):

$$\min \quad \left\{ \theta \times \sum_{k=1}^{g} \sum_{(\xi,\zeta)\in\Omega^k} \bar{R}_{\xi\zeta}^k \times z_{\xi\zeta}^k + (1-\theta) \times \sum_{k=1}^{g} \sum_{(\xi,\zeta)\in\Omega^k} ((\underline{R}_{\xi\zeta}^k + \bar{R}_{\xi\zeta}^k)/2) \times z_{\xi\zeta}^k \right\}$$

$$s.t. \quad \begin{cases} \sum_{k=1}^{g} \sum_{(\xi,\zeta)\in\Omega^k} \underline{R}_{\xi\zeta}^k \times \rho_{\zeta\xi}^k \geq \underline{a} \\ \sum_{k=1}^{g} \sum_{(\xi,\zeta)\in\Omega^k} \bar{R}_{\xi\zeta}^k \times \rho_{\zeta\xi}^k \geq \bar{a} \\ \rho_{\zeta\xi}^k + z_{\xi\zeta}^k \geq 0, \ (\xi,\zeta) \in \Omega^k; k = 1,2,\ldots,g \\ z_{\xi\zeta}^k \geq 0, \ (\xi,\zeta) \in \Omega^k; k = 1,2,\ldots,g \\ 0 \leq (\gamma_j^1)^+ \leq (\gamma_j^2)^+ \leq \cdots \leq (\gamma_j^{\#h})^+ \leq 1, \ j = 1,2,\ldots,n \end{cases}$$

(MOD-4.13)

Case 4.4 The hesitant fuzzy PIS is already known in advance. In this case, the variables need to be determined are just the weights of criteria, then the model (MOD-4.10) reduces to the following model (Zhang and Xu 2014):

$$\min \left\{ \theta \times \sum_{k=1}^{g} \sum_{(\xi,\zeta)\in\Omega^k} \underline{R}_{\xi\zeta}^k \times z_{\xi\zeta}^k + (1-\theta) \times \sum_{k=1}^{g} \sum_{(\xi,\zeta)\in\Omega^k} \left(\left(\underline{R}_{\xi\zeta}^k + \bar{R}_{\xi\zeta}^k \right) \Big/ 2 \right) \times z_{\xi\zeta}^k \right\}$$

$$s.t. \begin{cases} \sum_{k=1}^{g} \sum_{(\xi,\zeta)\in\Omega^k} \underline{R}_{\xi\zeta}^k \times \rho_{\zeta\xi}^k \geq \underline{a} \\ \sum_{k=1}^{g} \sum_{(\xi,\zeta)\in\Omega^k} \bar{R}_{\xi\zeta}^k \times \rho_{\zeta\xi}^k \geq \bar{a} \\ \rho_{\zeta\xi}^k + z_{\xi\zeta}^k \geq 0, \ (\xi,\zeta) \in \Omega^k; \ k = 1,2,\ldots,g \\ z_{\xi\zeta}^k \geq 0, (\xi,\zeta) \in \Omega^k; \ k = 1,2,\ldots,g \\ 0 \leq w_j \leq 1, \ \sum_{j=1}^{n} w_j = 1, j = 1,2,\ldots,n \end{cases}$$

(MOD-4.14)

Case 4.5 The weights of criteria are not completely unknown but are partially known, and the known weights of criteria are denoted by $\boldsymbol{\Delta}$. In this case, the model (MOD-4.10) is transformed into the following model (Zhang and Xu 2014):

$$\min \left\{ \theta \times \sum_{k=1}^{g} \sum_{(\xi,\zeta)\in\Omega^k} \bar{R}_{\xi\zeta}^k \times z_{\xi\zeta}^k + (1-\theta) \times \sum_{k=1}^{g} \sum_{(\xi,\zeta)\in\Omega^k} ((\underline{R}_{\xi\zeta}^k + \bar{R}_{\xi\zeta}^k)/2) \times z_{\xi\zeta}^k \right\}$$

$$s.t. \begin{cases} \sum_{k=1}^{g} \sum_{(\xi,\zeta)\in\Omega^k} \underline{R}_{\xi\zeta}^k \times \rho_{\zeta\xi}^k \geq \underline{a} \\ \sum_{k=1}^{g} \sum_{(\xi,\zeta)\in\Omega^k} \bar{R}_{\xi\zeta}^k \times \rho_{\zeta\xi}^k \geq \bar{a} \\ \rho_{\zeta\xi}^k + z_{\xi\zeta}^k \geq 0, \ (\xi,\zeta) \in \Omega^k; k = 1,2,\ldots,g \\ z_{\xi\zeta}^k \geq 0, \ (\xi,\zeta) \in \Omega^k; k = 1,2,\ldots,g \\ 0 \leq (\gamma_j^1)^+ \leq (\gamma_j^2)^+ \leq \cdots \leq (\gamma_j^{\#h})^+ \leq 1, j = 1,2,\ldots,n \\ w \in \Delta \end{cases}$$

(MOD-4.15)

It is apparent from the above analysis that the model (MOD-4.10) is extremely flexible, according to the different actual requirements, this model can be reduced to various special models for solving different decision making problems under different circumstances.

4.2.3 An Energy Project Selection Problem and the Analysis Process

Zhang and Xu (2014) modified the energy project selection problem introduced in Sect. 1.4.1 to demonstrate the implementation process of the proposed method. They also conducted a comparison analysis with the similar method to show the superiority of the proposed method.

4.2.3.1 Description of the Energy Project Selection Problem

According to the description of the energy project selection problem in Sect. 1.4.1, there are five energy projects $A_i (i = 1, 2, 3, 4, 5)$ to be invested, and four criteria to be considered: technological criterion C_1; environmental criterion C_2; socio-political criterion C_3; economic criterion C_4 (more details about them can be found in Kahraman and Kaya (2010). Three experts $e_k (k = 1, 2, 3)$ are invited to evaluate the performances of the five alternatives with respect to each criterion. The results evaluated by three experts are contained in hesitant fuzzy decision matrices, which are shown in Tables 4.4, 4.5 and 4.6 (Zhang and Xu 2014).

Meanwhile, the experts $e_k (k = 1, 2, 3)$ also provide their incomplete preference information through pair-wise comparisons over alternatives as follows (Zhang and Xu 2014):

$$\Omega^1 = \{\langle (A_1, A_2), [0.7, 0.8]\rangle, \langle (A_3, A_4), [0.8, 0.9]\rangle, \langle (A_5, A_3), [0.8, 0.9]\rangle\},$$

$$\Omega^2 = \left\{ \begin{array}{l} \langle (A_1, A_3), [0.6, 0.7]\rangle, \langle (A_3, A_4), [0.8, 0.9]\rangle, \\ \langle (A_4, A_2), [0.7, 0.8]\rangle, \langle (A_5, A_3), [0.7, 0.8]\rangle \end{array} \right\},$$

$$\Omega^3 = \{\langle (A_2, A_1), [0.7, 0.8]\rangle, \langle (A_3, A_1), [0.5, 0.6]\rangle, \langle (A_4, A_5), [0.9, 1.0]\rangle\}.$$

Table 4.4 Hesitant fuzzy decision matrix $\Re^1$

	C_1	C_2
A_1	H{0.3, 0.4, 0.5}	H{0.1, 0.7, 0.8, 0.9}
A_2	H{0.3, 0.5}	H{0.2, 0.5, 0.6, 0.7, 0.9}
A_3	H{0.6, 0.7}	H{0.6, 0.9}
A_4	H{0.3, 0.4, 0.7, 0.8}	H{0.2, 0.4, 0.7}
A_5	H{0.1, 0.3, 0.6, 0.7, 0.9}	H{0.4, 0.6, 0.7, 0.8}
	C_3	C_4
A_1	H{0.2, 0.4, 0.5}	H{0.3, 0.5, 0.6, 0.9}
A_2	H{0.1, 0.5, 0.6, 0.8}	H{0.3, 0.4, 0.7}
A_3	H{0.3, 0.5, 0.7}	H{0.4, 0.6}
A_4	H{0.1, 0.8}	H{0.6, 0.8, 0.9}
A_5	H{0.7, 0.8, 0.9}	H{0.3, 0.6, 0.7, 0.9}

Table 4.5 Hesitant fuzzy decision matrix $\Re^2$

	C_1	C_2
A_1	H{0.4, 0.6}	H{0.7, 0.8, 0.9}
A_2	H{0.3, 0.4, 0.5}	H{0.1, 0.5, 0.6, 0.7, 0.9}
A_3	H{0.2, 0.4, 0.5, 0.6, 0.7}	H{0.6, 0.9}
A_4	H{0.3, 0.7, 0.8}	H{0.2, 0.4, 0.7}
A_5	H{0.6, 0.7, 0.9}	H{0.4, 0.6, 0.7, 0.8}
	C_3	C_4
A_1	H{0.2, 0.4, 0.5}	H{0.3, 0.5, 0.6, 0.9}
A_2	H{0.4, 0.5, 0.6, 0.8}	H{0.3, 0.4, 0.7}
A_3	H{0.3, 0.5, 0.7}	H{0.3, 0.4, 0.5, 0.6}
A_4	H{0.4, 0.6, 0.8}	H{0.3, 0.5, 0.6, 0.8, 0.9}
A_5	H{0.5, 0.6, 0.7, 0.8, 0.9}	H{0.3, 0.8, 0.9}

Table 4.6 Hesitant fuzzy decision matrix $\Re^3$

	C_1	C_2
A_1	H{0.3, 0.4, 0.5, 0.6, 0.7}	H{0.7, 0.8}
A_2	H{0.3, 0.5}	H{0.5, 0.6, 0.7, 0.9}
A_3	H{0.5, 0.6, 0.7}	H{0.6, 0.9}
A_4	H{0.3, 0.4, 0.7, 0.8}	H{0.2, 0.4, 0.7}
A_5	H{0.7, 0.9}	H{0.4, 0.5, 0.6, 0.7, 0.8}
	C_3	C_4
A_1	H{0.2, 0.3, 0.4, 0.5}	H{0.5, 0.6, 0.9}
A_2	H{0.5, 0.6, 0.8}	H{0.4, 0.5, 0.6, 0.7}
A_3	H{0.3, 0.5, 0.7}	H{0.4, 0.5, 0.6, 0.7, 0.8}
A_4	H{0.1, 0.2, 0.7, 0.8}	H{0.6, 0.8, 0.9}
A_5	H{0.7, 0.8, 0.9}	H{0.3, 0.6, 0.7, 0.9}

where $\langle (A_1, A_3), [0.6, 0.7] \rangle$ in Ω^2 indicates that the expert e_2 moderately prefers A_1 to A_3, and the others have the similar meanings.

4.2.3.2 The Decision Making Process of the Proposed Method

In the following, we employ the interval programming model-based hesitant fuzzy LINMAP group decision method to solve the above energy project selection problem. It is easy to see from Tables 4.4, 4.5 and 4.6 that the numbers of values in different HFEs are different. In order to accurately calculate the distance between two HFEs, we should extend the shorter one until both of them have the same length. According to the regulations in Definition 1.3, we consider that the experts are risk-averse in the energy project selection problem, and normalize the hesitant fuzzy data by adding the minimal values as listed in Tables 4.7, 4.8 and 4.9 (Zhang and Xu 2014).

Table 4.7 Hesitant fuzzy normalized decision matrix $\tilde{\Re}^1$

	C_1	C_2
A_1	H{0.3, 0.3, 0.3, 0.4, 0.5}	H{0.1, 0.1, 0.7, 0.8, 0.9}
A_2	H{0.3, 0.3, 0.3, 0.3, 0.5}	H{0.2, 0.5, 0.6, 0.7, 0.9}
A_3	H{0.6, 0.6, 0.6, 0.6, 0.7}	H{0.6, 0.6, 0.6, 0.6, 0.9}
A_4	H{0.3, 0.3, 0.4, 0.7, 0.8}	H{0.2, 0.2, 0.2, 0.4, 0.7}
A_5	H{0.1, 0.3, 0.6, 0.7, 0.9}	H{0.4, 0.4, 0.6, 0.7, 0.8}
	C_3	C_4
A_1	H{0.2, 0.2, 0.2, 0.4, 0.5}	H{0.3, 0.3, 0.5, 0.6, 0.9}
A_2	H{0.1, 0.1, 0.5, 0.6, 0.8}	H{0.3, 0.3, 0.3, 0.4, 0.7}
A_3	H{0.3, 0.3, 0.3, 0.5, 0.7}	H{0.4, 0.4, 0.4, 0.4, 0.6}
A_4	H{0.1, 0.1, 0.1, 0.1, 0.8}	H{0.6, 0.6, 0.6, 0.8, 0.9}
A_5	H{0.7, 0.7, 0.7, 0.8, 0.9}	H{0.3, 0.3, 0.6, 0.7, 0.9}

Table 4.8 Hesitant fuzzy normalized decision matrix $\tilde{\Re}^2$

	C_1	C_2
A_1	H{0.4, 0.4, 0.4, 0.4, 0.6}	H{0.7, 0.7, 0.7, 0.8, 0.9}
A_2	H{0.3, 0.3, 0.3, 0.4, 0.5}	H{0.1, 0.5, 0.6, 0.7, 0.9}
A_3	H{0.2, 0.4, 0.5, 0.6, 0.7}	H{0.6, 0.6, 0.6, 0.6, 0.9}
A_4	H{0.3, 0.3, 0.3, 0.7, 0.8}	H{0.2, 0.2, 0.2, 0.4, 0.7}
A_5	H{0.6, 0.6, 0.6, 0.7, 0.9}	H{0.4, 0.4, 0.6, 0.7, 0.8}
	C_3	C_4
A_1	H{0.2, 0.2, 0.2, 0.4, 0.5}	H{0.3, 0.3, 0.5, 0.6, 0.9}
A_2	H{0.4, 0.4, 0.5, 0.6, 0.8}	H{0.3, 0.3, 0.3, 0.4, 0.7}
A_3	H{0.3, 0.3, 0.3, 0.5, 0.7}	H{0.3, 0.3, 0.4, 0.5, 0.6}
A_4	H{0.4, 0.4, 0.4, 0.6, 0.8}	H{0.3, 0.5, 0.6, 0.8, 0.9}
A_5	H{0.5, 0.6, 0.7, 0.8, 0.9}	H{0.3, 0.3, 0.3, 0.8, 0.9}

Table 4.9 Hesitant fuzzy normalized decision matrix $\tilde{\Re}^3$

	C_1	C_2
A_1	H{0.3, 0.4, 0.5, 0.6, 0.7}	H{0.7, 0.7, 0.7, 0.7, 0.8}
A_2	H{0.3, 0.3, 0.3, 0.3, 0.5}	H{0.5, 0.5, 0.6, 0.7, 0.9}
A_3	H{0.5, 0.5, 0.5, 0.6, 0.7}	H{0.6, 0.6, 0.6, 0.6, 0.9}
A_4	H{0.3, 0.3, 0.4, 0.7, 0.8}	H{0.2, 0.2, 0.2, 0.4, 0.7}
A_5	H{0.7, 0.7, 0.7, 0.7, 0.9}	H{0.4, 0.5, 0.6, 0.7, 0.8}
	C_3	C_4
A_1	H{0.2, 0.2, 0.3, 0.4, 0.5}	H{0.5, 0.5, 0.5, 0.6, 0.9}
A_2	H{0.5, 0.5, 0.5, 0.6, 0.8}	H{0.4, 0.4, 0.5, 0.6, 0.7}
A_3	H{0.3, 0.3, 0.3, 0.5, 0.7}	H{0.4, 0.5, 0.6, 0.7, 0.8}
A_4	H{0.1, 0.1, 0.2, 0.7, 0.8}	H{0.6, 0.6, 0.6, 0.8, 0.9}
A_5	H{0.7, 0.7, 0.7, 0.8, 0.9}	H{0.3, 0.3, 0.6, 0.7, 0.9}

Then, we proceed to solve the above problem for selecting the most appropriate energy policy, which involves the following two cases (i.e., Cases 4.6 and 4.7):

Case 4.6 (Zhang and Xu 2014). The information about the criteria weights is completely unknown.

For convenience, let

$$A^{+} = (h_1^{+}, h_2^{+}, h_3^{+}, h_4^{+}) = \begin{pmatrix} H\{h_{11}, h_{12}, h_{13}, h_{14}, h_{15}\}, H\{h_{21}, h_{22}, h_{23}, h_{24}, h_{25}\}, \\ H\{h_{31}, h_{32}, h_{33}, h_{34}, h_{35}\}, H\{h_{41}, h_{42}, h_{43}, h_{44}, h_{45}\} \end{pmatrix} \tag{4.15}$$

Then, we utilize the model (MOD-4.10) to construct the corresponding linear programming model (MOD-A.1) which is displayed in Appendix. Let $\theta = 0.5$ and $\varepsilon_{\tilde{a}} = [\underline{a}, \bar{a}] = [0.1, 0.1]$, the optimal weight vector $\boldsymbol{w}$ and the hesitant fuzzy PIS A^{+} can be obtained (Zhang and Xu 2014):

$$\begin{aligned} \boldsymbol{w} &= (0.2729, 0.1098, 0.1063, 0.5110)^{T}, \\ A^{+} &= (H\{0.0, 0.0, 0.8498, 0.8498, 0.8498\}, H\{0.0, 1.0, 1.0, 1.0, 1.0\}, \\ &\quad H\{1.0, 1.0, 1.0, 1.0, 1.0\}, H\{1.0, 1.0, 1.0, 1.0, 1.0\}). \end{aligned}$$

Based on the obtained weight vector and the hesitant fuzzy PIS, the distances between the alternatives $A_i (i = 1, 2, 3, 4, 5)$ and the hesitant fuzzy PIS A^{+} can be calculated by using Eq. (4.3), and the results are listed in Table 4.10. According to the increasing orders of the distances $D_i^k (i = 1, 2, 3, 4, 5,\ k = 1, 2, 3)$, the ranking order of the alternatives for each expert is obtained and is also listed in Table 4.10 (Zhang and Xu 2014).

At length, on the basis of the Borda's function (Hwang and Yoon 1981), the Borda's scores of alternatives $A_i (i = 1, 2, 3, 4, 5)$ are obtained and are shown in Table 4.11 (Zhang and Xu 2014). Therefore, the ranking order of alternatives can be easily obtained from Table 4.11 (Zhang and Xu 2014) as $A_2 \prec A_4 \prec A_1 \prec A_3 \prec A_5$, and the most desirable alternative is obtained as the energy project A_5.

Table 4.10 The distances and the ranking of alternatives

Experts	A_1	A_2	A_3	A_4	A_5	Ranking
e_1	0.3250	0.3511	0.2905	0.2904	0.2904	$A_2 \prec A_1 \prec A_3 \prec A_4 \sim A_5$
e_2	0.2830	0.3395	0.3363	0.2963	0.2835	$A_2 \prec A_3 \prec A_4 \prec A_5 \prec A_1$
e_3	0.2472	0.2472	0.2287	0.2715	0.2411	$A_4 \prec A_1 \sim A_2 \prec A_5 \prec A_3$

Table 4.11 Borda's scores of alternatives for each expert

Energy projects	Experts			Borda's scores
	e_1	e_2	e_3	
A_1	2	5	2	9
A_2	1	1	2	4
A_3	3	2	5	10
A_4	4	3	1	8
A_5	**4**	**4**	**4**	**12**

Case 4.7 (Zhang and Xu 2014). The information about the criteria weights is partially known and the known weight information is given as:

$$\Delta = \left\{ \begin{array}{l} 0.15 \leq w_1 \leq 0.2, 0.16 \leq w_2 \leq 0.18, 0.3 \leq w_3 \leq 0.35, \\ 0.3 \leq w_4 \leq 0.45, \sum_{j=1}^{4} w_j = 1 \end{array} \right\}.$$

Then, we utilize the model (MOD-4.15) to construct the corresponding linear programming model (MOD-A.2) which is displayed in Appendix. Taking $\theta = 0.5$ and $\varepsilon_{\tilde{a}} = [\underline{a}, \bar{a}] = [0.1, 0.1]$, we can obtain the optimal weight vector $\boldsymbol{w}$ and the hesitant fuzzy PIS A^+ as:

$$\boldsymbol{w} = (0.15, 0.16, 0.3, 0.39)^T$$

$$A^+ = \begin{pmatrix} H\{0.0123, 0.0123, 0.9086, 0.9086, 0.9086\}, \\ H\{0.0, 0.4716, 0.4716, 0.4716, 1.0\}, \\ H\{0.2836, 0.2836, 0.2836, 1.0, 1.0\}, \\ H\{1.0, 1.0, 1.0, 1.0, 1.0\} \end{pmatrix}.$$

Afterwards, the distances $D_i^k (i = 1, 2, 3, 4, 5,\ k = 1, 2, 3)$ of the alternatives A_i $(i = 1, 2, 3, 4, 5)$ to the hesitant fuzzy PIS A^+ can be derived by using Eq. (4.3), and the results are listed in Table 4.12 (Zhang and Xu 2014).

Furthermore, the ranking of alternatives for each expert $e_k (k = 1, 2, 3)$ can be obtained according to the increasing orders of the distances, which are also listed in Table 4.12. Finally, the Borda's scores of all alternatives $A_i (i = 1, 2, 3, 4, 5)$ can be obtained by using the Borda's function (Hwang and Yoon 1981), and the results are listed in Table 4.13 (Zhang and Xu 2014). Thus, the ranking order of the

Table 4.12 The distances and the ranking of alternatives

	A_1	A_2	A_3	A_4	A_5	Ranking
e_1	0.2964	0.3634	0.2626	0.2624	0.2625	$A_2 \prec A_1 \prec A_3 \prec A_5 \prec A_4$
e_2	0.2638	0.3168	0.2935	0.2473	0.2731	$A_2 \prec A_3 \prec A_5 \prec A_1 \prec A_4$
e_3	0.2290	0.2290	0.2144	0.2691	0.2327	$A_4 \prec A_5 \prec A_1 \sim A_2 \prec A_3$

Table 4.13 Borda's scores of alternatives for each expert

Energy projects	Experts			Borda's scores
	e_1	e_2	e_3	
A_1	2	4	3	9
A_2	1	1	3	5
A_3	3	2	5	10
$\mathbf{A_4}$	**5**	**5**	**1**	**11**
A_5	4	3	2	9

alternatives $A_i (i = 1, 2, 3, 4, 5)$ can be easily obtained as $A_2 \prec A_5 \sim A_1 \prec A_3 \prec A_4$, and the best alternative is the energy project A_4.

4.2.3.3 Comparison Analysis and Discussion

Zhang and Xu (2014) made a comparison with the intuitionistic fuzzy LINMAP (IF-LINMAP) method proposed by Li et al. (2010), which is the closest to Zhang and Xu (2014)'s method. In order to compare with the IF-LINMAP method, we consider the HFEs' envelopes, i.e., intuitionistic fuzzy data, and apply the IF-LINMAP method to solve the energy project selection problem under intuitionistic fuzzy environment.

Because the envelope of the HFE h is the IFN $I_{env}(h)$, we can transform hesitant fuzzy data in the energy project selection problem into intuitionistic fuzzy data as depicted in Tables 4.14, 4.15 and 4.16 (Zhang and Xu 2014).

Moreover, the IF-LINMAP method can only take into account pair-wise comparisons over alternatives with the crisp degree of truth 0 or 1, thus we also assume that the interval numbers $R_k(\xi, \zeta)$ in the above example reduces to the real number 0 or 1. Namely, the experts e_k $(k = 1, 2, 3)$ provide their comparison preference information over alternatives as (Zhang and Xu 2014):

$$\Omega^1 = \{(A_1, A_2), (A_3, A_4), (A_5, A_3)\},$$
$$\Omega^2 = \{(A_1, A_3), (A_3, A_4), (A_4, A_2), (A_5, A_3)\},$$
$$\Omega^3 = \{(A_2, A_1), (A_3, A_1), (A_4, A_5)\},$$

Table 4.14 Intuitionistic fuzzy decision matrix $\Re^1$

	C_1	C_2	C_3	C_4
A_1	$I(0.3, 0.5)$	$I(0.1, 0.1)$	$I(0.2, 0.5)$	$I(0.3, 0.1)$
A_2	$I(0.3, 0.5)$	$I(0.2, 0.1)$	$I(0.1, 0.2)$	$I(0.4, 0.3)$
A_3	$I(0.6, 0.3)$	$I(0.6, 0.1)$	$I(0.3, 0.3)$	$I(0.4, 0.4)$
A_4	$I(0.3, 0.2)$	$I(0.2, 0.3)$	$I(0.1, 0.2)$	$I(0.6, 0.1)$
A_5	$I(0.1, 0.1)$	$I(0.4, 0.2)$	$I(0.7, 0.1)$	$I(0.3, 0.1)$

Table 4.15 Intuitionistic fuzzy decision matrix $\Re^2$

	C_1	C_2	C_3	C_4
A_1	I(0.4, 0.4)	I(0.7, 0.1)	I(0.2, 0.5)	I(0.3, 0.1)
A_2	I(0.3, 0.5)	I(0.1, 0.1)	I(0.4, 0.2)	I(0.4, 0.3)
A_3	I(0.2, 0.3)	I(0.6, 0.1)	I(0.3, 0.3)	I(0.3, 0.4)
A_4	I(0.3, 0.2)	I(0.2, 0.3)	I(0.4, 0.2)	I(0.3, 0.1)
A_5	I(0.6, 0.1)	I(0.4, 0.2)	I(0.5, 0.1)	I(0.3, 0.1)

Table 4.16 Intuitionistic fuzzy decision matrix $\Re^3$

	C_1	C_2	C_3	C_4
A_1	I(0.3, 0.3)	I(0.7, 0.2)	I(0.2, 0.5)	I(0.5, 0.1)
A_2	I(0.3, 0.5)	I(0.5, 0.1)	I(0.5, 0.2)	I(0.4, 0.3)
A_3	I(0.5, 0.3)	I(0.6, 0.1)	I(0.3, 0.3)	I(0.4, 0.2)
A_4	I(0.3, 0.2)	I(0.2, 0.3)	I(0.1, 0.2)	I(0.6, 0.1)
A_5	I(0.7, 0.1)	I(0.4, 0.2)	I(0.7, 0.1)	I(0.3, 0.1)

where (A_1, A_2) in Ω^1 means that the expert e_1 prefers A_1 to A_2, and the others have the similar meanings.

Then, we employ the IF-LINMAP method (Li et al. 2010) to construct the corresponding optimal model and obtain its optimal solution as (where $\varepsilon_{\tilde{a}} = 0.1$):

$$\begin{aligned} \boldsymbol{w} &= (0.017, 0.013, 0.089, 0.881)^T, \\ \boldsymbol{u} &= (0.017, 0.01, 0.074, 0.4682), \\ \boldsymbol{v} &= (0, 0, 0.0149, 0.116). \end{aligned}$$

Using Eqs. (6), (61) and (62) in Li et al. (2010), the intuitionistic fuzzy PIS A^+ can be calculated as follows:

$$A^+ = (I(1.0, 0.0), I(1.0, 0.0), I(0.8315, 0.1674), I(0.5296, 0.1312)).$$

Therefore, the squares of the distances between the alternatives $A_i (i = 1, 2, 3, 4, 5)$ and the intuitionistic fuzzy PIS A^+ can be calculated using Eq. (6) in Li et al. (2010) as:

$$\begin{aligned} &d_1^1 = 0.0944, \quad d_2^1 = 0.0786, \quad d_3^1 = 0.0719, \quad d_4^1 = 0.0604, \quad d_5^1 = 0.0717, \\ &d_1^2 = 0.086, \quad d_2^2 = 0.05, \quad d_3^2 = 0.086, \quad d_4^2 = 0.0807, \quad d_5^2 = 0.071, \\ &d_1^3 = 0.0761, \quad d_2^3 = 0.0383, \quad d_3^3 = 0.0661, \quad d_4^3 = 0.0604, \quad d_5^3 = 0.0604. \end{aligned}$$

Comparing these distances, the rankings of the alternatives $A_i (i = 1, 2, 3, 4, 5)$ for the experts $e_k (k = 1, 2, 3)$ are generated as:

Table 4.17 Borda's scores of alternatives for each expert

Energy projects	Experts			Borda's scores
	e_1	e_2	e_3	
A_1	1	1	1	3
$\mathbf{A_2}$	**2**	**5**	**5**	**12**
A_3	3	1	2	6
A_4	5	3	3	11
A_5	4	4	3	11

$$A_1 \prec A_2 \prec A_3 \prec A_5 \prec A_4,$$
$$A_1 \sim A_3 \prec A_4 \prec A_5 \prec A_2,$$
$$A_1 \prec A_3 \prec A_5 \sim A_4 \prec A_2.$$

Similarly, the Borda's scores of the alternatives $A_i (i = 1, 2, 3, 4, 5)$ can be obtained as in Table 4.17 (Zhang and Xu 2014).

Then, the ranking of the alternatives $A_i (i = 1, 2, 3, 4, 5)$ can be easily obtained from Table 4.17 as $A_1 \prec A_3 \prec A_5 \sim A_4 \prec A_2$. Thus, the best alternative is the energy project A_2.

To provide a better view of the comparison results, we put the results of the ranking of alternatives obtained by Zhang and Xu (2014)'s method and by Li et al. (2010)'s method into Fig. 4.1 (Zhang and Xu 2014).

It is easy to see that the ranking of the alternatives obtained by Li et al. (2010)'s method is remarkably different from that obtained by Zhang and Xu (2014)'s method. The main reason is that Zhang and Xu (2014)'s method considers the hesitant fuzzy information which is represented by several possible values, not by a margin of error (as in IFNs), while if adopting Li et al. (2010)'s method, it needs to transform HFEs into IFNs, which gives rise to a difference in the accuracy of data in the two types. Moreover, Zhang and Xu (2014)'s method considered the situation that the set of all pair-wise comparison judgments from experts are not crisp but

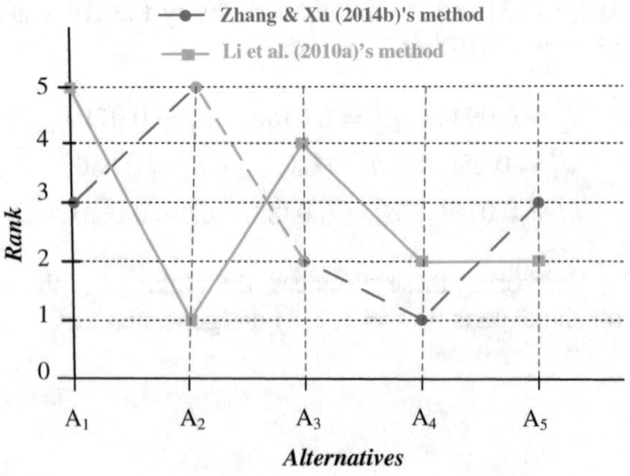

Fig. 4.1 The pictorial representation of the rankings of alternatives

interval numbers, while Li et al. (2010)'s method only considers pair-wise comparison judgments between alternatives with the crisp degree of truth 0 or 1, which is one of the most disadvantages in multidimensional analysis of preference (Sadi-Nezhad and Akhtari 2008). Thus, it is not hard to see that Zhang and Xu (2014)'s method has some desirable advantages over Li et al. (2010)'s method as follows:

(1) Zhang and Xu (2014)'s method, by extending the LINMAP to take into account the hesitant fuzzy assessments which are well-suited to handle the ambiguity and impreciseness inherent in the MCGDM problems, does not need to transform HFEs into IFNs but directly handle such problems, and thus obtains better final decision results.
(2) Zhang and Xu (2014)'s method can integrate more useful information into decision making process by considering the situation that all pair-wise comparison judgments from experts are interval numbers, and thus is capable of better modeling the real-life decision situations. In particular, when we meet some decision making situations where the decision makers have hesitancy in providing their preferences, Zhang and Xu (2014)'s method demonstrates its great superiority in handling this sort of decision making problems.

4.3 Hesitant Fuzzy Programming Model-Based LINMAP Method

To handle the hesitant fuzzy MCDM problem with incomplete weight information in which the pair-wise comparison judgments on alternatives are represented by HFEs, Zhang et al. (2015) further utilized the main structure of the LINMAP to develop a new hesitant fuzzy mathematical programming technique. We call it Zhang et al. (2015)'s method. Meanwhile, to address the incomplete and inconsistent preference structures of criteria weights, Zhang et al. (2015) also established a bi-objective nonlinear programming model with several deviation variables.

4.3.1 Hesitant Fuzzy Consistency and Inconsistency Indices

In a similar manner of the classical LINMAP method, Zhang et al. (2015)'s method requires the decision maker to provide not only the ratings of alternatives with respect to each criterion but also the pair-wise preference information over alternatives when evaluating the MCDM problems. Assume here that the decision maker gives the pair-wise comparison preference information over alternatives by a set of ordered pairs $\tilde{\Omega} = \{(\xi,\zeta)|A_\xi \succeq_{\tilde{R}(\xi,\zeta)} A_\zeta\}$ with hesitant fuzzy truth degrees $\tilde{R}(\xi,\zeta)$ $(\xi,\zeta \in \{1,2,\ldots,m\})$.

Then, the square of the distances between each pair alternatives $(\xi,\zeta)\in\tilde{\Omega}$ and the hesitant fuzzy PIS A^+ can be obtained by using Eq. (1.2), respectively, as follows:

$$D_\xi=\sum_{j=1}^{n} w_j d(h_{\xi j},h_j^+)^2=\sum_{j=1}^{n} w_j\left(\frac{1}{\#h}\sum_{\lambda=1}^{\#h}\left(\gamma_{\xi j}^{\lambda}-(\gamma_j^{\lambda})^+\right)^2\right), \tag{4.16}$$

and

$$D_\zeta=\sum_{j=1}^{n} w_j d(h_{\zeta j},h_j^+)^2=\sum_{j=1}^{n} w_j\left(\frac{1}{\#h}\sum_{\lambda=1}^{\#h}\left(\gamma_{\zeta j}^{\lambda}-(\gamma_j^{\lambda})^+\right)^2\right). \tag{4.17}$$

Remark 4.4 (Zhang et al. 2015). All HFEs in the decision matrix have the same length (if not, then add some values into the shorter one until they have the same length according to Definition 1.3), and we stipulate $\#h=\#h_{ij}=\#h_j^+$ $(i\in\{1,2,\ldots,m\},j\in\{1,2,\ldots,n\})$.

Usually, the decision maker for two given alternatives is presumed to prefer that alternative which is "closer" to the ideal alternative. Therefore, for each pair of alternatives $(\xi,\zeta)\in\tilde{\Omega}$, it is easily observed that:

(1) if $D_\xi>D_\zeta$, namely, the alternative A_ξ is farther from the hesitant fuzzy PIS A^+ than the alternative A_ζ, then the ranking order of the alternatives A_ξ and A_ζ determined by D_ξ and D_ζ based on $(\boldsymbol{w},A^+)$ is inconsistent with the ranking order obtained by the decision maker's preference relationship in $\tilde{\Omega}$.
(2) if $D_\xi\leq D_\zeta$, namely, the alternative A_ξ is closer to the hesitant fuzzy PIS A^+ than the alternative A_ζ, then the ranking order of the alternatives A_ξ and A_ζ determined by D_ξ and D_ζ based on $(\boldsymbol{w},A^+)$ is consistent with the ranking order obtained by the decision maker's preference relationship in $\tilde{\Omega}$.

Then, a hesitant fuzzy inconsistency index $(D_\zeta-D_\xi)^-$ is developed to measure the degree of inconsistency between the ranking order of the alternatives A_ξ and A_ζ in which one ranking order is determined by D_ξ and D_ζ based on $(\boldsymbol{w},A^+)$, and the other ranking order obtained by the decision maker's preference information in $\tilde{\Omega}$ as:

$$(D_\zeta-D_\xi)^-=\begin{cases}0, & (D_\zeta\geq D_\xi)\\ \tilde{R}(\xi,\zeta)\times(D_\xi-D_\zeta), & (D_\zeta<D_\xi)\end{cases} \tag{4.18}$$

Then, the hesitant fuzzy inconsistency index is rewritten as:

$$(D_\zeta-D_\xi)^-=\tilde{R}(\xi,\zeta)\max\{0,(D_\xi-D_\zeta)\} \tag{4.19}$$

The comprehensive hesitant fuzzy inconsistency index is defined as:

$$\begin{aligned}\tilde{B} &= \sum_{(\xi,\zeta)\in\tilde{\Omega}} (D_\zeta - D_\xi)^- \\ &= \sum_{(\xi,\zeta)\in\tilde{\Omega}} \tilde{R}(\xi,\zeta)\max\{0,(D_\xi - D_\zeta)\}\end{aligned} \tag{4.20}$$

On the other hand, a hesitant fuzzy consistency index $(D_\zeta - D_\xi)^+$ is proposed as follows:

$$(D_\zeta - D_\xi)^+ = \begin{cases} \tilde{R}(\xi,\zeta) \times (D_\zeta - D_\xi), & (D_\zeta \geq D_\xi) \\ 0, & (D_\zeta < D_\xi) \end{cases} \tag{4.21}$$

which is utilized to measure the degree of consistency between the ranking orders of the alternatives A_ξ and A_ζ in which one ranking order is determined by D_ξ and D_ζ based on $(\boldsymbol{w}, A^+)$, and the other ranking order obtained by the decision maker's preference information in $\tilde{\Omega}$.

The hesitant fuzzy consistency index in Eq. (4.21) can be also rewritten as:

$$(D_\zeta - D_\xi)^+ = \tilde{R}(\xi,\zeta)\max\{0,(D_\zeta - D_\xi)\} \tag{4.22}$$

and the comprehensive hesitant fuzzy consistency index is obtained as:

$$\begin{aligned}\tilde{G} &= \sum_{(\xi,\zeta)\in\tilde{\Omega}} (D_\zeta - D_\xi)^+ \\ &= \sum_{(\xi,\zeta)\in\tilde{\Omega}} \tilde{R}(\xi,\zeta)\max\{0,(D_\zeta - D_\xi)\}\end{aligned} \tag{4.23}$$

4.3.2 Optimization Model

According to the definitions of hesitant fuzzy consistent and inconsistency indices, it is easy to see that the smaller the hesitant fuzzy inconsistency index is, the better the final decision result is. Usually, the hesitant fuzzy consistency index $\tilde{G}$ should be bigger than the hesitant fuzzy inconsistency index $\tilde{B}$. We assume the hesitant fuzzy consistency index $\tilde{G}$ is not smaller than the hesitant fuzzy inconsistency index $\tilde{B}$ by ε_h. Here, ε_h is an arbitrary HFE given by the decision maker in advance. Then, we construct an optimal model to determine the weighting vector of criteria and the hesitant fuzzy PIS as follows (Zhang et al. 2015):

$$\min\ \{\tilde{B}\}$$
$$s.t.\ \begin{cases} \tilde{G} - \tilde{B} \geq \varepsilon_h \\ \boldsymbol{w} \in \boldsymbol{\Delta} \end{cases} \quad \text{(MOD-4.16)}$$

where $\boldsymbol{\Delta}$ is the preference structure of weights of criteria.

Using Eqs. (4.20) and (4.23), it can be easily derived that

$$\begin{aligned} \tilde{G} - \tilde{B} &= \sum_{(\xi,\zeta)\in\tilde{\Omega}} \left\{ (D_\zeta - D_\xi)^+ - (D_\zeta - D_\xi)^- \right\} \\ &= \sum_{(\xi,\zeta)\in\tilde{\Omega}} \left(\tilde{R}(\xi,\zeta) \times (D_\zeta - D_\xi) \right) \end{aligned} \tag{4.24}$$

Combining with Eqs. (4.16–4.17), the above Eq. (4.24) can be rewritten as follows:

$$\tilde{G} - \tilde{B} = \sum_{(\xi,\zeta)\in\tilde{\Omega}} \left\{ \tilde{R}(\xi,\zeta) \times \left\{ \begin{array}{l} \sum_{j=1}^{n} w_j \left(\frac{1}{\#h} \sum_{\lambda=1}^{\#h} \left((\gamma_{\zeta j}^{\lambda})^2 - (\gamma_{\xi j}^{\lambda})^2 \right) \right) \\ -2 \sum_{j=1}^{n} w_j \left(\frac{1}{\#h} \sum_{\lambda=1}^{\#h} (\gamma_j^{\lambda})^+ (\gamma_{\zeta j}^{\lambda} - \gamma_{\xi j}^{\lambda}) \right) \end{array} \right\} \right\} \tag{4.25}$$

Using Eqs. (4.20) and (4.24), the mathematical programming model (MOD-4.16) can be rewritten as (Zhang et al. 2015):

$$\min \left\{ \sum_{(\xi,\zeta)\in\tilde{\Omega}} \left\{ \tilde{R}(\xi,\zeta) \times \max\{0, (D_\xi - D_\zeta)\} \right\} \right\}$$
$$s.t. \begin{cases} \sum_{(\xi,\zeta)\in\tilde{\Omega}} \left\{ \tilde{R}(\xi,\zeta) \times (D_\zeta - D_\xi) \right\} \geq \varepsilon_h \\ \boldsymbol{w} \in \boldsymbol{\Delta} \end{cases} \quad \text{(MOD-4.17)}$$

For each pair of alternatives $(\xi,\zeta) \in \tilde{\Omega}$, let $z_{\xi\zeta} = \max\{0, (D_\xi - D_\zeta)\}$, then for each pair $(\xi,\zeta) \in \tilde{\Omega}$, $z_{\xi\zeta} \geq D_\xi - D_\zeta$, i.e., $z_{\xi\zeta} + D_\zeta - D_\xi \geq 0$ and $z_{\xi\zeta} \geq 0$. Thus, the mathematical programming model (MOD-4.17) can be converted into the following model (Zhang et al. 2015):

$$\min \left\{ \sum_{(\xi,\zeta)\in\tilde{\Omega}} z_{\xi\zeta} \tilde{R}(\xi,\zeta) \right\}$$
$$s.t. \begin{cases} \sum_{(\xi,\zeta)\in\tilde{\Omega}} \left(\tilde{R}(\xi,\zeta) \times (D_\zeta - D_\xi) \right) \geq \varepsilon_h \\ z_{\xi\zeta} - D_\xi + D_\zeta \geq 0 \quad ((\xi,\zeta) \in \tilde{\Omega}) \\ z_{\xi\zeta} \geq 0 \quad ((\xi,\zeta) \in \tilde{\Omega}) \\ \boldsymbol{w} \in \boldsymbol{\Delta} \end{cases} \quad \text{(MOD-4.18)}$$

Let $\rho_{\zeta\xi} = D_\zeta - D_\xi$ and $\vartheta_j^\lambda = w_j(\gamma_j^\lambda)^+$, since $0 \le (\gamma_j^1)^+ \le (\gamma_j^2)^+ \le \cdots \le (\gamma_j^{\#h})^+ \le 1$ and $0 \le w_j \le 1$, then we obtain $0 \le \vartheta_j^1 \le \cdots \le \vartheta_j^{\#h} \le w_j$ ($j \in \{1,2,\ldots,n\}$). By using Eqs. (4.16) and (4.17), we have

$$\begin{aligned}\rho_{\zeta\xi} &= \sum_{j=1}^{n} w_j\left(\frac{1}{\#h}\sum_{\lambda=1}^{\#h}\left(\left(\gamma_{\zeta j}^\lambda\right)^2 - \left(\gamma_{\xi j}^\lambda\right)^2\right)\right) - 2\sum_{j=1}^{n} w_j\left(\frac{1}{\#h}\sum_{\lambda=1}^{\#h}(\gamma_j^\lambda)^+\left(\gamma_{\zeta j}^\lambda - \gamma_{\xi j}^\lambda\right)\right)\\ &= \sum_{j=1}^{n} w_j\left(\frac{1}{\#h}\sum_{\lambda=1}^{\#h}\left(\left(\gamma_{\zeta j}^\lambda\right)^2 - \left(\gamma_{\xi j}^\lambda\right)^2\right)\right) - 2\sum_{j=1}^{n}\left(\frac{1}{\#h}\sum_{\lambda=1}^{\#h}\vartheta_j^\lambda(\gamma_{\zeta j}^\lambda - \gamma_{\xi j}^\lambda)\right)\end{aligned} \tag{4.26}$$

Consequently, the model (MOD-4.18) is further transformed into the following mathematical programming model (Zhang et al. 2015):

$$\min\left\{\sum_{(\xi,\zeta)\in\tilde{\Omega}} \tilde{R}(\xi,\zeta) \times z_{\xi\zeta}\right\}$$
$$s.t.\begin{cases}\sum_{(\xi,\zeta)\in\tilde{\Omega}} \left(\rho_{\zeta\xi} \times \tilde{R}(\xi,\zeta)\right) \ge \varepsilon_h\\ z_{\xi\zeta} + \rho_{\zeta\xi} \ge 0 \quad ((\xi,\zeta)\in\tilde{\Omega})\\ z_{\xi\zeta} \ge 0 \quad ((\xi,\zeta)\in\tilde{\Omega})\\ 0 \le \vartheta_j^1 \le \cdots \le \vartheta_j^{\#h} \le w_j \;\; (j \in \{1,2,\ldots,n\})\\ \boldsymbol{w} \in \boldsymbol{\Delta}\end{cases} \tag{MOD-4.19}$$

It is easy to notice that in the model (MOD-4.19), the objective function is an HFE and the right and left coefficients of the first constraint condition are also HFEs. This model (MOD-4.19) is called hesitant fuzzy mathematical programming model. Zhang et al. (2015) developed an effective method based on the ranking method of HFEs for solving the model (MOD-4.19).

According to the score-based ranking method of HFEs, the model (MOD-4.19) is transformed into the following programming model (MOD-4.20) (Zhang et al. 2015):

$$\min\left\{\sum_{(\xi,\zeta)\in\tilde{\Omega}} z_{\xi\zeta}s(\tilde{R}(\xi,\zeta))\right\}$$
$$s.t.\begin{cases}\sum_{(\xi,\zeta)\in\tilde{\Omega}} g_{\zeta\xi}s(\tilde{R}(\xi,\zeta)) \ge s(\varepsilon_h)\\ z_{\xi\zeta} + \rho_{\zeta\xi} \ge 0 \quad ((\xi,\zeta)\in\tilde{\Omega})\\ z_{\xi\zeta} \ge 0 \quad ((\xi,\zeta)\in\tilde{\Omega})\\ 0 \le \vartheta_j^1 \le \cdots \le \vartheta_j^{\#h} \le w_j \;\; (j \in \{1,2,\ldots,n\})\\ \boldsymbol{w} \in \boldsymbol{\Delta}\end{cases} \tag{MOD-4.20}$$

where $s(\tilde{R}(\xi,\zeta))$ and $s(\varepsilon_h)$ are the score functions of $\tilde{R}(\xi,\zeta)$ and ε_h, respectively.

It is observed that the model (MOD-4.20) is a crisp linear programming model which can be easily solved by using the Simplex method and needs very low time cost related to the non-linear programming model. Thus, the non-inferior solutions of the model (MOD-4.20), i.e., the criteria weighting vector $\boldsymbol{w} = (w_1, w_2, \ldots, w_n)^T$ and $\vartheta_j^\lambda (j \in \{1, 2, \ldots, n\}, \lambda \in \{1, 2, \ldots, \#h\})$, can be easily obtained. Based on the derived hesitant fuzzy PIS and the optimal weights of criteria, the distances between the alternatives and the hesitant fuzzy PIS are calculated and the best compromise alternative that has the shortest distance to the hesitant fuzzy PIS can be identified.

Remark 4.5 It is easy to see that in the model (MOD-4.20) there exist $\left(\left|\tilde{\Omega}\right| + n + nl\right)$ variables that need to be determined, including $\left|\tilde{\Omega}\right|$ variables $z_{\xi\zeta} \geq 0$ $\left((\xi, \zeta) \in \tilde{\Omega}\right)$, n weights of criteria w_j $(j \in \{1, 2, \ldots, n\})$, and $n \bullet \#h$ variables of ϑ_j^λ $(j \in \{1, 2, \ldots, n\},\ \ \lambda \in \{1, 2, \ldots, \#h\})$; and at the same time, we have $\left(\left|\tilde{\Omega}\right| + 2nl - n + 1\right)$ inequalities (excluding the nonnegative constraints for the variables and the incomplete weighed information $\boldsymbol{\Delta}$). In order to determine objectively these variables, the number $\left(\left|\tilde{\Omega}\right| + 2nl - n + 1\right)$ of inequalities should not be very small. In general, the larger the $\left|\tilde{\Omega}\right|$ is, the more precise and reliable the decision results $(\boldsymbol{w}, A^+)$ are.

4.3.3 Issues that Involve Inconsistent Preference Structure of Criteria Weights

In many practical decision making problems, the weight information of criteria provided by the decision maker may be inconsistent when using the five basic ranking forms introduced in Sect. 1.3.1 to express them, especially in some complex and uncertain environments. These inconsistent opinions on the importance degrees of the criteria may result in no feasible solutions that satisfy all conditions in $\boldsymbol{\Delta}$. Inspired by the idea of Chen (2014), Zhang et al. (2015) presented some non-negative deviation variables to relax the conditions in $\boldsymbol{\Delta}$, and constructed a bi-objective nonlinear programming model to address the problem with inconsistent weight information.

For convenience, we denote $\boldsymbol{\Delta}$ as follows:

$$
\begin{aligned}
\boldsymbol{\Delta} = \Big\{ (w_1, w_2, \ldots, w_n)^T \in \boldsymbol{\Delta_0} \big| & w_{j_1} \geq w_{j_2} \quad \textit{for all } j_1 \in \Lambda_{(1)1} \textit{ and } j_2 \in \Lambda_{(2)1} \\
& w_{j_1} - w_{j_2} \geq \tau^L_{j_1 j_2}, w_{j_1} - w_{j_2} \leq \tau^U_{j_1 j_2} \quad \textit{for all } j_1 \in \Lambda_{(1)2} \textit{ and } j_2 \in \Lambda_{(2)2} \\
& w_{j_1} - w_{j_2} - w_{j_3} + w_{j_4} \geq 0 \quad \textit{for all } j_1 \in \Lambda_{(1)3}, j_2 \in \Lambda_{(2)3}, j_3 \in \Lambda_{(3)3} \textit{ and } j_4 \in \Lambda_{(4)3} \\
& w_{j_1} / w_{j_2} \geq \tau_{j_1 j_2} \quad \textit{for all } j_1 \in \Lambda_{(1)4} \textit{ and } j_2 \in \Lambda_{(2)4} \\
& w_{j_1} \geq \tau^L_{j_1},\ w_{j_1} \leq \tau^U_{j_1} \quad \textit{for all } j_1 \in \Lambda_{(1)5} \Big\}
\end{aligned}
$$

where $j_1 \neq j_2 \neq j_3 \neq j_4$.

Now, several non-negative deviation variables are introduced to relax the conditions in $\mathbf{\Delta}$. This relaxed result is denoted by $\mathbf{\Delta}^{\dagger}$, as follows:

$$\mathbf{\Delta}^{\dagger} = \Big\{(w_1, w_2, \ldots, w_n) \in \mathbf{\Delta_0} \Big| w_{j_1} + o^-_{1j_1j_2} \geq w_{j_2} \quad \textit{for all } j_1 \in \Lambda_{(1)1} \textit{ and } j_2 \in \Lambda_{(2)1}$$
$$w_{j_1} - w_{j_2} + o^-_{2j_1j_2} \geq \tau^L_{j_1j_2}, w_{j_1} - w_{j_2} - o^+_{2j_1j_2} \leq \tau^U_{j_1j_2} \quad \textit{for all } j_1 \in \Lambda_{(1)2} \textit{ and } j_2 \in \Lambda_{(2)2}$$
$$w_{j_1} - w_{j_2} - w_{j_3} + w_{j_4} + o^-_{3j_1j_2j_3j_4} \geq 0 \quad \textit{for all } j_1 \in \Lambda_{(1)3}, j_2 \in \Lambda_{(2)3}, j_3 \in \Lambda_{(3)3} \textit{ and } j_4 \in \Lambda_{(4)3}$$
$$\frac{w_{j_1}}{w_{j_2}} + o^-_{4j_1j_2} \geq \tau_{j_1j_2} \quad \textit{for all } j_1 \in \Lambda_{(1)4} \textit{ and } j_2 \in \Lambda_{(2)4}$$
$$w_{j_1} + o^-_{5j_1} \geq \tau^L_{j_1},\ w_{j_1} - o^+_{5j_1} \leq \tau^U_{j_1} \quad \textit{for all } j_1 \in \Lambda_{(1)5}\Big\} \tag{4.27}$$

It is noted that all the deviation values $o^-_{1j_1j_2}, o^-_{2j_1j_2}, o^+_{2j_1j_2}, o^-_{3j_1j_2j_3j_4}, o^-_{4j_1j_2}, o^-_{5j_1}, o^+_{5j_1}$ $(j_1, j_2, j_3, j_4 \in \{1, 2, \ldots, n\})$ are non-negative real numbers and also are unknown in advance. To handle the inconsistent weight information, a bi-objective nonlinear programming model is constructed as follows (Zhang et al. 2015):

$$\min \left\{ \sum_{(\xi,\zeta)\in\tilde{\Omega}} z_{\xi\zeta} s\big(\tilde{R}(\xi,\zeta)\big) \right\}$$
$$\min \left\{ \sum_{j_1,j_2,j_3,j_4\in\{1,2,\ldots,n\}} \begin{pmatrix} o^-_{1j_1j_2} + o^-_{2j_1j_2} + o^+_{2j_1j_2} + o^-_{3j_1j_2j_3j_4} \\ + o^-_{4j_1j_2} + o^-_{5j_1} + o^+_{5j_1} \end{pmatrix} \right\}$$
$$s.t. \begin{cases} \sum_{(\xi,\zeta)\in\tilde{\Omega}} \big(\rho_{\zeta\xi} \times s(\tilde{R}(\xi,\zeta))\big) \geq s(\varepsilon_h) \\ z_{\xi\zeta} + \rho_{\zeta\xi} \geq 0 \quad ((\xi,\zeta) \in \tilde{\Omega}) \\ z_{\xi\zeta} \geq 0 \quad ((\xi,\zeta) \in \tilde{\Omega}) \\ 0 \leq \vartheta^1_j \leq \cdots \leq \vartheta^{\#h}_j \leq w_j \ (j \in \{1,2,\ldots,n\}) \\ (w_1, w_2, \ldots, w_n) \in \mathbf{\Delta}^{\dagger} \\ o^-_{1j_1j_2} \geq 0,\ j_1 \in \Lambda_{(1)1} \textit{ and } j_2 \in \Lambda_{(2)1} \\ o^-_{2j_1j_2} \geq 0, o^+_{2j_1j_2} \geq 0,\ j_1 \in \Lambda_{(1)2} \textit{ and } j_2 \in \Lambda_{(2)2} \\ o^-_{3j_1j_2j_3j_4} \geq 0,\ j_1 \in \Lambda_{(1)3}, j_2 \in \Lambda_{(2)3}, j_3 \in \Lambda_{(3)3} \textit{ and } j_4 \in \Lambda_{(4)3} \\ o^-_{4j_1j_2} \geq 0,\ j_1 \in \Lambda_{(1)4} \textit{ and } j_2 \in \Lambda_{(2)4} \\ o^-_{5j_1} \geq 0, o^+_{5j_1} \geq 0, j_1 \in \Lambda_{(1)5} \end{cases} \tag{MOD-4.21}$$

By using the min-max operator (Chen 2014), the bi-objective nonlinear programming model (MOD-4.21) can be converted into the single-objective nonlinear programming model (Zhang et al. 2015):

$$\min \{\chi\}$$
$$s.t.\begin{cases} \sum_{(\xi,\zeta)\in\tilde{\Omega}} z_{\xi\zeta} s(\tilde{R}(\xi,\zeta)) \le \chi \\ \sum_{j_1,j_2,j_3,j_4\in\{1,2,\ldots,n\}} \begin{pmatrix} o^-_{1j_1j_2} + o^-_{2j_1j_2} + o^+_{2j_1j_2} + o^-_{3j_1j_2j_3j_4} \\ + o^-_{4j_1j_2} + o^-_{5j_1} + o^+_{5j_1} \end{pmatrix} \le \chi \\ \sum_{(\xi,\zeta)\in\tilde{\Omega}} \left(\rho_{\zeta\xi} \times s(\tilde{R}(\xi,\zeta))\right) \ge s(\varepsilon_h) \\ z_{\xi\zeta} + \rho_{\zeta\xi} \ge 0 \quad ((\xi,\zeta)\in\tilde{\Omega}) \\ z_{\xi\zeta} \ge 0 \quad ((\xi,\zeta)\in\tilde{\Omega}) \\ 0 \le \vartheta^1_j \le \cdots \le \vartheta^{\#h}_j \le w_j \; (j\in\{1,2,\ldots,n\}) \\ (w_1,w_2,\ldots,w_n)\in\Delta^\dagger \\ o^-_{1j_1j_2} \ge 0,\, j_1\in\Lambda_{(1)1} \text{ and } j_2\in\Lambda_{(2)1} \\ o^-_{2j_1j_2} \ge 0,\, o^+_{2j_1j_2} \ge 0, j_1\in\Lambda_{(1)2} \text{ and } j_2\in\Lambda_{(2)2} \\ o^-_{3j_1j_2j_3j_4} \ge 0, j_1\in\Lambda_{(1)3}, j_2\in\Lambda_{(2)3}, j_3\in\Lambda_{(3)3} \text{ and } j_4\in\Lambda_{(4)3} \\ o^-_{4j_1j_2} \ge 0, j_1\in\Lambda_{(1)4} \text{ and } j_2\in\Lambda_{(2)4} \\ o^-_{5j_1} \ge 0,\, o^+_{5j_1} \ge 0, j_1\in\Lambda_{(1)5} \end{cases} \tag{MOD-4.22}$$

Apparently, the model (MOD-4.22) is a crisp linear programming model which can be easily solved by using the LINGO 11.0 or MATLAB 7.4. By solving this model, the optimal weighting vector of criteria $\boldsymbol{w} = (w_1, w_2, \ldots, w_n)^T$, the deviation values $o^-_{1j_1j_2}, o^-_{2j_1j_2}, o^+_{2j_1j_2}, o^-_{3j_1j_2j_3j_4}, o^-_{4j_1j_2}, o^-_{5j_1}, o^+_{5j_1}$ $(j_1, j_2, j_3, j_4 \in \{1, 2, \ldots, n\})$ and the hesitant fuzzy PIS $A^+ = (h^+_1, h^+_2, \ldots, h^+_n)$ are obtained, respectively. Analogously, based on the derived hesitant fuzzy PIS and the optimal weights of criteria, the distances between the alternatives and the hesitant fuzzy PIS can be calculated and the best compromise alternative that has the shortest distance to the hesitant fuzzy PIS can be identified.

Now, the algorithm (Algorithm 4.2) of the proposed approach for solving the aforementioned MCDM problem can be summarized as in Table 4.18 (Zhang et al. 2015).

4.3.4 Case Illustration and Discussions

To demonstrate the applicability and the implementation process of the proposed method, Zhang et al. (2015) presented an illustrating example involved with the supplier selection problem.

Table 4.18 The algorithm of the proposed decision method

Step 1:	Identify the evaluation criteria and the incomplete weight information structure
Step 2:	Express the pair-wise comparison preference judgments over alternatives with hesitant fuzzy truth degrees represented by $\tilde{\Omega} = \{(\xi, \zeta) \mid A_\xi \succeq_{\tilde{R}(\xi,\zeta)} A_\zeta\}$
Step 3:	Construct the hesitant fuzzy decision matrix $\Re$ and obtain the normalized decision matrix $\bar{\Re}$ by Eq. (2.11)
Step 4:	Calculate the hesitant fuzzy consistency and inconsistency indices by Eqs. (4.20) and (4.23), respectively
Step 5:	Construct the hesitant fuzzy programming model according to the model. (MOD-4.18), and transform the derived model into a linear programming model according to the model (MOD-4.20) if the weight information is incomplete and consistent, or a nonlinear programming model according to the model (MOD-4.22) if the weight information is incomplete and inconsistent
Step 6:	Get the optimal weight vector $\boldsymbol{w}$ and the hesitant fuzzy PIS A^+ through solving the model (MOD-4.20) or the model (MOD-4.22) by using the Simplex method
Step 7:	Calculate the relative distances $D_i (i \in \{1, 2, \ldots, m\})$ between each of the alternatives $A_i (i \in \{1, 2, \ldots, m\})$ and the hesitant fuzzy PIS A^+ using Eq. (1.2)
Step 8:	The ranking of all alternatives is generated according to the increasing order of the distances $D_i (i \in \{1, 2, \ldots, m\})$ and the best alternative from the alternative set $\boldsymbol{A}$ is determined

4.3.4.1 Description of the Supplier Selection Problem

With the increase of public awareness of the need to protect the environment, it is urgent for businesses to introduce and promote business practices that help ease the negative impacts of their actions on the environment (Wang and Chan 2013). In the automobile manufacturing industries, the manufacturers want to improve their environmental management practices, not only internally, but also with their suppliers. To this end, the automobile manufacturing company plans to find some environmentally and economically powerful suppliers as strategic partners, with whom the company intends to build long-term collaborative relationships. There are five qualified suppliers which are named, for our purposes, as A_1, A_2, A_3, A_4 and A_5. A decision organization including three experts (from the purchasing department, management department, environment department, respectively) is invited to evaluate these five suppliers and help the company choose an optimal supplier as its strategic partner. The supplier selection criteria have been determined by the decision organization as follows: (1) C_1 is the delivery capability; (2) C_2 is the environmental performance; (3) C_3 is the cost of product; (4) C_4 is the quality of product. It is noted that the criterion C_3 is the cost criterion and the others are the benefit criteria. The preference structure of criterion importance is also given as follows (Zhang et al. 2015):

$$\Delta = \left\{ (w_1, w_2, w_3, w_4) \in \Delta_0 \middle| \begin{array}{l} w_4 \geq w_1,\ w_2 \geq 2w_1,\ 0.15 \leq w_4 \leq 0.5, \\ 0.05 \leq w_2 - w_3 \leq 0.3,\ w_4 - w_3 \geq w_2 - w_1 \end{array} \right\}.$$

To get more reasonable evaluation results, in the real-world decision process the experts are usually required to give their evaluations anonymously. For the green supplier selection problem, the original assessments of suppliers on each criterion provided anonymously by the three experts are listed in Table 4.19. Although all of the experts provide their evaluation values of alternatives under each criterion, some of these values may be repeated. Considering the decision information provided anonymously by experts, thus we only collect all of the possible values for an alternative under a criterion, and each value provided only means that it is a possible value, while the times that the values repeated are negligible. Obviously, the HFE is just a tool to deal with such cases, and all possible evaluations for an alternative under each criterion can be considered as a HFE (Xu and Xia 2011). The collective opinions of the original assessments of suppliers with respect to criteria provided by the decision organization are taken as HFEs, listed in Table 4.19 (Zhang et al. 2015). For the element $H\ \{0.4, 0.5, 0.7\}$ in Table 4.19, it means that the decision organization has hesitancy among 0.4, 0.5 and 0.7 when providing the assessment of the alternative A_1 with respect to the criterion C_1, and the others have the similar meanings.

Moreover, the decision organization also provides the HFEs of ordered pairs for their preferences over the alternatives as follows (Zhang et al. 2015):

$$\tilde{\Omega} = \left\{ \begin{array}{l} \langle (1,2), \tilde{R}(1,2) \rangle, \langle (2,3), \tilde{R}(2,3) \rangle, \langle (2,4), \tilde{R}(2,4) \rangle, \langle (2,5), \tilde{R}(2,5) \rangle, \\ \langle (3,1), \tilde{R}(3,1) \rangle, \langle (3,4), \tilde{R}(3,4) \rangle, \langle (4,5), \tilde{R}(4,5) \rangle \end{array} \right\}$$

where the corresponding hesitant fuzzy truth degrees are as follows:

$$\begin{array}{ll} \tilde{R}(1,2) = H\{0.5, 0.6, 0.7\}, & \tilde{R}(2,3) = H\{0.6, 0.65, 0.7\} \\ \tilde{R}(2,4) = H\{0.8, 0.85, 0.9\}, & \tilde{R}(2,5) = H\{0.5, 0.7\}, \\ \tilde{R}(3,1) = H\{0.4, 0.5, 0.6\}, & \tilde{R}(3,4) = H\{0.6, 0.7, 0.95\}, \\ \tilde{R}(4,5) = H\{0.7, 0.9\}. & \end{array}$$

and $\tilde{R}(1,2) = H\{0.5, 0.6, 0.7\}$ means that the decision organization has hesitancy among the values 0.5, 0.6 and 0.7 when providing the degree to which the alternative A_1 is superior to A_2, and the others have the similar meanings.

Table 4.19 Ratings of the green suppliers by experts under various criteria

Suppliers	Criterion C_1				Criterion C_2			
	e_1	e_2	e_3	Collective opinion	e_1	e_2	e_3	Collective opinion
A_1	0.5	0.4	0.7	$H\{0.4, 0.5, 0.7\}$	0.7	0.7	0.7	$H\{0.7\}$
A_2	0.5	0.6	0.4	$H\{0.4, 0.5, 0.6\}$	0.8	0.7	0.9	$H\{0.7, 0.8, 0.9\}$
A_3	0.4	0.4	0.4	$H\{0.4\}$	0.6	0.9	0.6	$H\{0.6, 0.9\}$
A_4	0.4	0.6	0.3	$H\{0.3, 0.4, 0.6\}$	0.4	0.6	0.5	$H\{0.4, 0.5, 0.6\}$
A_5	0.7	0.8	0.6	$H\{0.6, 0.7, 0.8\}$	0.5	0.3	0.4	$H\{0.3, 0.4, 0.5\}$
Suppliers	Criterion C_3				Criterion C_4			
	e_1	e_2	e_3	Collective opinion	e_1	e_2	e_3	Collective opinion
A_1	0.5	0.4	0.55	$H\{0.4, 0.5, 0.55\}$	0.9	0.8	0.85	$H\{0.8, 0.85, 0.9\}$
A_2	0.55	0.4	0.45	$H\{0.4, 0.45, 0.55\}$	0.6	0.7	0.7	$H\{0.6, 0.7\}$
A_3	0.3	0.45	0.55	$H\{0.3, 0.45, 0.55\}$	0.6	0.8	0.7	$H\{0.6, 0.7, 0.8\}$
A_4	0.1	0.1	0.1	$H\{0.1\}$	0.3	0.25	0.15	$H\{0.15, 0.25, 0.3\}$
A_5	0.5	0.6	0.5	$H\{0.5, 0.6\}$	0.7	0.45	0.4	$H\{0.4, 0.45, 0.7\}$

4.3.4.2 Illustration of the Proposed Approach

In the following, we employ the proposed method to solve the above green supplier evaluation problem. Firstly, we normalize the hesitant fuzzy decision data provided by the decision organization in Table 4.19, and the normalized results are listed in Table 4.20 (Zhang et al. 2015).

Taking $\varepsilon_h = H\{0.01\}$ and $\varepsilon_w = 0.01$, we utilize the model (MOD-4.19) to construct the hesitant fuzzy programming model (MOD-A.3) which is displayed in Appendix. Then, the derived model (MOD-A.3) is converted into the crisp linear programming model based on the ranking approach of HFEs [i.e., using the model (MOD-4.20)]. By solving the corresponding linear programming model using the LINGO 11.0, the components of the optimal solution can be obtained:

Table 4.20 The hesitant fuzzy normalized decision matrix

	C_1	C_2	C_3	C_4
A_1	$H\{0.4, 0.5, 0.7\}$	$H\{0.7, 0.7, 0.7\}$	$H\{0.45, 0.5, 0.6\}$	$H\{0.8, 0.85, 0.9\}$
A_2	$H\{0.4, 0.5, 0.6\}$	$H\{0.7, 0.8, 0.9\}$	$H\{0.45, 0.55, 0.6\}$	$H\{0.6, 0.6, 0.7\}$
A_3	$H\{0.4, 0.4, 0.4\}$	$H\{0.6, 0.6, 0.9\}$	$H\{0.45, 0.55, 0.7\}$	$H\{0.6, 0.7, 0.8\}$
A_4	$H\{0.3, 0.4, 0.6\}$	$H\{0.4, 0.5, 0.6\}$	$H\{0.9, 0.9, 0.9\}$	$H\{0.15, 0.25, 0.3\}$
A_5	$H\{0.6, 0.7, 0.8\}$	$H\{0.3, 0.4, 0.5\}$	$H\{0.4, 0.4, 0.5\}$	$H\{0.4, 0.45, 0.7\}$

$$
\begin{aligned}
& w=(w_1,w_2,w_3,w_4)^{\mathrm{T}}=(0.1217,0.2434,0.1934,0.4415)^{\mathrm{T}},\\
& \vartheta_1^1=0, \vartheta_1^2=0.0017, \quad \vartheta_1^3=0.1217, \quad \vartheta_2^1=0.1921,\\
& \vartheta_2^2=0.2434, \quad \vartheta_2^3=0.2434, \quad \vartheta_3^1=0.1594, \quad \vartheta_3^2=0.1594,\\
& \vartheta_3^3=0.1594, \quad \vartheta_4^1=0, \quad \vartheta_4^2=0.3824, \quad \vartheta_4^3=0.3824.
\end{aligned}
$$

Using the equation $\gamma_{h_j^+}^{\lambda}=\vartheta_j^{\lambda}/w_j$, we can further obtain the hesitant fuzzy PIS A^+ as:

$$
A^+=\left(h_1^+,h_2^+,h_3^+,h_4^+\right)=\begin{pmatrix} H\{0.0,0.014,1.0\}, H\{0.7892,1.0,1.0\},\\ H\{0.8242,0.8242,0.8242\},\\ H\{0.0,0.8661,0.8661\} \end{pmatrix}.
$$

The distances between the alternatives $A_i (i=1,2,3,4,5)$ and the hesitant fuzzy PIS A^+ are calculated by utilizing Eq. (1.2) as below:

$$
\begin{aligned}
& d(A_1,A^+)=0.1484, \quad d(A_2,A^+)=0.1162, \quad d(A_3,A^+)=0.1164,\\
& d(A_4,A^+)=0.1692, \quad d(A_5,A^+)=0.1852.
\end{aligned}
$$

By comparing the distances $d(A_i,A^+)$ $(i=1,2,3,4,5)$, we obtain the ranking order of alternatives as: $A_5 \prec A_4 \prec A_1 \prec A_3 \prec A_2$. Thus, the best optimal alternative is the green supplier A_2.

4.3.4.3 Sensitivity Analysis

The sensitivity analysis of the parameter ε_h was performed by modifying its values. Usually, by increasing or decreasing the values of the parameter ε_h, we recalculate the optimal weighting vector w and the hesitant fuzzy PIS A^+, and obtain the ranking orders of alternatives.

As far as the green supplier selection problem is concerned, we change the values of the parameter ε_h from $H\{0.0001\}$ to $H\{0.1\}$, and obtain the calculation results of the optimal weighting vector w, the hesitant fuzzy PIS A^+ and the ranking order of alternatives. These calculation results are listed in Table 4.21 (Zhang et al. 2015).

From the sensitivity analysis results presented in Table 4.21, we notice that both the optimal weighting vector w and the hesitant fuzzy PIS A^+ are not sensitive to the parameter ε_h, and in spite of the alteration in the value of ε_h, the obtained ranking orders of alternatives are consistent. In general, the value of the parameter ε_h is given by the decision maker a priori and ε_h should not be too big which usually takes the value less than $H\{0.1\}$.

Table 4.21 The sensitivity analysis results with different values of ε_h

ε_h	$w = (w_1, w_2, w_3, w_4)^T$	$A^+ = (h_1^+, h_2^+, h_3^+, h_4^+)$	The ranking results of alternatives
$H\{0.0001\}$	(0.1155,0.231, 0.181, 0.4725)	($H\{0, 0.014, 1\}$, $H\{0.7892, 1, 1\}$, $H\{0.8242, 0.8242, 0.8242\}$, $H\{0, 0.8661, 0.8661\}$)	$A_5 \prec A_4 \prec A_1 \prec A_3 \prec A_2$
$H\{0.001\}$	(0.1155, 0.231, 0.181, 0.4725)	($H\{0, 0.014,1\}$, $H\{0.7892,1,1\}$, $H\{0.8242, 0.8242, 0.8242\}$, $H\{0, 0.8661, 0.8661\}$)	$A_5 \prec A_4 \prec A_1 \prec A_3 \prec A_2$
$H\{0.01\}$	(0.1217, 0.2434, 0.1934, 0.4415)	($H\{0, 0.014,1\}$, $H\{0.7892,1,1\}$, $H\{0.8242, 0.8242, 0.8242\}$, $H\{0, 0.8661, 0.8661\}$)	$A_5 \prec A_4 \prec A_1 \prec A_3 \prec A_2$
$H\{0.02\}$	(0.1241, 0.2481, 0.1981, 0.4297)	($H\{0, 0.2119,1\}$, $H\{0.8759, 0.8759,1\}$, $H\{0.8445, 0.8445, 0.8445\}$, $H\{0, 0.8797, 0.8797\}$)	$A_5 \prec A_4 \prec A_1 \prec A_3 \prec A_2$
$H\{0.05\}$	(0.1216, 0.2432, 0.1932, 0.442)	($H\{0, 0.6686,1\}$, $H\{0.757, 0.757, 0.9996\}$, $H\{1, 1, 1\}$, $H\{0.2269, 0.8649, 0.8649\}$)	$A_5 \prec A_4 \prec A_1 \prec A_3 \prec A_2$
$H\{0.08\}$	(0.1142, 0.2284, 0.1784, 0.479)	($H\{0, 0.3835, 0.9378\}$, $H\{1, 1, 1\}$, $H\{1, 1, 1\}$, $H\{0.0635, 0.9344, 0.9344\}$)	$A_5 \prec A_4 \prec A_1 \prec A_3 \prec A_2$
$H\{0.1\}$	(0.1156, 0.2312, 0.1812, 0.472)	($H\{0, 0.4721, 0.7852\}$, $H\{1, 1, 1\}$, $H\{1, 1, 1\}$, $H\{0.1297, 0.9252, 0.9252\}$)	$A_5 \prec A_4 \prec A_1 \prec A_3 \prec A_2$

4.3.4.4 Discussion of the Inconsistent Weights of Criteria

In some practical decision making processes, the decision maker might express the inconsistent opinions on the criterion importance when he/she employs the five basic ranking forms of criteria weights introduced in Chap. 1. In the above green supplier selection problem, the five basic ranking forms are expressed as follows:

$$\begin{aligned}
\mathbf{\Delta_1} &= \{(w_1, w_2, w_3, w_4) \in \mathbf{\Delta_0} | w_4 \geq w_1\}, \\
\mathbf{\Delta_2} &= \{(w_1, w_2, w_3, w_4) \in \mathbf{\Delta_0} | 0.05 \leq w_2 - w_3 \leq 0.3\}, \\
\mathbf{\Delta_3} &= \{(w_1, w_2, w_3, w_4) \in \mathbf{\Delta_0} | w_4 - w_3 \geq w_2 - w_1\}, \\
\mathbf{\Delta_4} &= \{(w_1, w_2, w_3, w_4) \in \mathbf{\Delta_0} | w_2 \geq 2w_1\}, \\
\mathbf{\Delta_5} &= \{(w_1, w_2, w_3, w_4) \in \mathbf{\Delta_0} | 0.15 \leq w_4 \leq 0.5\}.
\end{aligned}$$

Thus, $\Delta = \mathbf{\Delta_1} \cup \mathbf{\Delta_2} \cup \mathbf{\Delta_3} \cup \mathbf{\Delta_4} \cup \mathbf{\Delta_5}$. Now, we assume that a new condition $w_3 \geq w_2$ is added to the set $\mathbf{\Delta}_1$, then the set $\mathbf{\Delta}_1$ is updated as follows:

$$\boldsymbol{\Delta}_1^{(new)} = \{(w_1, w_2, w_3, w_4) \in \boldsymbol{\Delta}_0 | w_4 \geq w_1, w_3 \geq w_2\},$$

and Δ is updated as follows:

$$\Delta^{(new)} = \left\{ (w_1, w_2, w_3, w_4) \in \Delta_0 \left| \begin{array}{l} w_4 \geq w_1, \;\; w_3 \geq w_2, \;\; 0.05 \leq w_2 - w_3 \leq 0.3, \\ w_4 - w_3 \geq w_2 - w_1, \\ w_2 \geq 2w_1, \;\; 0.15 \leq w_4 \leq 0.5 \end{array} \right. \right\}.$$

It is easy to see that the condition of $w_3 \geq w_2$ in $\boldsymbol{\Delta}_1^{(new)}$ conflicts with the condition of $0.05 \leq w_2 - w_3 \leq 0.3$ in $\boldsymbol{\Delta}_2$. In other words, the weights in $\boldsymbol{\Delta}^{(new)}$ exist partially inconsistent which may result in no feasible solutions that satisfy all conditions in $\boldsymbol{\Delta}^{(new)}$. To handle the inconsistent weight information, some deviation variables are provided to relax the conditions in $\boldsymbol{\Delta}^{(new)}$ into $\boldsymbol{\Delta}^{\dagger(new)}$, as follows (Zhang et al. 2015):

$$\boldsymbol{\Delta}^{\dagger(new)} = \left\{ (w_1, w_2, \ldots, w_n) \in \boldsymbol{\Delta}_0 \left| \begin{array}{l} w_4 + o_{141}^- \geq w_1, \quad w_3 + o_{132}^- \geq w_2, \\ w_2 - w_3 \; + o_{223}^- \geq 0.05, \\ w_4 + o_{54}^- \geq 0.15, \quad w_2 - w_3 - o_{223}^+ \leq 0.3, \\ w_4 - w_3 - w_2 + w_1 + o_{34321}^- \geq 0, \\ w_2/w_1 + o_{421}^- \geq 2, \quad w_4 - o_{54}^+ \leq 0.5 \end{array} \right. \right\}$$

where all deviation variables $o_{141}^-, o_{132}^-, o_{223}^-, o_{223}^+, o_{34321}^-, o_{421}^-, o_{54}^+$ and o_{54}^- are non-negative real numbers.

According to the model (MOD-4.21) proposed in Sect. 4.3.3, a bi-objective nonlinear programming model (MOD-A.4) is constructed, which is listed in Appendix. Based on the model (MOD-4.22) proposed in Sect. 4.3.3, the model (MOD-A.4) is converted into the corresponding optimal model (MOD-A.5) which is also listed in Appendix.

By solving the model (MOD-A.5) using the Simplex method, the components of the optimal solution can be obtained. By combining with the equation $\gamma_{h_j^+}^{\lambda} = \vartheta_j^{\lambda}/w_j$, we can obtain the optimal weight vector $\boldsymbol{w}$ and the hesitant fuzzy PIS A^+ as follows:

$$\begin{aligned} \boldsymbol{w} &= (w_1, w_2, w_3, w_4)^{\mathrm{T}} = (0.133, 0.267, 0.233, 0.367)^T, \quad \chi = 0.05, \\ A^+ &= \left(h_1^+, h_2^+, h_3^+, h_4^+\right) \\ &= \begin{pmatrix} H\{0,1,1\}, H\{0,1,1\}, H\{0.7712, 0.7712, \\ 0.7712\}, H\{0.0825, 1, 1\} \end{pmatrix}, \end{aligned}$$

$$o_{141}^- = o_{223}^+ = o_{34321}^- = o_{421}^- = o_{54}^+ = o_{54}^- = 0, \;\; o_{132}^- = 0.0335, \;\; o_{223}^- = 0.0165.$$

Then, the distances $d(A_i, A^+)(i = 1, 2, 3, 4, 5)$ of the alternatives $A_i(i = 1, 2, 3, 4, 5)$ to the hesitant fuzzy PIS A^+ are calculated by using Eq. (1.2) as follows:

$$d(A_1, A^+) = 0.1658, \quad d(A_2, A^+) = 0.1468, \quad d(A_3, A^+) = 0.1471,$$
$$d(A_4, A^+) = 0.2109, \quad d(A_5, A^+) = 0.1681.$$

By comparing these relative distances, we can obtain the ranking of the alternatives as: $A_4 \prec A_5 \prec A_1 \prec A_3 \prec A_2$. It is easy to see that the ranking of alternatives determined based on inconsistent weights is remarkably different from that based on the consistent weights. The main reason is that the weight distribution among these four criteria under the inconsistent weight information is different from those using the consistent information.

4.4 Conclusions and Future Research Directions

In this chapter, we have introduced the hesitant fuzzy LINMAP group decision method with interval programming models proposed by Zhang and Xu (2014) and the hesitant fuzzy programming model-based LINMAP method developed by Zhang et al. (2015). The former is mainly used to solve the MCGDM problems in which the ratings of alternatives with respect to criteria are taken as HFEs, and all pair-wise comparison judgments are represented by interval numbers. The most two important advantages of the former are that (1) it can take sufficiently into account the experts' hesitancy in expressing their assessment information for criteria values by using HFEs and (2) it can simultaneously consider the uncertainty of preference information over alternatives by using interval numbers. Compared with the IF-LINMAP method (Li et al. 2010), this decision method integrates more useful information into the decision making process, and does not need to transform HFEs into IFNs but directly deals with the MCGDM problems with hesitant fuzzy information, and thus obtains better final decision results.

While the latter is mainly utilized to address the MCDM problems with incomplete weights in which the ratings of alternatives with each criterion are taken as HFEs and the incomplete judgments on pair-wise comparisons of alternatives with hesitant degrees are also represented by HFEs. The latter makes several contributions to the hesitant fuzzy theory and its applications. First, the concept of hesitant fuzzy programming model in which both the objective function and some constraints' coefficients take the form of HFEs has been proposed. Second, an effective approach has been presented to solve the derived model. Third, a bi-objective programming model has been constructed to address the issues of incomplete and inconsistent weights of the criteria.

In future studies, we will develop the corresponding decision support systems based on these two decision making methods to solve the real-world decision problems in case of considering the decision maker's hesitation. We will also focus on some additional experimental studies with different sizes of randomly generated problems and discuss how the components of the optimal weight vector are obtained. Furthermore, the potential of combining the proposed decision methods with other useful MCDM techniques within the environment of HFEs will also be taken into consideration in the future.

Appendix

$$
\min \left\{ \begin{array}{l}
(0.05\theta+0.75)\times z_{12}^{1}+(0.05\theta+0.85)\times z_{34}^{1}+(0.05\theta+0.85)\times z_{53}^{1}+(0.05\theta+0.65)\times z_{13}^{2}+ \\
(0.05\theta+0.85)\times z_{34}^{2}+(0.05\theta+0.75)\times z_{42}^{2}+(0.05\theta+0.75)\times z_{53}^{2}+(0.05\theta+0.75)\times z_{21}^{3}+ \\
(0.05\theta+0.55)\times z_{31}^{3}+(0.05\theta+0.95)\times z_{45}^{3}
\end{array} \right\}
$$

$$
s.t. \left\{ \begin{array}{l}
0.014w_1+0.002w_2-0.148w_3+0.136w_4+0.04w_3\times(3h_{31}+2h_{32}+3h_{33}-h_{34}h_{35})-0.04w_1\times h_{12}- \\
0.04w_2\times(h_{22}+h_{23}-4h_{24}-h_{25})-0.08w_4\times(h_{41}+h_{42}+h_{43})+z_{12}^{1}\geq 0 \\
0.092w_1+0.296w_2+0.066w_3-0.306w_4-0.04w_1\times(2h_{13}-h_{12}-h_{11}+3h_{14}+3h_{15})+0.04w_4\times(3h_{41}+4h_{42}+ \\
2h_{43}+2h_{44}+2h_{45})-0.04w_3\times(4h_{32}-h_{31}+2h_{33}+2h_{34}+2h_{35})-0.08w_2\times(h_{12}+h_{22}+2h_{23}+2h_{24}+2h_{25})+z_{34}^{1}\geq 0 \\
0.382w_3-0.088w_2-0.034w_1+0.168w_4+0.04w_4\times(3h_{41}+3h_{42}+2h_{43}-h_{44}-h_{45})+0.04w_2\times(h_{12}- \\
h_{22}+2h_{24}+2h_{25})-0.04w_3\times(2h_{31}+3h_{32}+4h_{33}+4h_{34}+4h_{35})-0.04w_1\times(2h_{11}+h_{12}-3h_{14}-5h_{15})+z_{53}^{1}\geq 0 \\
0.134w_2-0.06w_1-0.096w_3+0.13w_4+0.04w_1(h_{11}+2h_{12}+h_{13}-2h_{15})-0.04w_2\times(2h_{22}+h_{23}+h_{24}+h_{25})- \\
w_4\times(0.12h_{41}+0.04h_{42}+0.04h_{43})+0.04w_3\times(2h_{31}+h_{32}+h_{33}+h_{34}+h_{35})+z_{13}^{2}\geq 0 \\
0.296w_2-0.02w_1-0.094w_3-0.24w_4+0.04w_4\times(3h_{41}+3h_{42}+2h_{43}+2h_{44})+0.04w_1\times(h_{11}+h_{12}- \\
2h_{13}-h_{14}+h_{15})-0.08w_2\times(h_{12}+h_{22}+2h_{23}+2h_{24}+2h_{25})+0.04w_3\times(h_{31}+h_{32}+h_{33}+h_{34}+h_{35})+z_{34}^{2}\geq 0 \\
0.144w_1-0.23w_2-0.018w_3+0.246w_4-0.04w_4\times(2h_{41}+4h_{42}+3h_{43}+2h_{44})+0.04w_3\times h_{33}+ \\
0.04w_2\times(2h_{12}+3h_{22}+4h_{23}+3h_{24}-h_{25})-0.12w_1\times(h_{11}+h_{12})+z_{42}^{2}\geq 0 \\
0.216w_1-0.088w_2+0.308w_3+0.154w_4+0.04w_2\times(h_{12}-h_{22}+2h_{24}+2h_{25})-0.04w_3\times(2h_{31}+3h_{32}+ \\
4h_{33}+3h_{34}+2h_{35})-0.04w_4\times(3h_{41}+3h_{42}-h_{43})-0.04w_1\times(2h_{11}+h_{12}+h_{13}+2h_{14}+4h_{15})+z_{53}^{2}\geq 0 \\
0.234w_3-0.088w_2-0.148w_1-0.1w_4+0.04w_1\times(2h_{11}+3h_{12}+2h_{13}+h_{14})+0.04w_2\times(h_{23}-h_{12}+ \\
2h_{24}+2h_{25})-0.04w_3\times(3h_{31}+2h_{32}+2h_{33}+3h_{34}+3h_{35})+0.04w_4\times(2h_{41}+2h_{44}+2h_{45})+z_{21}^{3}\geq 0 \\
0.05w_1-0.07w_2+0.086w_3-0.004w_4+0.04w_2\times(h_{22}-h_{21}+h_{23}+h_{24}+h_{25})-w_1\times(0.04h_{14}+0.08h_{15})+ \\
0.04w_4\times(h_{41}-h_{42}-h_{43}+h_{45})-0.04w_3\times(2h_{31}+h_{32}+h_{34}+h_{35})+z_{31}^{3}\geq 0 \\
0.138w_4-0.208w_2-0.448w_3-0.058w_1-0.04w_4\times(h_{42}+3h_{44}+3h_{45})+0.04w_1(h_{11}+2h_{13}-2h_{15})+ \\
0.04w_2\times(h_{12}+3h_{22}+4h_{23}+2h_{24}+2h_{25})+0.04w_3\times(h_{31}+7h_{32}+6h_{33}+6h_{34}+6h_{35})+z_{45}^{3}\geq 0 \\
0.0258w_2-0.145w_1-0.1558w_3-0.2352w_4+0.028w_1h_{11}-0.028w_1h_{12}+0.004w_1h_{13}+ \\
0.028w_2h_{12}+0.092w_1h_{14}+0.204w_1h_{15}+0.06w_2h_{22}+0.004w_2h_{23}-0.212w_2h_{24}+0.012w_2h_{25}+ \\
0.016w_3h_{31}+0.028w_3h_{32}-0.024w_3h_{33}+0.16w_3h_{34}+0.128w_3h_{35}+0.112w_4h_{41}+ \\
0.236w_4h_{42}+0.108w_4h_{43}-0.028w_4h_{44}-0.044w_4h_{45}\geq \underline{a} \\
0.0236w_2-0.1352w_1-0.1422w_3-0.2191w_4+0.026w_1h_{11}-0.026w_1h_{12}+0.004w_1h_{13}+ \\
0.026w_2h_{12}+0.086w_1h_{14}+0.19w_1h_{15}+0.056w_2h_{22}+0.004w_2h_{23}-0.198w_2h_{24}+ \\
0.012w_2h_{25}+0.014w_3h_{31}+0.024w_3h_{32}-0.024w_3h_{33}+0.148w_3h_{34}+0.118w_3h_{35}+ \\
0.104w_4h_{41}+0.22w_4h_{42}+0.1w_4h_{43}-0.026w_4h_{44}-0.04w_4h_{45}\geq \bar{a} \\
z_{12}^{1},z_{34}^{1},z_{53}^{1},z_{13}^{2},z_{34}^{2},z_{42}^{2},z_{53}^{2},z_{21}^{3},z_{31}^{3},z_{45}^{3}\geq 0 \\
0\leq h_{11}\leq h_{12}\leq h_{13}\leq h_{14}\leq h_{15}\leq 1,\quad 0\leq h_{21}\leq h_{22}\leq h_{23}\leq h_{24}\leq h_{25}\leq 1 \\
0\leq h_{31}\leq h_{32}\leq h_{33}\leq h_{34}\leq h_{35}\leq 1,\quad 0\leq h_{41}\leq h_{42}\leq h_{43}\leq h_{44}\leq h_{45}\leq 1 \\
0\leq w_j\leq 1(j=1,2,3,4),\quad w_1+w_2+w_3+w_4=1
\end{array} \right.
\tag{MOD-A.1}
$$

$$\min\left\{\begin{array}{l}(0.05\theta+0.75)\times z_{12}^{1}+(0.05\theta+0.85)\times z_{34}^{1}+(0.05\theta+0.85)\times z_{53}^{1}+(0.05\theta+0.65)\times z_{13}^{2}+\\(0.05\theta+0.85)\times z_{34}^{2}+(0.05\theta+0.75)\times z_{42}^{2}+(0.05\theta+0.75)\times z_{53}^{2}+(0.05\theta+0.75)\times z_{21}^{3}+\\(0.05\theta+0.55)\times z_{31}^{3}+(0.05\theta+0.95)\times z_{45}^{3}\end{array}\right\}$$

$$s.t.\left\{\begin{array}{l}0.014w_1+0.002w_2-0.148w_3+0.136w_4+0.04w_3\times(3h_{31}+2h_{32}+3h_{33}-h_{34}h_{35})-0.04w_1\times h_{12}-\\0.04w_2\times(h_{22}+h_{23}-4h_{24}-h_{25})-0.08w_4\times(h_{41}+h_{42}+h_{43})+z_{12}^{1}\geq 0\\0.092w_1+0.296w_2+0.066w_3-0.306w_4-0.04w_1\times(2h_{13}-h_{12}-h_{11}+3h_{14}+3h_{15})+0.04w_4\times(3h_{41}+4h_{42}+\\2h_{43}+2h_{44}+2h_{45})-0.04w_3\times(4h_{32}-h_{31}+2h_{33}+2h_{34}+2h_{35})-0.08w_2\times(h_{12}+h_{22}+2h_{23}+2h_{24}+2h_{25})+z_{34}^{1}\geq 0\\0.382w_3-0.088w_2-0.034w_1+0.168w_4+0.04w_4\times(3h_{41}+3h_{42}+2h_{43}-h_{44}-h_{45})+0.04w_2\times(h_{12}-\\h_{22}+2h_{24}+2h_{25})-0.04w_3\times(2h_{31}+3h_{32}+4h_{33}+4h_{34}+4h_{35})-0.04w_1\times(2h_{11}+h_{12}-3h_{14}-5h_{15})+z_{53}^{1}\geq 0\\0.134w_2-0.06w_1-0.096w_3+0.13w_4+0.04w_1(h_{11}+2h_{12}+h_{13}-2h_{15})-0.04w_2\times(2h_{22}+h_{23}+h_{24}+h_{25})-\\w_4\times(0.12h_{41}+0.04h_{42}+0.04h_{43})+0.04w_3\times(2h_{31}+h_{32}+h_{33}+h_{34}+h_{35})+z_{13}^{2}\geq 0\\0.296w_2-0.02w_1-0.094w_3-0.24w_4+0.04w_4\times(3h_{41}+3h_{42}+2h_{43}+2h_{44})+0.04w_1\times(h_{11}+h_{12}-\\2h_{13}-h_{14}+h_{15})-0.08w_2\times(h_{12}+h_{22}+2h_{23}+2h_{24}+2h_{25})+0.04w_3\times(h_{31}+h_{32}+h_{33}+h_{34}+h_{35})+z_{34}^{2}\geq 0\\0.144w_1-0.23w_2-0.018w_3+0.246w_4-0.04w_4\times(2h_{41}+4h_{42}+3h_{43}+2h_{44})+0.04w_3\times h_{33}+\\0.04w_2\times(2h_{12}+3h_{22}+4h_{23}+3h_{24}-h_{25})-0.12w_1\times(h_{11}+h_{12})+z_{42}^{2}\geq 0\\0.216w_1-0.088w_2+0.308w_3+0.154w_4+0.04w_2\times(h_{12}-h_{22}+2h_{24}+2h_{25})-0.04w_3\times(2h_{31}+3h_{32}+\\4h_{33}+3h_{34}+2h_{35})-0.04w_4\times(3h_{41}+3h_{42}-h_{43})-0.04w_1\times(2h_{11}+h_{12}+h_{13}+2h_{14}+4h_{15})+z_{53}^{2}\geq 0\\0.234w_3-0.088w_2-0.148w_1-0.1w_4+0.04w_1\times(2h_{11}+3h_{12}+2h_{13}+h_{14})+0.04w_2\times(h_{23}-h_{12}+\\2h_{24}+2h_{25})-0.04w_3\times(3h_{31}+2h_{32}+2h_{33}+3h_{34}+3h_{35})+0.04w_4\times(2h_{41}+2h_{44}+2h_{45})+z_{21}^{3}\geq 0\\0.05w_1-0.07w_2+0.086w_3-0.004w_4+0.04w_2\times(h_{22}-h_{21}+h_{23}+h_{24}+h_{25})-w_1\times(0.04h_{14}+0.08h_{15})+\\0.04w_4\times(h_{41}-h_{42}-h_{43}+h_{45})-0.04w_3\times(2h_{31}+h_{32}+h_{34}+h_{35})+z_{31}^{3}\geq 0\\0.138w_4-0.208w_2-0.448w_3-0.058w_1-0.04w_4\times(h_{42}+3h_{44}+3h_{45})+0.04w_1(h_{11}+2h_{13}-2h_{15})+\\0.04w_2\times(h_{12}+3h_{22}+4h_{23}+2h_{24}+2h_{25})+0.04w_3\times(h_{31}+7h_{32}+6h_{33}+6h_{34}+6h_{35})+z_{45}^{3}\geq 0\\0.0258w_2-0.145w_1-0.1558w_3-0.2352w_4+0.028w_1h_{11}-0.028w_1h_{12}+0.004w_1h_{13}+\\0.028w_2h_{12}+0.092w_1h_{14}+0.204w_1h_{15}+0.06w_2h_{22}+0.004w_2h_{23}-0.212w_2h_{24}+0.012w_2h_{25}+\\0.016w_3h_{31}+0.028w_3h_{32}-0.024w_3h_{33}+0.16w_3h_{34}+0.128w_3h_{35}+0.112w_4h_{41}+\\0.236w_4h_{42}+0.108w_4h_{43}-0.028w_4h_{44}-0.044w_4h_{45}\geq \underline{a}\\0.0236w_2-0.1352w_1-0.1422w_3-0.2191w_4+0.026w_1h_{11}-0.026w_1h_{12}+0.004w_1h_{13}+\\0.026w_2h_{12}+0.086w_1h_{14}+0.19w_1h_{15}+0.056w_2h_{22}+0.004w_2h_{23}-0.198w_2h_{24}+\\0.012w_2h_{25}+0.014w_3h_{31}+0.024w_3h_{32}-0.024w_3h_{33}+0.148w_3h_{34}+0.118w_3h_{35}+\\0.104w_4h_{41}+0.22w_4h_{42}+0.1w_4h_{43}-0.026w_4h_{44}-0.04w_4h_{45}\geq \bar{a}\\z_{12}^{1},z_{34}^{1},z_{53}^{1},z_{13}^{2},z_{34}^{2},z_{42}^{2},z_{53}^{2},z_{21}^{3},z_{31}^{3},z_{45}^{3}\geq 0\\0\leq h_{11}\leq h_{12}\leq h_{13}\leq h_{14}\leq h_{15}\leq 1,\quad 0\leq h_{21}\leq h_{22}\leq h_{23}\leq h_{24}\leq h_{25}\leq 1\\0\leq h_{31}\leq h_{32}\leq h_{33}\leq h_{34}\leq h_{35}\leq 1,\quad 0\leq h_{41}\leq h_{42}\leq h_{43}\leq h_{44}\leq h_{45}\leq 1\\0.15\leq w_1\leq 0.2,0.16\leq w_2\leq 0.18,0.3\leq w_3\leq 0.35,0.3\leq w_4\leq 0.45,w_1+w_2+w_3+w_4=1\end{array}\right.$$

(MOD-A.2)

$$
\min\left\{\begin{array}{l}
\{0.5,0.6,0.7\}\otimes z_{12}+\{0.6,0.65,0.7\}\otimes z_{23}+\{0.8,0.85,0.9\}\otimes z_{24}+\{0.5,0.7\}\otimes z_{25}\\
+\{0.4,0.5,0.6\}\otimes z_{31}+\{0.6,0.7,0.95\}\otimes z_{34}+\{0.7,0.9\}\otimes z_{45}
\end{array}\right\}
$$

$$
s.t.\left\{\begin{array}{l}
-0.00667w_1+0.157w_2+0.0175w_3-0.321w_4+0.0667\vartheta_1^1-0.2\vartheta_1^2-0.0667\vartheta_2^2+\\
0.0333\vartheta_2^3+0.133\vartheta_4^1+0.167\vartheta_4^2+0.133\vartheta_4^3+z_{12}\geq 0\\
-0.133w_1-0.137w_2+0.0433w_3+0.0933w_4+0.133\vartheta_1^1+0.133\vartheta_1^2+0.133\vartheta_2^2+\\
0.0667\vartheta_2^3-0.0667\vartheta_3^1-0.0667\vartheta_4^1-0.0667\vartheta_4^2+z_{23}\geq 0\\
-0.09w_1-0.39w_2+0.522w_3-0.345w_4+0.333\vartheta_1^2+0.0667\vartheta_1^3+0.2\vartheta_2^2+0.2\vartheta_2^3-\\
0.2\vartheta_3^1-0.233\vartheta_3^2-0.3\vartheta_3^3+0.267\vartheta_4^1+0.233\vartheta_4^2+0.3\vartheta_4^3+z_{24}\geq 0\\
0.203w_1-0.48w_2-0.0983w_3-0.095w_4-0.133\vartheta_1^1+0.2\vartheta_1^2-0.133\vartheta_1^3+0.267\vartheta_2^2+\\
0.267\vartheta_2^3+0.0667\vartheta_3^1+0.1\vartheta_3^2+0.0333\vartheta_3^3-0.0333\vartheta_4^1+0.1\vartheta_4^2+0.133\vartheta_4^3+z_{25}\geq 0\\
0.14w_1-0.02w_2-0.608w_3+0.288w_4-0.2\vartheta_1^1-0.0667\vartheta_1^2-0.0667\vartheta_2^2-0.0667\vartheta_2^3+\\
0.0667\vartheta_3^1+0.0333\vartheta_3^2-0.0667\vartheta_4^1-0.1\vartheta_4^2-0.133\vartheta_4^3+z_{31}\geq 0\\
0.0433w_1-0.253w_2+0.478w_3-0.438w_4-0.133\vartheta_1^1+0.2\vartheta_1^2+0.0667\vartheta_1^3-0.0667\vartheta_2^2+\\
0.133\vartheta_2^3-0.133\vartheta_3^1-0.233\vartheta_3^2-0.3\vartheta_3^3+0.333\vartheta_4^1+0.3\vartheta_4^2+0.3\vartheta_4^3+z_{34}\geq 0\\
0.293w_1-0.09w_2-0.63w_3+0.25w_4-0.133\vartheta_1^1-0.133\vartheta_1^2+0.2\vartheta_1^3+0.0667\vartheta_2^2+\\
0.0667\vartheta_2^3+0.267\vartheta_3^1+0.333\vartheta_3^2+0.0333\vartheta_3^3-0.3\vartheta_4^1-0.133\vartheta_4^2-0.167\vartheta_4^3+z_{45}\geq 0\\
\{0.5,0.6,0.7\}\otimes(-0.00667w_1+0.157w_2+0.0175w_3-0.321w_4+0.0667\vartheta_1^1-\\
0.2\vartheta_1^2-0.0667\vartheta_2^2+0.0333\vartheta_2^3+0.133\vartheta_4^1+0.167\vartheta_4^2+0.133\vartheta_4^3)+\\
\{0.6,0.65,0.7\}\otimes(-0.133w_1-0.137w_2+0.0433w_3+0.0933w_4+0.133\vartheta_1^1+\\
0.133\vartheta_1^2+0.133\vartheta_2^2+0.0667\vartheta_2^3-0.0667\vartheta_3^1-0.0667\vartheta_4^1-0.0667\vartheta_4^2)+\\
\{0.8,0.85,0.9\}\otimes(-0.09w_1-0.39w_2+0.522w_3-0.345w_4+0.333\vartheta_1^2+0.0667\vartheta_1^3+\\
0.2\vartheta_2^2+0.2\vartheta_2^3-0.2\vartheta_3^1-0.233\vartheta_3^2-0.3\vartheta_3^3+0.267\vartheta_4^1+0.233\vartheta_4^2+0.3\vartheta_4^3)+\\
\{0.5,0.7\}\otimes(0.203w_1-0.48w_2-0.0983w_3-0.095w_4-0.133\vartheta_1^1+0.2\vartheta_1^2-0.133\vartheta_1^3+\\
0.267\vartheta_2^2+0.267\vartheta_2^3+0.0667\vartheta_3^1+0.1\vartheta_3^2+0.0333\vartheta_3^3-0.0333\vartheta_4^1+0.1\vartheta_4^2+0.133\vartheta_4^3)+\\
\{0.4,0.5,0.6\}\otimes(0.14w_1-0.02w_2-0.608w_3+0.288w_4-0.2\vartheta_1^1-0.0667\vartheta_1^2-\\
0.0667\vartheta_2^2-0.0667\vartheta_2^3+0.0667\vartheta_3^1+0.0333\vartheta_3^2-0.0667\vartheta_4^1-0.1\vartheta_4^2-0.133\vartheta_4^3)+\\
\{0.6,0.7,0.95\}\otimes(0.0433w_1-0.253w_2+0.478w_3-0.438w_4-0.133\vartheta_1^1+0.2\vartheta_1^2+0.0667\vartheta_1^3-\\
0.0667\vartheta_2^2+0.133\vartheta_2^3-0.133\vartheta_3^1-0.233\vartheta_3^2-0.3\vartheta_3^3+0.333\vartheta_4^1+0.3\vartheta_4^2+0.3\vartheta_4^3)+\\
\{0.7,0.9\}\otimes(0.293w_1-0.09w_2-0.63w_3+0.25w_4-0.133\vartheta_1^1-0.133\vartheta_1^2+0.2\vartheta_1^3+0.0667\vartheta_2^2+\\
0.0667\vartheta_2^3+0.267\vartheta_3^1+0.333\vartheta_3^2+0.0333\vartheta_3^3-0.3\vartheta_4^1-0.133\vartheta_4^2-0.167\vartheta_4^3)\succeq\{0.01\}\\
w_4\geq w_1,\ 0.05\leq w_2-w_2\leq 0.3,\ \ w_4-w_3\geq w_2-w_1,\ w_2\geq 2w_1,\ 0.15\leq w_4\leq 0.5\\
w_1+w_2+w_3+w_4=1,\ 0.01\leq w_j\leq 1\quad (j=1,2,3,4)\\
z_{12},z_{23},z_{24},z_{31},z_{34},z_{25},z_{45}\geq 0\\
00\leq\vartheta_j^1\leq\vartheta_j^2\leq\vartheta_j^3\leq w_j\ (j=1,2,3,4)
\end{array}\right.
\qquad\text{(MOD-A.3)}
$$

$$
\min\left\{\begin{array}{l} H\{0.5,0.6,0.7\}\otimes z_{12}+H\{0.6,0.65,0.7\}\otimes z_{23}+H\{0.8,0.85,0.9\}\otimes z_{24}+H\{0.5,0.7\}\otimes z_{25} \\ +H\{0.4,0.5,0.6\}\otimes z_{31}+H\{0.6,0.7,0.95\}\otimes z_{34}+H\{0.7,0.9\}\otimes z_{45} \end{array}\right\}
$$

$$
\min\{o_{141}^{-}+o_{132}^{-}+o_{223}^{-}+o_{223}^{+}+o_{34321}^{-}+o_{421}^{-}+o_{54}^{+}+o_{54}^{-}\}
$$

$$
s.t.\left\{\begin{array}{l}
-0.00667w_1+0.157w_2+0.0175w_3-0.321w_4+0.0667\vartheta_1^1-0.2\vartheta_1^2-0.0667\vartheta_2^2 \\
+0.0333\vartheta_2^3+0.133\vartheta_4^1+0.167\vartheta_4^2+0.133\vartheta_4^3+z_{12}\geq 0 \\
-0.133w_1-0.137w_2+0.0433w_3+0.0933w_4+0.133\vartheta_1^1+0.133\vartheta_1^2+0.133\vartheta_2^2+0.0667\vartheta_2^3 \\
-0.0667\vartheta_3^1-0.0667\vartheta_4^1-0.0667\vartheta_4^2+z_{23}\geq 0 \\
-0.09w_1-0.39w_2+0.522w_3-0.345w_4+0.333\vartheta_1^2+0.0667\vartheta_1^3+0.2\vartheta_2^2+0.2\vartheta_2^3 \\
-0.2\vartheta_3^1-0.233\vartheta_3^2-0.3\vartheta_3^3+0.267\vartheta_4^1+0.233\vartheta_4^2+0.3\vartheta_4^3+z_{24}\geq 0 \\
0.203w_1-0.48w_2-0.0983w_3-0.095w_4-0.133\vartheta_1^1+0.2\vartheta_1^2-0.133\vartheta_1^3+0.267\vartheta_2^2+0.267\vartheta_2^3 \\
+0.0667\vartheta_3^1+0.1\vartheta_3^2+0.0333\vartheta_3^3-0.0333\vartheta_4^1+0.1\vartheta_4^2+0.133\vartheta_4^3+z_{25}\geq 0 \\
0.14w_1-0.02w_2-0.608w_3+0.288w_4-0.2\vartheta_1^1-0.0667\vartheta_1^2-0.0667\vartheta_2^2-0.0667\vartheta_2^3 \\
+0.0667\vartheta_3^1+0.0333\vartheta_3^2-0.0667\vartheta_4^1-0.1\vartheta_4^2-0.133\vartheta_4^3+z_{31}\geq 0 \\
0.0433w_1-0.253w_2+0.478w_3-0.438w_4-0.133\vartheta_1^1+0.2\vartheta_1^2+0.0667\vartheta_1^3-0.0667\vartheta_2^2 \\
+0.133\vartheta_2^3-0.133\vartheta_3^1-0.233\vartheta_3^2-0.3\vartheta_3^3+0.333\vartheta_4^1+0.3\vartheta_4^2+0.3\vartheta_4^3+z_{34}\geq 0 \\
0.293w_1-0.09w_2-0.63w_3+0.25w_4-0.133\vartheta_1^1-0.133\vartheta_1^2+0.2\vartheta_1^3+0.0667\vartheta_2^2+0.0667\vartheta_2^3 \\
+0.267\vartheta_3^1+0.333\vartheta_3^2+0.0333\vartheta_3^3-0.3\vartheta_4^1-0.133\vartheta_4^2-0.167\vartheta_4^3+z_{45}\geq 0 \\
\{0.5,0.6,0.7\}\otimes(-0.00667w_1+0.157w_2+0.0175w_3-0.321w_4+0.0667\vartheta_1^1-0.2\vartheta_1^2 \\
-0.0667\vartheta_2^2+0.0333\vartheta_2^3+0.133\vartheta_4^1+0.167\vartheta_4^2+0.133\vartheta_4^3) \\
+\{0.6,0.65,0.7\}\otimes(-0.133w_1-0.137w_2+0.0433w_3+0.0933w_4+0.133\vartheta_1^1 \\
+0.133\vartheta_1^2+0.133\vartheta_2^2+0.0667\vartheta_2^3-0.0667\vartheta_3^1-0.0667\vartheta_4^1-0.0667\vartheta_4^2)+ \\
\{0.8,0.85,0.9\}\otimes(-0.09w_1-0.39w_2+0.522w_3-0.345w_4+0.333\vartheta_1^2+0.0667\vartheta_1^3 \\
+0.2\vartheta_2^2+0.2\vartheta_2^3-0.2\vartheta_3^1-0.233\vartheta_3^2-0.3\vartheta_3^3+0.267\vartheta_4^1+0.233\vartheta_4^2+0.3\vartheta_4^3) \\
+\{0.5,0.7\}\otimes(0.203w_1-0.48w_2-0.0983w_3-0.095w_4-0.133\vartheta_1^1+0.2\vartheta_1^2-0.133\vartheta_1^3 \\
+0.267\vartheta_2^2+0.267\vartheta_2^3+0.0667\vartheta_3^1+0.1\vartheta_3^2+0.0333\vartheta_3^3-0.0333\vartheta_4^1+0.1\vartheta_4^2+0.133\vartheta_4^3) \\
+\{0.4,0.5,0.6\}\otimes(0.14w_1-0.02w_2-0.608w_3+0.288w_4-0.2\vartheta_1^1-0.0667\vartheta_1^2 \\
-0.0667\vartheta_2^2-0.0667\vartheta_2^3+0.0667\vartheta_3^1+0.0333\vartheta_3^2-0.0667\vartheta_4^1-0.1\vartheta_4^2-0.133\vartheta_4^3) \\
+\{0.6,0.7,0.95\}\otimes(0.0433w_1-0.253w_2+0.478w_3-0.438w_4-0.133\vartheta_1^1+0.2\vartheta_1^2+0.0667\vartheta_1^3 \\
-0.0667\vartheta_2^2+0.133\vartheta_2^3-0.133\vartheta_3^1-0.233\vartheta_3^2-0.3\vartheta_3^3+0.333\vartheta_4^1+0.3\vartheta_4^2+0.3\vartheta_4^3) \\
+\{0.7,0.9\}\otimes(0.293w_1-0.09w_2-0.63w_3+0.25w_4-0.133\vartheta_1^1-0.133\vartheta_1^2+0.2\vartheta_1^3+0.0667\vartheta_2^2 \\
+0.0667\vartheta_2^3+0.267\vartheta_3^1+0.333\vartheta_3^2+0.0333\vartheta_3^3-0.3\vartheta_4^1-0.133\vartheta_4^2-0.167\vartheta_4^3)\succeq H\{0.01\} \\
z_{12},z_{23},z_{24},z_{31},z_{34},z_{25},z_{45}\geq 0 \\
0\leq\vartheta_j^1\leq\vartheta_j^2\leq\vartheta_j^3\leq w_j\ (j=1,2,3,4) \\
w_4+o_{141}^{-}\geq w_1,w_3+o_{132}^{-}\geq w_2,w_2-w_3\ +o_{223}^{-}\geq 0.05,w_2-w_3-o_{223}^{+}\leq 0.3, \\
w_4-w_3-w_2+w_1+o_{34321}^{-}\geq 0,w_2/w_1+o_{421}^{-}\geq 2,\ w_4-o_{54}^{+}\leq 0.5,\ w_4+o_{54}^{-}\geq 0.15 \\
w_1+w_2+w_3+w_4=1,\ \ 0.01\leq w_j\leq 1\quad (j=1,2,3,4) \\
o_{141}^{-},o_{132}^{-},o_{223}^{-},o_{223}^{+},o_{34321}^{-},o_{421}^{-},o_{54}^{+},o_{54}^{-}\geq 0
\end{array}\right.
$$

(MOD-A.4)

$$
\min \{y\}
$$

$$
s.t.\left\{\begin{array}{l}
0.6z_{12}+0.65z_{23}+0.85z_{24}+0.6z_{25}+0.5z_{31}+0.75z_{34}+0.8z_{45}\leq y\\
o_{141}^{-}+o_{132}^{-}+o_{223}^{-}+o_{223}^{+}+o_{34321}^{-}+o_{421}^{-}+o_{54}^{+}+o_{54}^{-}\leq y\\
-0.00667w_1+0.157w_2+0.0175w_3-0.321w_4+0.0667\vartheta_1^1-0.2\vartheta_1^2-0.0667\vartheta_2^2\\
+0.0333\vartheta_2^3+0.133\vartheta_4^1+0.167\vartheta_4^2+0.133\vartheta_4^3+z_{12}\geq 0\\
-0.133w_1-0.137w_2+0.0433w_3+0.0933w_4+0.133\vartheta_1^1+0.133\vartheta_1^2+0.133\vartheta_2^2+0.0667\vartheta_2^3\\
-0.0667\vartheta_3^1-0.0667\vartheta_4^1-0.0667\vartheta_4^2+z_{23}\geq 0\\
-0.09w_1-0.39w_2+0.522w_3-0.345w_4+0.333\vartheta_1^2+0.0667\vartheta_1^3+0.2\vartheta_2^2+0.2\vartheta_2^3\\
-0.2\vartheta_3^1-0.233\vartheta_3^2-0.3\vartheta_3^3+0.267\vartheta_4^1+0.233\vartheta_4^2+0.3\vartheta_4^3+z_{24}\geq 0\\
0.203w_1-0.48w_2-0.0983w_3-0.095w_4-0.133\vartheta_1^1+0.2\vartheta_1^2-0.133\vartheta_1^3+0.267\vartheta_2^2+0.267\vartheta_2^3\\
+0.0667\vartheta_3^1+0.1\vartheta_3^2+0.0333\vartheta_3^3-0.0333\vartheta_4^1+0.1\vartheta_4^2+0.133\vartheta_4^3+z_{25}\geq 0\\
0.14w_1-0.02w_2-0.608w_3+0.288w_4-0.2\vartheta_1^1-0.0667\vartheta_1^2-0.0667\vartheta_2^2-0.0667\vartheta_2^3\\
+0.0667\vartheta_3^1+0.0333\vartheta_3^2-0.0667\vartheta_4^1-0.1\vartheta_4^2-0.133\vartheta_4^3+z_{31}\geq 0\\
0.0433w_1-0.253w_2+0.478w_3-0.438w_4-0.133\vartheta_1^1+0.2\vartheta_1^2+0.0667\vartheta_1^3-0.0667\vartheta_2^2\\
+0.133\vartheta_2^3-0.133\vartheta_3^1-0.233\vartheta_3^2-0.3\vartheta_3^3+0.333\vartheta_4^1+0.3\vartheta_4^2+0.3\vartheta_4^3+z_{34}\geq 0\\
0.293w_1-0.09w_2-0.63w_3+0.25w_4-0.133\vartheta_1^1-0.133\vartheta_1^2+0.2\vartheta_1^3+0.0667\vartheta_2^2+0.0667\vartheta_2^3\\
+0.267\vartheta_3^1+0.333\vartheta_3^2+0.0333\vartheta_3^3-0.3\vartheta_4^1-0.133\vartheta_4^2-0.167\vartheta_4^3+z_{45}\geq 0\\
0.6\times(-0.00667w_1+0.157w_2+0.0175w_3-0.321w_4+0.0667\vartheta_1^1-0.2\vartheta_1^2\\
-0.0667\vartheta_2^2+0.0333\vartheta_2^3+0.133\vartheta_4^1+0.167\vartheta_4^2+0.133\vartheta_4^3)\\
+0.65\times(-0.133w_1-0.137w_2+0.0433w_3+0.0933w_4+0.133\vartheta_1^1\\
+0.133\vartheta_1^2+0.133\vartheta_2^2+0.0667\vartheta_2^3-0.0667\vartheta_3^1-0.0667\vartheta_4^1-0.0667\vartheta_4^2)+\\
0.85\times(-0.09w_1-0.39w_2+0.522w_3-0.345w_4+0.333\vartheta_1^2+0.0667\vartheta_1^3\\
+0.2\vartheta_2^2+0.2\vartheta_2^3-0.2\vartheta_3^1-0.233\vartheta_3^2-0.3\vartheta_3^3+0.267\vartheta_4^1+0.233\vartheta_4^2+0.3\vartheta_4^3)\\
+0.6\times(0.203w_1-0.48w_2-0.0983w_3-0.095w_4-0.133\vartheta_1^1+0.2\vartheta_1^2-0.133\vartheta_1^3\\
+0.267\vartheta_2^2+0.267\vartheta_2^3+0.0667\vartheta_3^1+0.1\vartheta_3^2+0.0333\vartheta_3^3-0.0333\vartheta_4^1+0.1\vartheta_4^2+0.133\vartheta_4^3)\\
+0.5\times(0.14w_1-0.02w_2-0.608w_3+0.288w_4-0.2\vartheta_1^1-0.0667\vartheta_1^2\\
-0.0667\vartheta_2^2-0.0667\vartheta_2^3+0.0667\vartheta_3^1+0.0333\vartheta_3^2-0.0667\vartheta_4^1-0.1\vartheta_4^2-0.133\vartheta_4^3)\\
+0.75\times(0.0433w_1-0.253w_2+0.478w_3-0.438w_4-0.133\vartheta_1^1+0.2\vartheta_1^2+0.0667\vartheta_1^3\\
-0.0667\vartheta_2^2+0.133\vartheta_2^3-0.133\vartheta_3^1-0.233\vartheta_3^2-0.3\vartheta_3^3+0.333\vartheta_4^1+0.3\vartheta_4^2+0.3\vartheta_4^3)\\
+0.8\times(0.293w_1-0.09w_2-0.63w_3+0.25w_4-0.133\vartheta_1^1-0.133\vartheta_1^2+0.2\vartheta_1^3+0.0667\vartheta_2^2\\
+0.0667\vartheta_2^3+0.267\vartheta_3^1+0.333\vartheta_3^2+0.0333\vartheta_3^3-0.3\vartheta_4^1-0.133\vartheta_4^2-0.167\vartheta_4^3)\geq 0.01\\
z_{12},z_{23},z_{24},z_{31},z_{34},z_{25},z_{45}\geq 0\\
w_4+o_{141}^{-}\geq w_1,\ w_3+o_{132}^{-}\geq w_2,\ w_2-w_3+o_{223}^{-}\geq 0.05,\ w_2-w_3-o_{223}^{+}\leq 0.3,\\
w_4-w_3-w_2+w_1+o_{34321}^{-}\geq 0,\ w_2/w_1+o_{421}^{-}\geq 2,\ w_4-o_{54}^{+}\leq 0.5,\ w_4+o_{54}^{-}\geq 0.15\\
w_1+w_2+w_3+w_4=1,\ \ 0\leq w_j\leq 1\quad (j=1,2,3,4)\\
0\leq\vartheta_j^1\leq\vartheta_j^2\leq\vartheta_j^3\leq w_j\ (j=1,2,3,4)\\
o_{141}^{-},o_{132}^{-},o_{223}^{-},o_{223}^{+},o_{34321}^{-},o_{421}^{-},o_{54}^{+},o_{54}^{-}\geq 0
\end{array}\right.
\tag{MOD-A.5}
$$

References

Chen, T. Y. (2013). An interval-valued intuitionistic fuzzy LINMAP method with inclusion comparison possibilities and hybrid averaging operations for multiple criteria group decision making. *Knowledge-Based Systems, 45*, 134–146.

Chen, T. Y. (2014). Interval-valued intuitionistic fuzzy QUALIFLEX method with a likelihood-based comparison approach for multiple criteria decision analysis. *Information Sciences, 261*, 149–169.

Hwang, C. L., & Yoon, K. (1981). *Multiple Attribute Decision Making Methods and Applications*. Berlin, Heidelberg: Springer.

Ishibuchi, H., & Tanaka, H. (1990). Multiobjective programming in optimization of the interval objective function. *European Journal of Operational Research, 48*, 219–225.

Kahraman, C., & Kaya, İ. (2010). A fuzzy multicriteria methodology for selection among energy alternatives. *Expert Systems with Applications, 37*, 6270–6281.

Lai, K. K., Wang, S., Xu, J., Zhu, S., & Fang, Y. (2002). A class of linear interval programming problems and its application to portfolio selection. *IEEE Transactions on Fuzzy Systems, 10*, 698–704.

Li, D. F., Chen, G. H., & Huang, Z.-G. (2010). Linear programming method for multiattribute group decision making using IF sets. *Information Sciences, 180*, 1591–1609.

Li, D. F., & Wan, S. P. (2014). Fuzzy heterogeneous multiattribute decision making method for outsourcing provider selection. *Expert Systems with Applications, 41*, 3047–3059.

Li, D. F., & Yang, J. B. (2004). Fuzzy linear programming technique for multiattribute group decision making in fuzzy environments. *Information Sciences, 158*, 263–275.

Sadi-Nezhad, S., & Akhtari, P. (2008). Possibilistic programming approach for fuzzy multidimensional analysis of preference in group decision making. *Applied Soft Computing, 8*, 1703–1711.

Srinivasan, V., & Shocker, A. D. (1973). Linear programming techniques for multidimensional analysis of preferences. *Psychometrika, 38*, 337–369.

Tong, S. (1994). Interval number and fuzzy number linear programmings. *Fuzzy Sets and Systems, 66*, 301–306.

Wan, S. P., & Li, D. F. (2013). Fuzzy LINMAP approach to heterogeneous MADM considering comparisons of alternatives with hesitation degrees. *Omega, 41*, 925–940.

Wang, Z. J., & Li, K. W. (2012). An interval-valued intuitionistic fuzzy multiattribute group decision making framework with incomplete preference over alternatives. *Expert Systems with Applications, 39*, 13509–13516.

Xu, Z. S., & Xia, M. M. (2011). Distance and similarity measures for hesitant fuzzy sets. *Information Sciences, 181*, 2128–2138.

Zhang, X. L., & Xu, Z. S. (2014). Interval programming method for hesitant fuzzy multi-attribute group decision making with incomplete preference over alternatives. *Computers & Industrial Engineering, 75*, 217–229.

Zhang, X. L., Xu, Z. S., & Xing, X. M. (2015). Hesitant fuzzy programming techniques for multidimensional analysis of hesitant fuzzy preferences. *OR Spectrum*. doi:10.1007/s00291-015-0420-0.

Zhu, B., & Xu, Z. S. (2014). Consistency measures for hesitant fuzzy linguistic preference relations. *IEEE Transactions on Fuzzy Systems, 22*(1), 35–45.

Chapter 5
Consensus Model-Based Hesitant Fuzzy Multiple Criteria Group Decision Analysis

Abstract The MCGDM is an important research topic in decision theory. Many useful methods have been proposed to solve various MCGDM problems, but very few methods simultaneously take them into account from the perspectives of both the ranking and the magnitude of decision data, especially for the hesitant fuzzy situations. The purpose of this chapter is to develop a consensus model-based hesitant fuzzy group decision making method to handle hesitant fuzzy MCGDM problems by simultaneously taking into account such two aspects of hesitant fuzzy decision data. Firstly, an ordinal consensus index is presented for measuring the consensus among individual experts' opinions from the perspective of the ranking of decision information. Then, a cardinal consensus index is presented for measuring the consensus among individual experts' opinions from the perspective of the magnitude of decision information. Afterwards, a linear programming model to derive the weights of experts is constructed according to the idea that the expert with a larger consensus should be assigned a larger weight. Finally, the relative closeness indices of alternatives are determined and the ranking of alternatives is identified.

The MCGDM refers to the process that multiple experts make evaluations with their respective knowledge, experience and preference for a set of alternatives over multiple criteria, and the decision results from each expert are aggregated to form an overall ranking result for these alternatives. The MCGDM as an important research topic in decision theory has broad applications in the fields of natural science, social science, economy and management, etc. In the practical MCGDM process, the experts usually come from various research areas and may have many differences in knowledge structure, express abilities, evaluation levels and individual preferences as well as practical experience. These differences imply that different experts should be assigned different weights which reflect the corresponding importance in the group. However, in several group decision making literature, the weights of experts have usually been ignored or the weights of experts are determined only from the perspective of the magnitudes of decision data, but relatively very few approaches simultaneously take into account the decision data from the perspectives of both the ranking and the magnitude, especially in hesitant fuzzy environment.

X. Zhang and Z. Xu, *Hesitant Fuzzy Methods for Multiple Criteria Decision Analysis*, Studies in Fuzziness and Soft Computing 345,
DOI 10.1007/978-3-319-42001-1_5

The purpose of this chapter is to develop a consensus model-based hesitant fuzzy group decision making method to handle the hesitant fuzzy MCGDM problems by simultaneously taking into account the decision data from the perspectives of both the ranking and the magnitude, which is motivated by the literature (Zhang and Xu 2014, 2015). Firstly, an ordinal consensus index is presented for measuring the consensus among individual experts' opinion from the perspective of the ranking of decision information. Then, a cardinal consensus index is presented for measuring the consensus among individual experts' opinion from the perspective of the magnitude of decision information. A linear programming model to derive the weights of experts is constructed based on the idea that the expert with a larger consensus should be assigned a larger weight. Afterwards, we calculate the relative closeness indices of alternatives to identify the ranking of alternatives. At length, an example is given to verify the proposed approach.

5.1 Maximizing Consensus Model for Deriving the Weights of Experts

In this chapter, we explore the hesitant fuzzy MCGDM problem in which the weights of criteria are completely known but the weights of experts are partially known or completely unknown. This hesitant fuzzy MCGDM problem is expressed by the group decision making matrix $\Re^k = (h_{ij}^k)_{m\times n}(k = 1, 2, \ldots, g)$, where h_{ij}^k is a HFE indicating the criterion value of the alternative A_i with respect to the criterion C_j provided by the expert e_k.

In what follows, we first establish a maximizing consensus model to objectively determine the weights of experts under hesitant fuzzy environments.

5.1.1 *Group Ordinal Consensus Index*

As previously mentioned, most of existing methods to determine the experts' weights only consider the decision information from the perspective of the magnitude but fail to consider the decision data from the perspective of the ranking. In fact, this sort of information is very important to determine the experts' weights. For this reason, by borrowing the idea from the literature (Pang and Liang 2012; Zhang and Xu 2014; 2015) we introduce a concept of group ordinal consensus (abbreviated as *GOC*) to measure the consensus degree among the individual expert's opinions from the perspective of the ranking of decision data.

In a given MCGDM problem with HFEs, for $\forall A_\xi, A_\zeta \in \boldsymbol{A}$, it can be said for the expert $e_k \in \boldsymbol{E}$ that the alternative $A_\xi \in \boldsymbol{A}$ dominates $A_\zeta \in \boldsymbol{A}$ with respect to the criterion $C_j \in \boldsymbol{C}$ if $s(h_{\xi j}^k) \geq s(h_{\zeta j}^k)$, and denoted by $A_\xi (R_{C_j}^{e_k})^{\geq} A_\zeta$. The dominance relation $(R_{C_j}^{e_k})^{\geq}$ can be expressed as:

$$\left(R_{C_j}^{e_k}\right)^{\geq} = \left\{(A_\xi, A_\zeta) \in \boldsymbol{A} \times \boldsymbol{A} \middle| s(h_{\xi j}^k) \geq s(h_{\zeta j}^k)\right\} \tag{5.1}$$

where $s(h_{\xi j}^k)$ and $s(h_{\zeta j}^k)$ are the scores of $h_{\xi j}^k$ and $h_{\zeta j}^k$, respectively.

For the expert $e_k \in \boldsymbol{E}$, the dominance class of the alternative $A_\xi \in \boldsymbol{A}$ on the criterion $C_j \in \boldsymbol{C}$ is defined as follows:

$$[A_\xi]_{C_j}^{e_k \geq} = \left\{A_\zeta \in \boldsymbol{A} \middle| s(h_{\xi j}^k) \geq s(h_{\zeta j}^k)\right\} \tag{5.2}$$

Then, the family set of the dominance classes under the expert $e_k \in \boldsymbol{E}$ for the alternative set $\boldsymbol{A}$ with respect to the criterion $C_j \in \boldsymbol{C}$ is defined as follows:

$$\boldsymbol{A}/\left(R_{C_j}^{e_k}\right)^{\geq} = \left\{[A_1]_{C_j}^{e_k \geq}, [A_2]_{C_j}^{e_k \geq}, \ldots, [A_m]_{C_j}^{e_k \geq}\right\} \tag{5.3}$$

The ordinal consensus index with respect to the criterion $C_j \in \boldsymbol{C}$ between the expert $e_k \in \boldsymbol{E}$ and the expert $e_l \in \boldsymbol{E}(k \neq l)$ is defined as:

$$OC_k^l(C_j) = \frac{1}{m}\sum_{\xi=1}^{m}\left(\frac{\left|[A_\xi]_{C_j}^{e_k \geq} \cap [A_\xi]_{C_j}^{e_l \geq}\right|}{\left|[A_\xi]_{C_j}^{e_k \geq} \cup [A_\xi]_{C_j}^{e_l \geq}\right|}\right) \tag{5.4}$$

Apparently, $1/m \leq OC_k^l(C_j) \leq 1$.

Furthermore, the weighted ordinal consensus index between the expert e_k and the expert $e_l(k \neq l)$ is calculated by the following equation:

$$OC_k^l = \frac{1}{m}\sum_{j=1}^{n}\sum_{\xi=1}^{m} w_j\left(\frac{\left|[A_\xi]_{C_j}^{e_k \geq} \cap [A_\xi]_{C_j}^{e_l \geq}\right|}{\left|[A_\xi]_{C_j}^{e_k \geq} \cup [A_\xi]_{C_j}^{e_l \geq}\right|}\right) \tag{5.5}$$

and the weighted ordinal consensus index between the expert e_k and all the other experts is calculated by the following equation:

$$OC_k = \frac{1}{m}\sum_{l=1, l\neq k}^{g}\sum_{j=1}^{n}\sum_{\xi=1}^{m} w_j\left(\frac{\left|[A_\xi]_{C_j}^{e_k \geq} \cap [A_\xi]_{C_j}^{e_l \geq}\right|}{\left|[A_\xi]_{C_j}^{e_k \geq} \cup [A_\xi]_{C_j}^{e_l \geq}\right|}\right) \tag{5.6}$$

The sum of OC_k is normalized into a unit as:

$$O\bar{C}_k = \frac{\sum_{l=1, l\neq k}^{g}\sum_{j=1}^{n}\sum_{\xi=1}^{m} w_j\left(\frac{\left|[A_\xi]_{C_j}^{e_k \geq} \cap [A_\xi]_{C_j}^{e_l \geq}\right|}{\left|[A_\xi]_{C_j}^{e_k \geq} \cup [A_\xi]_{C_j}^{e_l \geq}\right|}\right)}{\sum_{k=1}^{g}\sum_{l=1, l\neq k}^{g}\sum_{j=1}^{n}\sum_{\xi=1}^{m} w_j\left(\frac{\left|[A_\xi]_{C_j}^{e_k \geq} \cap [A_\xi]_{C_j}^{e_l \geq}\right|}{\left|[A_\xi]_{C_j}^{e_k \geq} \cup [A_\xi]_{C_j}^{e_l \geq}\right|}\right)} \tag{5.7}$$

Thus, the group ordinal consensus index is defined as:

$$GOC = \frac{\sum_{k=1}^{g} \varpi_k \sum_{l=1,l\neq k}^{g} \sum_{j=1}^{n} \sum_{\xi=1}^{m} w_j \left(\left| \frac{[A_\xi]_{C_j}^{e_k \geq} \bigcap [A_\xi]_{C_j}^{e_l \geq}}{[A_\xi]_{C_j}^{e_k \geq} \bigcup [A_\xi]_{C_j}^{e_l \geq}} \right| \right)}{\sum_{k=1}^{g} \sum_{l=1,l\neq k}^{g} \sum_{j=1}^{n} \sum_{\xi=1}^{m} w_j \left(\left| \frac{[A_\xi]_{C_j}^{e_k \geq} \bigcap [A_\xi]_{C_j}^{e_l \geq}}{[A_\xi]_{C_j}^{e_k \geq} \bigcup [A_\xi]_{C_j}^{e_l \geq}} \right| \right)} \tag{5.8}$$

5.1.2 Group Cardinal Consensus Index

In this subsection, we will analyze the consensus situation among the experts' opinions from the perspective of the magnitude of decision information and give the definition of group cardinal consensus (abbreviated as *GCC*). We employ the similarity measure of HFEs to calculate the group cardinal consensus index.

Hence, for the alternative $A_\xi \in \boldsymbol{A}$, the cardinal consensus index with respect to the criterion $C_j \in \boldsymbol{C}$ between the expert e_k and the expert $e_l (k \neq l)$ is defined as:

$$CC_k^l(A_\xi, C_j) = 1 - \sqrt{\frac{1}{\#h} \sum_{\lambda=1}^{\#h} \left((\gamma_{\xi j}^{\lambda})^k - (\gamma_{\xi j}^{\lambda})^l \right)^2} \tag{5.9}$$

For all criteria, the weighted cardinal consensus index for the alternative $A_\xi \in \boldsymbol{A}$ between the expert $e_k \in \boldsymbol{E}$ and the expert $e_l \in \boldsymbol{E} (k \neq l)$ is defined as:

$$CC_k^l(A_\xi) = \sum_{j=1}^{n} \left(w_j \times \left(1 - \sqrt{\frac{1}{\#h} \sum_{\lambda=1}^{\#h} \left((\gamma_{\xi j}^{\lambda})^k - (\gamma_{\xi j}^{\lambda})^l \right)^2} \right) \right) \tag{5.10}$$

Then, for all criteria and alternatives, the weighted cardinal consensus index between the expert e_k and the expert $e_l (k \neq l)$ is defined as:

$$CC_k^l = \frac{1}{m} \sum_{\xi=1}^{m} \sum_{j=1}^{n} \left(w_j \times \left(1 - \sqrt{\frac{1}{\#h} \sum_{\lambda=1}^{\#h} \left((\gamma_{\xi j}^{\lambda})^k - (\gamma_{\xi j}^{\lambda})^l \right)^2} \right) \right) \tag{5.11}$$

Furthermore, the weighted cardinal consensus index between the expert e_k and all the other experts is defined as:

$$CC_k = \sum_{l=1,l\neq k}^{g} \left(\frac{1}{m} \sum_{\xi=1}^{m} \sum_{j=1}^{n} \left(w_j \times \left(1 - \sqrt{\frac{1}{\#h} \sum_{\lambda=1}^{\#h} \left((\gamma_{\xi j}^{\lambda})^k - (\gamma_{\xi j}^{\lambda})^l \right)^2} \right) \right) \right) \tag{5.12}$$

and the sum of CC_k is normalized into a unit as:

$$C\bar{C}_k = \frac{\sum_{l=1,l\neq k}^{g}\left(\frac{1}{m}\sum_{\xi=1}^{m}\sum_{j=1}^{n}\left(w_j \times \left(1-\sqrt{\frac{1}{\#h}\sum_{\lambda=1}^{\#h}\left((\gamma_{\xi j}^{\lambda})^k-(\gamma_{\xi j}^{\lambda})^l\right)^2}\right)\right)\right)}{\sum_{k=1}^{g}\left(\sum_{l=1,l\neq k}^{g}\left(\frac{1}{m}\sum_{\xi=1}^{m}\sum_{j=1}^{n}\left(w_j \times \left(1-\sqrt{\frac{1}{\#h}\sum_{\lambda=1}^{\#h}\left((\gamma_{\xi j}^{\lambda})^k-(\gamma_{\xi j}^{\lambda})^l\right)^2}\right)\right)\right)\right)} \tag{5.13}$$

Thus, the group cardinal consensus index is obtained as:

$$GCC = \frac{\sum_{k=1}^{g}\left(\varpi_k\sum_{l=1,l\neq k}^{g}\left(\frac{1}{m}\sum_{\xi=1}^{m}\sum_{j=1}^{n}\left(w_j \times \left(1-\sqrt{\frac{1}{\#h}\sum_{\lambda=1}^{\#h}\left((\gamma_{\xi j}^{\lambda})^k-(\gamma_{\xi j}^{\lambda})^l\right)^2}\right)\right)\right)\right)}{\sum_{k=1}^{g}\left(\sum_{l=1,l\neq k}^{g}\left(\frac{1}{m}\sum_{\xi=1}^{m}\sum_{j=1}^{n}\left(w_j \times \left(1-\sqrt{\frac{1}{\#h}\sum_{\lambda=1}^{\#h}\left((\gamma_{\xi j}^{\lambda})^k-(\gamma_{\xi j}^{\lambda})^l\right)^2}\right)\right)\right)\right)} \tag{5.14}$$

5.1.3 The Model for Determining the Experts' Weights

Based on the aforementioned analysis, we have to choose the experts' weight vector $\varpi \in \Gamma$ to maximize group consensus index for all criteria and alternatives from the perspectives of both the ranking and magnitude of decision information.

To do so, we can construct a bi-objective programming model to determine the experts' weights as:

$$\begin{aligned} &\max\{GOC, GCC\} \\ &s.t. \quad \varpi \in \Gamma \end{aligned} \tag{MOD-5.1}$$

Substituting GOC and GCC in the model (MOD-5.1) by Eqs. (5.8) and (5.14), respectively, then this model (MOD-5.1) can be rewritten as:

$$\begin{aligned} &\max\left\{\begin{aligned} &\frac{\sum_{k=1}^{g}\varpi_k\sum_{l=1,l\neq k}^{g}\sum_{j=1}^{n}\sum_{\xi=1}^{m}w_j\left(\left|[A_\xi]_{C_j}^{e_k\geq}\cap[A_\xi]_{C_j}^{e_l\geq}\right|\Big/\left|[A_\xi]_{C_j}^{e_k\geq}\cup[A_\xi]_{C_j}^{e_l\geq}\right|\right)}{\sum_{k=1}^{g}\sum_{l=1,l\neq k}^{g}\sum_{j=1}^{n}\sum_{\xi=1}^{m}w_j\left(\left|[A_\xi]_{C_j}^{e_k\geq}\cap[A_\xi]_{C_j}^{e_l\geq}\right|\Big/\left|[A_\xi]_{C_j}^{e_k\geq}\cup[A_\xi]_{C_j}^{e_l\geq}\right|\right)}, \\ &\frac{\sum_{k=1}^{g}\left(\varpi_k\sum_{l=1,l\neq k}^{g}\left(\frac{1}{m}\sum_{\xi=1}^{m}\sum_{j=1}^{n}\left(w_j\times\left(1-\sqrt{\frac{1}{\#h}\sum_{\lambda=1}^{\#h}\left((\gamma_{\xi j}^{\lambda})^k-(\gamma_{\xi j}^{\lambda})^l\right)^2}\right)\right)\right)\right)}{\sum_{k=1}^{g}\left(\sum_{l=1,l\neq k}^{g}\left(\frac{1}{m}\sum_{\xi=1}^{m}\sum_{j=1}^{n}\left(w_j\times\left(1-\sqrt{\frac{1}{\#h}\sum_{\lambda=1}^{\#h}\left((\gamma_{\xi j}^{\lambda})^k-(\gamma_{\xi j}^{\lambda})^l\right)^2}\right)\right)\right)\right)} \end{aligned}\right\} \\ &s.t. \quad \varpi \in \Gamma \end{aligned} \tag{MOD-5.2}$$

Obviously, the model (MOD-5.2) is a bi-objective linear programming model on the decision variables. There exist several solution methods for this model, and in this chapter we focus on developing a weight average approach for solving the model (MOD-5.2) in the sense of Pareto optimality.

By the weighted average approach, the model (MOD-5.2) can be transformed to the parametric linear programming model (MOD-5.3) as below:

$$
\begin{aligned}
&\max\left\{\begin{array}{l}
\theta \times \dfrac{\sum_{k=1}^{g} \varpi_k \sum_{l=1,l\neq k}^{g} \sum_{j=1}^{n} \sum_{\xi=1}^{m} w_j \left(\left| [A_\xi]_{C_j}^{e_k \geq} \bigcap [A_\xi]_{C_j}^{e_l \geq} \right| \Big/ \left| [A_\xi]_{C_j}^{e_k \geq} \bigcup [A_\xi]_{C_j}^{e_l \geq} \right| \right)}{\sum_{k=1}^{g} \sum_{l=1,l\neq k}^{g} \sum_{j=1}^{n} \sum_{\xi=1}^{m} w_j \left(\left| [A_\xi]_{C_j}^{e_k \geq} \bigcap [A_\xi]_{C_j}^{e_l \geq} \right| \Big/ \left| [A_\xi]_{C_j}^{e_k \geq} \bigcup [A_\xi]_{C_j}^{e_l \geq} \right| \right)} \\
+ (1-\theta) \times \dfrac{\sum_{k=1}^{g} \left(\varpi_k \sum_{l=1,l\neq k}^{g} \left(\frac{1}{m} \sum_{\xi=1}^{m} \sum_{j=1}^{n} \left(w_j \times \left(1 - \sqrt{\frac{1}{\#h} \sum_{\lambda=1}^{\#h} \left((\gamma_{\xi j}^{\lambda})^k - (\gamma_{\xi j}^{\lambda})^l \right)^2} \right) \right) \right) \right)}{\sum_{k=1}^{g} \left(\sum_{l=1,l\neq k}^{g} \left(\frac{1}{m} \sum_{\xi=1}^{m} \sum_{j=1}^{n} \left(w_j \times \left(1 - \sqrt{\frac{1}{\#h} \sum_{\lambda=1}^{\#h} \left((\gamma_{\xi j}^{\lambda})^k - (\gamma_{\xi j}^{\lambda})^l \right)^2} \right) \right) \right) \right)}
\end{array}\right\} \\
&s.t. \quad \boldsymbol{\varpi} \in \Gamma
\end{aligned}
\tag{MOD-5.3}
$$

where $0 \leq \theta \leq 1$ is a parameter, which is determined by the experts in advance.

The solution to the above maximization model (MOD-5.3) could be found by MATLAB with optimization toolbox or Lingo software package.

5.2 An Approach Based on TOPSIS for Solving the MCGDM Problem

Once the weights of experts are determined, it is necessary to develop a decision making method to identify the ranking of alternatives and select the optimal alternative. In the following, we employ the hesitant fuzzy TOPSIS method developed by Xu and Zhang (2013) to rank all alternatives.

The hesitant fuzzy TOPSIS method first needs to identify the hesitant fuzzy PIS A^+ and the hesitant fuzzy NIS A^- by employing the following equations:

$$
A^+ = (h_1^+, h_2^+, \ldots, h_n^+) = \left(H \left\{ \begin{array}{ll} \max_{i=1}^{m} {}_{k=1}^{g} (\gamma_{ij}^1)^k, & \max_{i=1}^{m} {}_{k=1}^{g} (\gamma_{ij}^2)^k, \\ \ldots, & \max_{i=1}^{m} {}_{k=1}^{g} (\gamma_{ij}^{\#h_{ij}})^k \end{array} \right\} \right) \tag{5.15}
$$

$$
A^- = (h_1^-, h_2^-, \ldots, h_n^-) = \left(H \left\{ \begin{array}{ll} \min_{i=1}^{m} {}_{k=1}^{g} (\gamma_{ij}^1)^k, & \min_{i=1}^{m} {}_{k=1}^{g} (\gamma_{ij}^2)^k, \\ \ldots, & \min_{i=1}^{m} {}_{k=1}^{g} (\gamma_{ij}^{\#h_{ij}})^k \end{array} \right\} \right) \tag{5.16}
$$

Furthermore, the distances for each expert between each of the alternatives $A_i (i \in \{1, 2, \ldots, m\})$ and the hesitant fuzzy PIS A^+ as well as the hesitant fuzzy NIS A^-, respectively, are calculated by the following equations.

$$D_i^{k+} = \sum_{j=1}^{n} w_j d_E\left(h_{ij}^k, h_j^+\right) = \sum_{j=1}^{n} w_j \sqrt{\frac{1}{\#h}\sum_{\lambda=1}^{\#h}\left((\gamma_{ij}^{\lambda})^k - (\gamma_j^{\lambda})^+\right)^2} \tag{5.17}$$

$$D_i^{k-} = \sum_{j=1}^{n} w_j d_E\left(h_{ij}^k, h_j^-\right) = \sum_{j=1}^{n} w_j \sqrt{\frac{1}{\#h}\sum_{\lambda=1}^{\#h}\left((\gamma_{ij}^{\lambda})^k - (\gamma_j^{\lambda})^-\right)^2} \tag{5.18}$$

Afterwards, the distances for all experts between each of the alternatives $A_i (i \in \{1, 2, \ldots, m\})$ and the hesitant fuzzy PIS A^+ as well as the hesitant fuzzy NIS A^-, are obtained by the following equations, respectively.

$$D_i^+ = \sum_{k=1}^{g} \left(\varpi_k \times D_i^{k+}\right) = \sum_{k=1}^{g}\sum_{j=1}^{n} \varpi_k w_j \sqrt{\frac{1}{\#h}\sum_{\lambda=1}^{\#h}\left((\gamma_{ij}^{\lambda})^k - (\gamma_j^{\lambda})^+\right)^2} \tag{5.19}$$

$$D_i^- = \sum_{k=1}^{g} \left(\varpi_k \times D_i^{k-}\right) = \sum_{k=1}^{g}\sum_{j=1}^{n} \varpi_k w_j \sqrt{\frac{1}{\#h}\sum_{\lambda=1}^{\#h}\left((\gamma_{ij}^{\lambda})^k - (\gamma_j^{\lambda})^-\right)^2} \tag{5.20}$$

The relative closeness indices $CI_i (i \in \{1, 2, \ldots, m\})$ of the alternatives $A_i (i \in \{1, 2, \ldots, m\})$ to the hesitant fuzzy PIS A^+ are calculated by using the following formula:

$$CI(A_i) = \frac{D_i^-}{D_i^+ + D_i^-} \tag{5.21}$$

It is easily seen that $0 \leq CI(A_i) \leq 1 (i \in \{1, 2, \ldots, m\})$.

In general, the bigger the relative closeness index $CI\ (A_i)$ is, and the better the alternative A_i is. The alternative A_i is closer to the hesitant fuzzy PIS A^+ and farther from the hesitant fuzzy NIS A^- as the CI_i approaches 1. If $CI(A_i) = 0$, then it means that the alternative A_i is farthest from the hesitant fuzzy PIS A^+ and is closest to the hesitant fuzzy NIS A^-. Conversely, if $CI(A_i) = 1$, then it means that the alternative A_i is closest to the hesitant fuzzy PIS A^+ and is farthest from the hesitant fuzzy NIS A^-. Thus, the alternative with the maximal relative closeness index is the best alternative, namely,

$$A^* = \left\{A_i : \left\langle i : CI(A_i) = \max_{1 \leq l \leq m} CI(A_l)\right\rangle\right\} \tag{5.22}$$

Based on the above models and analysis, an algorithm (Algorithm 5.1) is presented for solving the hesitant fuzzy MCGDM problems where the weights of criteria are completely known but the weights of experts are partially known. The Algorithm 5.1 involves the following steps:

Step 1. For a MCGDM problem with HFEs, we construct the decision matrices $\Re^k = (h_{ij}^k)_{m\times n} (k = 1, 2, \ldots, g)$, where the arguments $h_{ij}^k (i \in \{1, 2, \ldots, m\}, j \in \{1, 2, \ldots, n\}, k \in \{1, 2, \ldots, g\})$ are HFEs and indicate the ratings of the alternative $A_i \in \boldsymbol{A}$ with respect to the criterion $C_j \in \boldsymbol{C}$ given by the expert $e_k \in \boldsymbol{E}$.

Step 2. Employ Eqs. (5.8) and (5.14) to calculate the *GOC* and *GCC*, respectively.

Step 3. Determine the optimal weight vector $\boldsymbol{\varpi}^* = (\varpi_1^*, \varpi_2^*, \ldots, \varpi_g^*)^T$ of the experts by employing the model (MOD-5.3).

Step 4. Identify the hesitant fuzzy PIS A^+ and the hesitant fuzzy NIS A^- by employing Eqs. (5.15) and (5.16), respectively.

Step 5. Utilize Eqs. (5.17) and (5.18) to calculate the distances for each expert between each of the alternatives $A_i (i \in \{1, 2, \ldots, m\})$ and the hesitant fuzzy PIS A^+ as well as the hesitant fuzzy NIS A^-, respectively.

Step 6. Use Eqs. (5.19) and (5.20) to aggregate the distances for all experts between each of the alternatives $A_i (i \in \{1, 2, \ldots, m\})$ and the hesitant fuzzy PIS A^+ as well as the hesitant fuzzy NIS A^-, respectively.

Step 7. Calculate the relative closeness indices $CI(A_i) (i \in \{1, 2, \ldots, m\})$ of the alternatives $A_i (i \in \{1, 2, \ldots, m\})$ to the hesitant fuzzy PIS A^+ by using Eq. (5.21).

Step 8. Rank the alternatives and select the best alternative by using Eq. (5.22).

5.3 Illustrative Example

In this section, we consider an illustrative example modified from Liao and Kao (2011), Zhang and Xu (2015) to demonstrate the implementation process of the proposed method.

The company Formosa Watch Co., Ltd. is a large, well-known manufacturer that sells watches in its own chain stores in Asia. For developing new products, its board of directors wishes to select material suppliers to purchase key components in order to achieve the competitive advantage in the market. The decision committee including four experts $\{e_1, e_2, e_3, e_4\}$ has been formed to select a supplier from five qualified suppliers $\{A_1, A_2, A_3, A_4, A_5\}$ according to the following four criteria: (1) C_1 is the quality of product; (2) C_2 is the relationship closeness; (3) C_3 is the delivery capabilities; (4) C_4 is the experience time. The weighted vector of the criteria is given as $\boldsymbol{w} = (w_1, w_2, w_3, w_4)^T = (0.1, 0.3, 0.1, 0.5)^T$. The information about the experts' weights is also given as:

$$\Gamma = \begin{cases} \varpi_2 \leq 0.26, \varpi_2 + \varpi_3 \geq 0.45, \varpi_3 \geq 0.9\varpi_2 \\ \varpi_2 + \varpi_4 \leq 0.52, \varpi_1 + \varpi_2 + \varpi_3 + \varpi_4 = 1 \\ \varpi_k \geq 0, k = 1, 2, 3, 4 \end{cases}$$

The experts $e_k (k = 1,2,3,4)$ represent, respectively, the ratings of five possible alternatives $A_i (i = 1,2,3,4,5)$ by HFEs $h_{ij}^k (i = 1,2,3,4,5, j = 1,2,3,4)$ with respect to the criteria $C_j (j = 1,2,3,4)$, which are listed in Tables 5.1, 5.2, 5.3 and 5.4 (i.e., hesitant fuzzy decision matrices $\Re^k = (h_{ij}^k)_{5\times4} (k = 1,2,3,4)$).

In what follows, we employ the proposed method to help the company select the optimal supplier. Firstly, according to the score function-based ranking method of HFEs, it is easy to see that

Table 5.1 Hesitant fuzzy decision matrix $\Re^1$

	C_1	C_2	C_3	C_4
A_1	H{0.3, 0.4, 0.6}	H{0.4, 0.6, 0.7}	H{0.3, 0.4, 0.6}	H{0.4, 0.6}
A_2	H{0.5, 0.6, 0.7}	H{0.8}	H{0.5, 0.6}	H{0.5, 0.6}
A_3	H{0.4, 0.6}	H{0.8, 0.9}	H{0.75}	H{0.3, 0.4}
A_4	H{0.55}	H{0.6, 0.7, 0.8}	H{0.6, 0.7}	H{0.4, 0.7, 0.8}
A_5	H{0.2, 0.3, 0.4}	H{0.5, 0.6}	H{0.5, 0.7}	H{0.6, 0.7}

Table 5.2 Hesitant fuzzy decision matrix $\Re^2$

	C_1	C_2	C_3	C_4
A_1	H{0.2, 0.4, 0.5}	H{0.4, 0.6, 0.7}	H{0.4, 0.6}	H{0.7, 0.9}
A_2	H{0.5, 0.7}	H{0.7}	H{0.4, 0.5, 0.7}	H{0.5, 0.6, 0.7}
A_3	H{0.65}	H{0.6, 0.9}	H{0.75}	H{0.4, 0.5}
A_4	H{0.6, 0.8}	H{0.3, 0.7, 0.8}	H{0.6, 0.7, 0.8}	H{0.6, 0.7, 0.8}
A_5	H{0.2, 0.3, 0.5}	H{0.6, 0.7}	H{0.5, 0.7}	H{0.65}

Table 5.3 Hesitant fuzzy decision matrix $\Re^3$

	C_1	C_2	C_3	C_4
A_1	H{0.3, 0.4, 0.7}	H{0.6, 0.7}	H{0.6}	H{0.4, 0.6}
A_2	H{0.5}	H{0.6, 0.8}	H{0.5, 0.6}	H{0.5, 0.6, 0.8}
A_3	H{0.5, 0.6}	H{0.5, 0.8, 0.9}	H{0.7, 0.8}	H{0.3, 0.5}
A_4	H{0.5, 0.8}	H{0.4, 0.5}	H{0.6, 0.7}	H{0.55}
A_5	H{0.5, 0.7}	H{0.6}	H{0.5, 0.7, 0.8}	H{0.6, 0.7}

Table 5.4 Hesitant fuzzy decision matrix $\Re^4$

	C_1	C_2	C_3	C_4
A_1	H{0.4}	H{0.5, 0.6, 0.7}	H{0.2, 0.3, 0.4}	H{0.5, 0.6}
A_2	H{0.5, 0.6}	H{0.7, 0.8}	H{0.3, 0.6}	H{0.5, 0.7, 0.8}
A_3	H{0.4, 0.5, 0.6}	H{0.6, 0.9}	H{0.6, 0.7, 0.8}	H{0.5}
A_4	H{0.5, 0.7}	H{0.4, 0.7, 0.8}	H{0.75}	H{0.6, 0.7, 0.8}
A_5	H{0.4, 0.5}	H{0.7}	H{0.6, 0.7}	H{0.6, 0.7}

$$s(h^1_{51}) = 0.3 < s(h^1_{11}) = 0.43 < s(h^1_{31}) = 0.5 < s(h^1_{41}) = 0.55 < s(h^1_{21}) = 0.6.$$

Thus, the dominance classes of the alternatives $A_i (i = 1, 2, \ldots, m)$ with respect to the criterion C_1 for the expert e_1 are obtained by using Eq. (5.2) as follows:

$$[A_1]^{e_1 \geq}_{C_1} = \{A_1, A_5\}, \quad [A_2]^{e_1 \geq}_{C_1} = \{A_1, A_2, A_3, A_4, A_5\}, \quad [A_3]^{e_1 \geq}_{C_1} = \{A_1, A_3, A_5\},$$
$$[A_4]^{e_1 \geq}_{C_1} = \{A_1, A_3, A_4, A_5\}, \quad [A_5]^{e_1 \geq}_{C_1} = \{A_5\}.$$

Then, the dominance class of the set of alternatives with respect to the criterion C_1 for the expert e_1 is obtained by Eq. (5.3) as:

$$\begin{aligned} \boldsymbol{A}/\left(R^{e_1}_{C_1}\right)^{\geq} &= \left\{[A_1]^{e_1 \geq}_{C_1}, [A_2]^{e_1 \geq}_{C_1}, [A_3]^{e_1 \geq}_{C_1}, [A_4]^{e_1 \geq}_{C_1}, [A_5]^{e_1 \geq}_{C_1}\right\} \\ &= \{\{A_1, A_5\}, \{A_1, A_2, A_3, A_4, A_5\}, \{A_1, A_3, A_5\}, \{A_1, A_3, A_4, A_5\}, \{A_5\}\}. \end{aligned}$$

Analogously, the dominance class of the set of alternatives with respect to the criterion C_1 for the expert e_2 is obtained by Eq. (5.3) as:

$$\begin{aligned} \boldsymbol{A}/\left(R^{e_2}_{C_1}\right)^{\geq} &= \left\{[A_1]^{e_2 \geq}_{C_1}, [A_2]^{e_2 \geq}_{C_1}, [A_3]^{e_2 \geq}_{C_1}, [A_4]^{e_2 \geq}_{C_1}, [A_5]^{e_2 \geq}_{C_1}\right\} \\ &= \{\{A_1, A_5\}, \{A_1, A_2, A_5\}, \{A_1, A_2, A_3, A_5\}, \{A_1, A_2, A_3, A_4, A_5\}, \{A_5\}\}. \end{aligned}$$

Hence, the ordinal consensus index with respect to the criterion C_1 between the expert e_1 and the expert e_2 is obtained by Eq. (5.4) as follows:

$$OC^1_2(C_1) = \frac{1}{5}\sum_{i=1}^{m} \frac{\left|[A_i]^{e_1 \geq}_{C_1} \cap [A_i]^{e_2 \geq}_{C_1}\right|}{\left|[A_i]^{e_1 \geq}_{C_1} \cup [A_i]^{e_2 \geq}_{C_1}\right|} = \frac{1}{5}\left(\frac{2}{2} + \frac{3}{5} + \frac{3}{4} + \frac{4}{5} + \frac{1}{1}\right) = 0.83.$$

Likewise, by Eqs. (5.1)–(5.4) it is easy to obtain the following calculation results in Table 5.5.

Table 5.5 The ordinal consensus indices for all criteria and all alternatives

$OC^l_k(\cdot)$	Criteria				$\sum_{j=1}^{n} w_j OC^l_k(C_j)$
	C_1	C_2	C_3	C_4	
$OC^1_2(\cdot)$	0.83	0.7	1.0	0.733	0.7595
$OC^1_3(\cdot)$	0.59	0.7	0.7	0.9	0.7890
$OC^1_4(\cdot)$	0.72	0.563	0.92	0.83	0.7479
$OC^2_3(\cdot)$	0.6	0.7	0.7	0.65	0.6650
$OC^2_4(\cdot)$	0.7	0.83	0.92	0.74	0.7810
$OC^3_4(\cdot)$	0.8	0.547	0.61	0.84	0.7251

Furthermore, from the data in Table 5.5 we can calculate the normalized ordinal consensus indices for the experts $e_k(k = 1, 2, 3, 4)$ by Eqs. (5.6) and (5.7) as follows:

$$O\bar{C}_1 = 0.2570, \quad O\bar{C}_2 = 0.2468, \quad O\bar{C}_3 = 0.2439, \quad O\bar{C}_4 = 0.2523.$$

Thus, the group ordinal consensus index is obtained by Eq. (5.8) as follows:

$$GOC = 0.257\varpi_1 + 0.2468\varpi_2 + 0.2439\varpi_3 + 0.2523\varpi_4.$$

On the other hand, we need to calculate the consensus degree among individual experts' opinions from the perspective of the magnitudes of decision data. We here employ the hesitant fuzzy similarity measure to calculate the consensus degree. First, we normalize the decision matrices by Definition 1.2 based on the assumption that the experts are risk-averse. The normalized decision matrices are displayed in Tables 5.6, 5.7, 5.8 and 5.9.

By using Eqs. (5.9)–(5.11), it is easy to obtain the calculation results in Table 5.10. Furthermore, the normalized cardinal consensus indices for the experts $e_k(k = 1, 2, 3, 4)$ are obtained by Eqs. (5.12)–(5.13) as:

Table 5.6 Hesitant fuzzy normalized decision matrix $\bar{\mathfrak{R}}^1$

	C_1	C_2	C_3	C_4
A_1	H{0.3, 0.4, 0.6}	H{0.4, 0.6, 0.7}	H{0.3, 0.4, 0.6}	H{0.4, 0.4, 0.6}
A_2	H{0.5, 0.6, 0.7}	H{0.8, 0.8, 0.8}	H{0.5, 0.5, 0.6}	H{0.5, 0.5, 0.6}
A_3	H{0.4, 0.4, 0.6}	H{0.8, 0.8, 0.9}	H{0.75, 0.75, 0.75}	H{0.3, 0.3, 0.4}
A_4	H{0.55, 0.55, 0.55}	H{0.6, 0.7, 0.8}	H{0.6, 0.6, 0.7}	H{0.4, 0.7, 0.8}
A_5	H{0.2, 0.3, 0.4}	H{0.5, 0.5, 0.6}	H{0.5, 0.5, 0.7}	H{0.6, 0.6, 0.7}

Table 5.7 Hesitant fuzzy normalized decision matrix $\bar{\mathfrak{R}}^2$

	C_1	C_2	C_3	C_4
A_1	H{0.2, 0.4, 0.5}	H{0.4, 0.6, 0.7}	H{0.4, 0.4, 0.6}	H{0.7, 0.7, 0.9}
A_2	H{0.5, 0.5, 0.7}	H{0.7, 0.7, 0.7}	H{0.4, 0.5, 0.7}	H{0.5, 0.6, 0.7}
A_3	H{0.65, 0.65, 0.65}	H{0.6, 0.6, 0.9}	H{0.75, 0.75, 0.75}	H{0.4, 0.4, 0.5}
A_4	H{0.6, 0.6, 0.8}	H{0.3, 0.7, 0.8}	H{0.6, 0.7, 0.8}	H{0.6, 0.7, 0.8}
A_5	H{0.2, 0.3, 0.5}	H{0.6, 0.6, 0.7}	H{0.5, 0.5, 0.7}	H{0.65, 0.65, 0.65}

Table 5.8 Hesitant fuzzy normalized decision matrix $\bar{\mathfrak{R}}^3$

	C_1	C_2	C_3	C_4
A_1	H{0.3, 0.4, 0.7}	H{0.6, 0.6, 0.7}	H{0.6, 0.6, 0.6}	H{0.4, 0.4, 0.6}
A_2	H{0.5, 0.5, 0.5}	H{0.6, 0.6, 0.8}	H{0.5, 0.5, 0.6}	H{0.5, 0.6, 0.8}
A_3	H{0.5, 0.5, 0.6}	H{0.5, 0.8, 0.9}	H{0.7, 0.7, 0.8}	H{0.3, 0.3, 0.5}
A_4	H{0.5, 0.5, 0.8}	H{0.4, 0.4, 0.5}	H{0.6, 0.6, 0.7}	H{0.55, 0.55, 0.55}
A_5	H{0.5, 0.5, 0.7}	H{0.6, 0.6, 0.6}	H{0.5, 0.7, 0.8}	H{0.6, 0.6, 0.7}

Table 5.9 Hesitant fuzzy normalized decision matrix $\Re^4$

	C_1	C_2	C_3	C_4
A_1	H{0.4, 0.4, 0.4}	H{0.5, 0.6, 0.7}	H{0.2, 0.3, 0.4}	H{0.5, 0.5, 0.6}
A_2	H{0.5, 0.5, 0.6}	H{0.7, 0.7, 0.8}	H{0.3, 0.3, 0.6}	H{0.5, 0.7, 0.8}
A_3	H{0.4, 0.5, 0.6}	H{0.6, 0.6, 0.9}	H{0.6, 0.7, 0.8}	H{0.5, 0.5, 0.5}
A_4	H{0.5, 0.5, 0.7}	H{0.4, 0.7, 0.8}	H{0.8, 0.8, 0.8}	H{0.6, 0.7, 0.8}
A_5	H{0.4, 0.4, 0.5}	H{0.7, 0.7, 0.7}	H{0.6, 0.6, 0.7}	H{0.6, 0.6, 0.7}

Table 5.10 The cardinal consensus indices for all criteria and all alternatives

CC_k^l	Alternatives					
	A_1	A_2	A_3	A_4	A_5	$\frac{1}{5}\sum_{i=1}^{5} CC_k^l(A_i)$
$CC_2^1(\cdot)$	0.8361	0.9152	0.8804	0.8671	0.9392	0.8876
$CC_3^1(\cdot)$	0.9388	0.8736	0.9060	0.8091	0.9355	0.8926
$CC_4^1(\cdot)$	0.9148	0.8694	0.8491	0.8855	0.9257	0.8889
$CC_3^2(\cdot)$	0.7861	0.9214	0.9029	0.8228	0.9210	0.8708
$CC_4^2(\cdot)$	0.8334	0.9219	0.9325	0.9631	0.9294	0.9161
$CC_4^3(\cdot)$	0.8925	0.9245	0.8681	0.8228	0.9459	0.8907

$$\bar{CC}_1 = 0.2496, \quad \bar{CC}_2 = 0.2501, \quad \bar{CC}_3 = 0.2482, \quad \bar{CC}_4 = 0.2521.$$

Thus, the group cardinal consensus index is obtained by Eq. (5.14) as:

$$GCC = 0.2496\varpi_1 + 0.2501\varpi_2 + 0.2482\varpi_3 + 0.2521\varpi_4.$$

According to the model (MOD-5.3), the following optimal model is established to determine the weights of experts:

$$\max\left\{\begin{array}{c} \theta(0.257\varpi_1 + 0.2468\varpi_2 + 0.2439\varpi_3 + 0.2523\varpi_4) \\ +(1-\theta)(0.2496\varpi_1 + 0.2501\varpi_2 + 0.2482\varpi_3 + 0.2521\varpi_4) \end{array}\right\}$$

$$s.t.\begin{cases} \varpi_2 \leq 0.26 \\ \varpi_2 + \varpi_3 \geq 0.45 \\ \varpi_3 \geq 0.9\varpi_2 \\ \varpi_2 + \varpi_4 \leq 0.52 \\ \varpi_1 + \varpi_2 + \varpi_3 + \varpi_4 = 1 \\ \varpi_k \geq 0, k = 1, 2, 3, 4 \end{cases} \qquad \text{(MOD-5.4)}$$

Let $\theta = 0.5$ and by solving the model (MOD-5.4), the optimal weight vector of the experts is obtained as:

$$\boldsymbol{\varpi} = (\varpi_1, \varpi_2, \varpi_3, \varpi_4)^T = (0.26, 0.22, 0.198, 0.322)^T$$

Afterwards, we employ Eqs. (5.15) and (5.16) to identify the hesitant fuzzy PIS A^+ and the hesitant fuzzy NIS A^-, respectively:

$$A^+ = \begin{pmatrix} H\{0.5,0.5,0.8\}, & H\{0.8,0.8,0.9\}, \\ H\{0.8,0.8,0.8\}, & H\{0.7,0.7,0.9\} \end{pmatrix},$$
$$A^- = \begin{pmatrix} H\{0.2,0.3,0.4\}, & H\{0.4,0.4,0.4\}, \\ H\{0.2,0.3,0.4\}, & H\{0.3,0.3,0.4\} \end{pmatrix}.$$

Furthermore, the distances for each expert between each of the alternatives $A_i (i \in \{1, 2, \ldots, m\})$ and the hesitant fuzzy PIS A^+ as well as the hesitant fuzzy NIS A^-, respectively, are calculated by Eqs. (5.17) and (5.18). The calculation results are displayed in Table 5.11.

By Eqs. (5.19) and (5.20), the distances for all experts between each of the alternatives $A_i (i \in \{1, 2, \ldots, 5\})$ and the hesitant fuzzy PIS A^+ as well as the hesitant fuzzy NIS A^-, respectively, are obtained as:

Table 5.11 Distances for each expert between alternatives and A^+ as well as A^-

Experts	Alternatives	D_i^{k+}	D_i^{k-}
e_1	A_1	0.2909	0.1480
	A_2	0.1716	0.2647
	A_3	0.2371	0.1831
	A_4	0.1660	0.3070
	A_5	0.2170	0.2071
e_2	A_1	0.1447	0.2924
	A_2	0.1642	0.2675
	A_3	0.2373	0.2165
	A_4	0.1519	0.3373
	A_5	0.1872	0.2529
e_3	A_1	0.2441	0.1810
	A_2	0.1671	0.2723
	A_3	0.2717	0.1958
	A_4	0.2555	0.1756
	A_5	0.1662	0.2660
e_4	A_1	0.2656	0.1649
	A_2	0.1485	0.2999
	A_3	0.2162	0.2315
	A_4	0.1251	0.3311
	A_5	0.1496	0.2790

$$D_1^+ = 0.2413,\quad D_2^+ = 0.1617,\quad D_3^+ = 0.2373,\quad D_4^+ = 0.1674,\quad D_5^+ = 0.1787,$$
$$D_1^- = 0.1917,\quad D_2^- = 0.2782,\quad D_3^- = 0.2085,\quad D_4^- = 0.2954,\quad D_5^- = 0.2520.$$

By Eq. (5.21), the relative closeness indices $CI(A_i)(i \in \{1,2,\ldots,5\})$ of the alternatives $A_i(i \in \{1,2,\ldots,5\})$ to the hesitant fuzzy PIS A^+ are calculated as:

$$CI(A_1) = 0.4427,\quad CI(A_2) = 0.6325,\quad CI(A_3) = 0.4678,$$
$$CI(A_4) = 0.6383,\quad CI(A_5) = 0.5851.$$

Obviously, the ranking of the alternatives based on their relative closeness indices is obtained as:

$$A_4 \succ A_2 \succ A_5 \succ A_3 \succ A_1$$

Thus, the optimal alternative with the maximal relative closeness index is A_4.

5.4 Conclusions

In group decision making, the experts may have vague knowledge about the preference degree of one alternative with respect to a criterion and cannot estimate their preferences with exact numerical values. It is more suitable to provide their preferences by means of HFEs when they have hesitation in the decision making process. In this chapter, we have developed a consensus model-based hesitant fuzzy group decision making method for solving the MCGDM problems in which the criteria values take the form of HFEs, and the weights of criteria are known in advance but the weights of experts are partially known or unknown. The proposed method first defines the consensus index from the perspectives of the ranking and the magnitude of decision information and derives the experts' weights on the basis of the idea that the consensus indices among the individual experts' opinions should be maximized, then utilizes the extended TOPSIS to rank all alternatives. The prominent characteristic of the developed approach is that it cannot only take into account the experts' weights and reduce the influence of unjust arguments on the decision results, but also take full advantage of the decision information from both the perspective of ranking and the angle of the sizes of values. In addition, the developed approach can be extended to other MCGDM problems where the decision information is in other forms, such as interval numbers, linguistic variables, and can be applied to many other practical fields.

References

Liao, C. N., & Kao, H. P. (2011). An integrated fuzzy TOPSIS and MCGP approach to supplier selection in supply chain management. *Expert Systems with Applications, 38*, 10803–10811.

Pang, J., & Liang, J. (2012). Evaluation of the results of multi-attribute group decision-making with linguistic information. *Omega, 40*, 294–301.

Xu, Z. S., & Zhang, X. L. (2013). Hesitant fuzzy multi-attribute decision making based on TOPSIS with incomplete weight information. *Knowledge-Based Systems, 52*, 53–64.

Zhang, X. L., & Xu, Z. S. (2014). Deriving experts' weights based on consistency maximization in intuitionistic fuzzy group decision making. *Journal of Intelligent and Fuzzy Systems, 27*, 221–233.

Zhang, X. L., & Xu, Z. S. (2015). Soft computing based on maximizing consensus and fuzzy TOPSIS approach to interval-valued intuitionistic fuzzy group decision making. *Applied Soft Computing, 26*, 42–56.

References

Chapter 6
Heterogeneous Multiple Criteria Group Decision Analysis

Abstract The purpose of this chapter is to develop the deviation modeling method to deal with the heterogeneous MCGDM problems with incomplete weight information in which the decision information is expressed as real numbers, interval numbers, linguistic variables, IFNs, HFEs and HFLTSs. The most characteristic of the deviation modeling method is that it does not unify the heterogeneous information but directly calculates the distances to the PIS and NIS. Compared with Li et al. (2010b)'s method, the advantages of the developed method are that (1) it can accommodate more complicated decision data, including IFNs, HFEs and HFLTSs; (2) it utilizes the maximization deviation model to determine objectively the weights of the criteria for each expert, which avoids the subjective randomness of selecting the weights; (3) it can consider fully the consistency between the opinions of the individual experts and the group, thus the final decision results derived by it are more persuasive.

The MCGDM problems with multiple formats of decision data are usually called heterogeneous MCGDM problems which have broad applications in the fields of natural science, social science, economy and management, etc. Owing to the multiformity of criteria values, it is difficult for calculation in this sort of heterogeneous MCGDM problems. To this end, many useful and valuable methods have been proposed to solve such a type of heterogeneous decision making problems, which can be roughly divided into the following two categories (Zhang et al. 2015). The first type of methods to deal with heterogeneous decision making problems is the methods based on the distances to the ideal solution. By investigating a heterogeneous MCGDM problem in which the decision data take the forms of linguistic variables, TFNs, interval numbers and crisp numbers, Li et al. (2010) developed a systematical approach in which it computes the distances to the ideal solution and the alternatives are ranked according to the proposed relative closeness degrees. By unifying the heterogeneous information into TFNs, Zhang and Lu (2003) developed a closeness coefficient-based method in which it calculates the distances between the alternatives and both the positive and negative ideal solutions and a closeness coefficient is obtained to rank the alternatives. Ma et al. (2010) developed a fuzzy MCGDM process model based on distances to the ideal solution to deal with the

X. Zhang and Z. Xu, *Hesitant Fuzzy Methods for Multiple Criteria Decision Analysis*, Studies in Fuzziness and Soft Computing 345,
DOI 10.1007/978-3-319-42001-1_6

heterogeneous information including real numbers, Boolean values and linguistic variables. Wan and Li (2014) presented an intuitionistic fuzzy programming method based on the distances to the ideal solution for solving the heterogeneous MCGDM problems with intuitionistic fuzzy truth degree whose decision data are expressed in real numbers, interval numbers, TrFNs and IFNs. The second type of methods to deal with the heterogeneous decision making problems is the methods by unifying heterogeneous information into 2-tuple linguistic variables. For example, Herrera et al. (2005) proposed a method that converts all the heterogeneous information into the 2-tuple linguistic information for solving the heterogeneous group decision making problems whose decision information is expressed as real numbers, interval numbers, linguistic variables. To handle the heterogeneous information (i.e., real numbers, interval numbers and linguistic variables) in engineering evaluation processes, Martínez et al. (2007) defined different fuzzy functions which can transform the heterogeneous information into 2-tuple linguistic variable.

In the practical heterogeneous MCGDM process, the weights of both experts and criteria are partially known or completely unknown, but the aforementioned methods fail to solve this issue. Moreover, with the increasing complexity of the real-world decision making problem, the experts often hesitate among several values to provide their assessments in the evaluation process. For instance, when the criteria are quite quantitative because of their nature, HFEs are usually used for the experts to express their preferences provoked by hesitation. When the decision criteria are quite qualitative, the HFLTSs (Rodríguez et al. 2012) are employed for the experts to represent their assessments. Considering the fact that the hesitant situations are very common in the real-world decision problems, Zhang et al. (2015) developed a deviation modeling method to deal with the heterogeneous hesitant fuzzy MCGDM problems with incomplete weight information in which the decision data are expressed as real numbers, interval numbers, linguistic variables, IFNs, HFEs and HFLTSs. There are three key issues being addressed in this approach, the first one is to construct a maximizing deviation optimal model in order to determine the optimal weights of criteria for each expert. Borrowing the idea of TOPSIS, the second one is to calculate the relative closeness indices of the alternatives for each expert. The third one is to establish a minimizing deviation optimal model based on the idea that the opinion of the individual expert should be consistent with that of the group to the greatest extent, which is used to determine the weights of experts and identify the optimal alternative.

6.1 Heterogeneous Type of Decision Information

In this chapter, heterogeneous information mainly includes crisp numbers, interval numbers, linguistic variables, IFNs, HFEs and HFLTSs. Because the concepts of interval numbers, IFNs and HFEs have been introduced in the previous chapters, we now review the main concepts and the distance measures of linguistic variables and HFLTSs, respectively.

6.1.1 Linguistic Variables and Fuzzy Numbers

The linguistic variables proposed by Zadeh (1975) are established by the linguistic descriptors and their semantics, which are often used to express the experts' preferences in qualitative situations. The linguistic term set based on the ordered structure approach (Yager 1995) can be defined as follows:

Definition 6.1 (Yager 1995). Let $L = \{l_0, l_1, \ldots, l_q\}$ be a finite and totally ordered discrete label set. Any label, $l_i \in L$, represents a possible value for a linguistic variable, and it must have the following characteristics: (1) the set L is ordered, i.e., $l_i \geq l_j$, if $i \geq j$; (2) there exists a negation operator: $Neg(l_i) = l_j$, such that $j = q - i$; (3) there exist a maximization operator and a minimization operator: $Max(l_i, l_j) = l_i$ and $Min(l_i, l_j) = l_j$ if $i \geq j$.

For example, a linguistic term set L with seven-point rating scales could be:

$$L = \left\{ \begin{array}{l} l_0 : \textit{definitely low},\ l_1 : \textit{very low},\ l_2 : \textit{low},\ l_3 : \textit{medium}, \\ l_4 : \textit{high},\ l_5 : \textit{very high},\ l_6 : \textit{definitely high} \end{array} \right\} \tag{6.1}$$

Due to the fact that the linguistic terms provided by the experts are the approximate assessments, several authors (Cheng et al. 2003; Delgado et al. 1998) advised the decision maker to employ the TrFNs to capture and represent the uncertainty and vagueness of such linguistic assessments. The concept of TrFNs is introduced in Chap. 2. The Hamming distance measure for TrFNs is introduced as follows (Chen 2000):

$$d(\tilde{\alpha}_1, \tilde{\alpha}_2) = \frac{1}{4}(|a_1 - a_2| + |b_1 - b_2| + |c_1 - c_2| + |d_1 - d_2|) \tag{6.2}$$

A special case of TrFN $\tilde{\alpha} = Tr(a, b, c, d)$ is the TFN $\tilde{\alpha} = T(a, b, d)$ if $b = c$. For convenience, we denote a TFN as $\tilde{\beta} = T(a, b, c)$. The linguistic term sets with seven-point rating scales and their semantics captured by TFNs are described in Fig. 6.1.

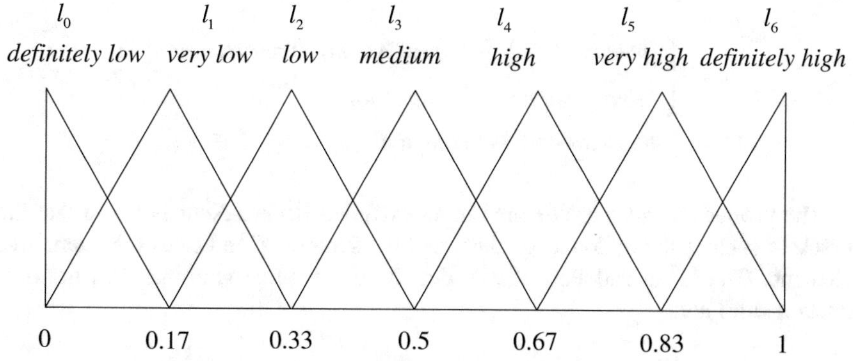

Fig. 6.1 The linguistic term sets with seven-point rating scales and their semantics

The Hamming distance between TFNs can be given as follows (Chen 2000):

$$d(\tilde{\beta}_1, \tilde{\beta}_2) = \frac{1}{3}(|a_1 - a_2| + |b_1 - b_2| + |c_1 - c_2|) \tag{6.3}$$

6.1.2 Hesitant Fuzzy Linguistic Term Sets

In order to capture the hesitancy of the decision maker under qualitative situations, Rodríguez et al. (2012) proposed the concept of HFLTS. Its definition is presented as below:

Definition 6.2 (Rodriguez et al. 2012). Let $L = \{l_0, l_1, \ldots, l_q\}$ be a linguistic term set, a HFLTS hl is an ordered finite subset of consecutive linguistic terms of L. Generally, the HFLTS hl is denoted by $hl = HL\{l_i, l_{i+1}, \ldots, l_j\}$ where $l_k \in L$ $(k = i, i+1, \ldots, j)$.

Following the previous example, two different HFLTSs hl_1 and hl_2 might be given as:

$$hl_1 = HL\{l_0 : \textit{definitely low}, l_1 : \textit{very low}, l_2 : \textit{low}\},$$
$$hl_2 = HL\{l_1 : \textit{very low}, l_2 : \textit{low}, l_3 : \textit{medium}\}.$$

A special case of the HFLTS hl is the linguistic variable if the HFLTS hl only contains a single linguistic term. Because the HFLTSs are not similar to human beings' way of thinking and reasoning, Rodríguez et al. (2012) proposed the concept of comparative linguistic expressions that are close to human beings' cognitive processes to be semantically represented by HFLTSs and generated by context-free grammars. The definitions of comparative linguistic expressions and context-free grammars are reviewed as:

Definition 6.3 (Rodriguez et al. 2012). Let G_H be a context-free grammar and $L = \{l_0, l_1, \ldots, l_q\}$ be a linguistic term set. The elements of $G_H = (V_N, V_T, I, D)$ are defined as:

$$V_N = \left\{ \begin{array}{l} \langle \textit{primary term} \rangle, \langle \textit{composite term} \rangle, \langle \textit{unary relation} \rangle, \\ \langle \textit{binary relation} \rangle, \langle \textit{conjunction} \rangle \end{array} \right\};$$
$$V_T = \{\textit{atmost}, \textit{atleast}, \textit{between}, \textit{and}, l_0, \ldots, l_g\}; I \in V_N.$$

The production rules are defined in an extended Backus-Naur Form so that the brackets enclose the optional elements and the symbol "|" indicates the alternative elements (Bordogna and Pasi 1993). For the context-free grammar G_H, the production rules are:

$$D = \{I ::= \langle primary\ term \rangle | \langle composite\ term \rangle; \langle primary\ term \rangle ::= l_0 | l_1 | \ldots | l_g;$$
$$\langle unary\ relation \rangle ::= atleast | atmost; \langle binary\ relation \rangle ::= between;$$
$$\langle conjunction \rangle ::= and; \langle composite\ term \rangle ::$$
$$= \langle unary\ relation \rangle \langle primary\ term \rangle | \langle binary\ relation \rangle \langle primary\ term \rangle \langle conjunction \rangle$$
$$\langle primary\ term \rangle \}.$$

For example, let $L = \{l_0, l_1, l_2, l_3, l_4, l_5, l_6\}$ be a linguistic term set, two comparative linguistic expressions obtained by the context-free grammar G_{HL} might be represented as:

$$ll_1 = \{atmost\, l_2\}, ll_2 = \{between\, l_1\, and\, l_3\}$$

Then, Rodríguez et al. (2012) presented a transformation function $E_{G_{HL}}$ to transform the comparative linguistic expressions generated by G_{HL} into HFLTSs as:

$$E_{G_{HL}}(l_i) = HL\{l_i | l_i \in L\}, E_{G_{HL}}(atmost\, l_i) = HL\{l_j | l_j \leq l_i\, and\, l_i, l_j \in L\},$$
$$E_{G_{HL}}(atleast\, l_i) = HL\{l_j | l_j \geq l_i\, and\, l_i, l_j \in L\},$$
$$E_{G_{HL}}(between\, l_i\, and\, l_j) = HL\{l_k | l_i \leq l_k \leq l_j\, and\, l_i, l_k, l_j \in L\}.$$

Following the previous example, two HFLTSs hl_1 and hl_2 can be obtained by using the transformation function E_{G_H} as:

$$hl_1 = E_{G_{HL}}(ll_1) = HL\{l_0, l_1, l_2\}, \quad hl_2 = E_{G_{HL}}(ll_2) = HL\{l_1, l_2, l_3\}.$$

For further applications of HFLTSs in decision making, different models based on the different envelopes of HFLTSs, such as the envelope of interval linguistic variables (Rodriguez et al. 2012, 2013) and the envelope of TrFNs (Liu and Rodríguez 2014), are proposed to handle the decision making problems with HFLTSs. In this chapter, we employ the fuzzy envelope of HFLTSs proposed by Liu and Rodríguez (2014), i.e., the TrFNs, to carry out the computing with words processes. The general process to derive this fuzzy envelope of HFLTSs is reviewed as follows:

Let $hl = HL\{l_i, l_{i+1}, \ldots, l_j\}$ be a HFLTS, where $l_k \in L\ (k = i, i+1, \ldots, j)$. Due to the fact that all the linguistic terms $l_k \in L$ can be usually captured by TFNs $\tilde{\beta}_k = T(a_k, b_k, c_k)$, it is logical to aggregate the set of points of TFNs of linguistic terms in the HFLTS to obtain the fuzzy envelope for the HFLTS. The set of elements is given as $T = \{a_k, b_k, b_{k+1}, \ldots, b_j, c_j\}$ for $c_{k-1} = b_k = a_{k+1}$. Thus, its fuzzy envelope $\tilde{\alpha} = Tr(a, b, c, d)$ is calculated as:

$$
\begin{aligned}
a &= \min\{a_k, b_k, b_{k+1}, \ldots, b_j, c_j\} = a_i, \\
d &= \max\{a_k, b_k, b_{k+1}, \ldots, b_j, c_j\} = c_j, \\
b &= OWA_{w^{\xi}}(b_i, b_{i+1}, \ldots, b_j), \\
c &= OWA_{w^{\zeta}}(b_i, b_{i+1}, \ldots, b_j),
\end{aligned}
$$

where the OWA weighting vectors for computing b and c being in the forms of ξ and $\zeta(\xi, \zeta = 1, 2;\ \xi = \zeta \ or \ \xi \neq \zeta)$, respectively, can be determined by the following equations with a parameter $\upsilon \in [0, 1]$:

$$
\begin{aligned}
w^1 &= \left(w_i^1, w_{i+1}^1, w_{i+2}^1 \ldots, w_{j-1}^1, w_j^1\right)^T \\
&= \left(\upsilon, \upsilon(1-\upsilon), \upsilon(1-\upsilon)^2, \ldots, \upsilon(1-\upsilon)^{j-2}, (1-\upsilon)^{j-1}\right)^T
\end{aligned}
\tag{6.4}
$$

$$
\begin{aligned}
w^2 &= \left(w_i^2, w_{i+1}^2, w_{i+2}^2 \ldots, w_{j-1}^2, w_j^2\right)^T \\
&= \left(\upsilon^{j-1}, (1-\upsilon)\upsilon^{j-2}, (1-\upsilon)\upsilon^{j-3}, \ldots, (1-\upsilon)\upsilon, (1-\upsilon)\right)^T
\end{aligned}
\tag{6.5}
$$

The detailed process to obtain the fuzzy envelope (TrFNs) for comparative linguistic expressions can be found in Liu and Rodríguez (2014).

6.2 Description of the Heterogeneous MCGDM Problems

The MCGDM is the process that a group consisting of $g(g \geq 2)$ experts has to select an optimal alternative from $m(m \geq 2)$ potential alternatives or rank all potential alternatives based on $n(n \geq 2)$ criteria. We denote the alternative set by $\boldsymbol{A} = \{A_1, A_2, \ldots, A_m\}$, the criteria set by $\boldsymbol{C} = \{C_1, C_2, \ldots, C_n\}$ and the expert set by $\boldsymbol{E} = \{e_1, e_2, \ldots, e_g\}$. Because the evaluation information of alternatives with respect to criteria provided by experts depends on the natures of criteria, in this chapter, we consider six distinct forms of evaluation information such as crisp numbers, interval numbers, linguistic variables, IFNs, HFEs and HFLTSs. In general, the criterion C_j in the criteria set $\boldsymbol{C}$ is evaluated using only one of the aforementioned six distinct information forms. Thus, the criteria set $\boldsymbol{C}$ is divided into the six subsets $\boldsymbol{C}_\xi$ ($\xi =$ 1, 2, 3, 4, 5, 6) in which the criteria values are expressed as crisp numbers, interval numbers, linguistic variables, IFNs, HFEs and HFLTSs, respectively. Let

$$
\begin{aligned}
&\boldsymbol{C}_1 = \{C_1, C_2, \ldots, C_{j_1}\}, && \boldsymbol{C}_2 = \{C_{j_1+1}, C_{j_1+2}, \ldots, C_{j_2}\}, \\
&\boldsymbol{C}_3 = \{C_{j_2+1}, C_{j_2+2}, \ldots, C_{j_3}\}, && \boldsymbol{C}_4 = \{C_{j_3+1}, C_{j_3+2}, \ldots, C_{j_4}\}, \\
&\boldsymbol{C}_5 = \{C_{j_4+1}, C_{j_4+2}, \ldots, C_{j_5}\}, && \boldsymbol{C}_6 = \{C_{j_5+1}, C_{j_5+2}, \ldots, C_n\}
\end{aligned}
$$

where $1 \leq j_1 \leq j_2 \leq j_3 \leq j_4 \leq j_5 \leq n$, $\boldsymbol{C}_\xi \bigcap \boldsymbol{C}_\zeta = \varnothing \quad (\xi, \zeta = 1, 2, \ldots, 6;\ \xi \neq \zeta)$ and $\bigcup_{\xi=1}^{6} \boldsymbol{C}_\xi = \boldsymbol{C}$.

Let the ratings of the alternative $A_i \in \boldsymbol{A}$ on the criterion $C_j \in \boldsymbol{C}$ given by the expert $e_k \in \boldsymbol{E}$ be expressed as x_{ij}^k (Zhang et al. 2015).

(1) If $j \in \{1, 2, \ldots, j_1\}$, then $x_{ij}^k = a_{ij}^k$ is expressed as a crisp number;
(2) If $j \in \{j_1+1, j_1+2, \ldots, j_2\}$, then $x_{ij}^k = [\underline{a}_{ij}^k, \bar{a}_{ij}^k]$ is expressed as an interval number;
(3) If $j \in \{j_2+1, j_2+2, \ldots, j_3\}$, then $x_{ij}^k = l_{ij}^k \in L$ is a linguistic term which can be usually captured by a TFN $l_{ij}^k = T(a_{ij}^k, b_{ij}^k, c_{ij}^k)$;
(4) If $j \in \{j_3+1, j_3+2, \ldots, j_4\}$, then $x_{ij}^k = I(\mu_{ij}^k, v_{ij}^k)$ is expressed as an IFN;
(5) If $j \in \{j_4+1, j_4+2, \ldots, j_5\}$, then $x_{ij}^k = H\{(h_{ij}^k)^1, (h_{ij}^k)^2, \ldots, (h_{ij}^k)^{\#h_{ij}^k}\}$ is expressed as a HFE;
(6) If $j \in \{j_5+1, j_5+2, \ldots, n\}$, then $x_{ij}^k = HL\{(l_{ij}^k)^1, (l_{ij}^k)^2, \ldots, (l_{ij}^k)^{\#l_{ij}^k}\}(l_{ij}^k \in L)$ is expressed as a HFLTS, we here employ its fuzzy envelope (i.e., the TrFN) instead of it to carry out the computing with words processes. That is to say, the semantic of HFLTS is captured by a TrFN $x_{ij}^k = Tr(a_{ij}^k, b_{ij}^k, c_{ij}^k, d_{ij}^k)$.

For convenience of understanding, the six cases of x_{ij}^k are rewritten as the form in Eq. (6.6) (Zhang et al. 2015).

$$x_{ij}^k = \begin{cases} a_{ij}^k, & \text{if } j \in \{1, 2, \ldots, j_1\} \\ [\underline{a}_{ij}^k, \bar{a}_{ij}^k], & \text{if } j \in \{j_1+1, j_1+2, \ldots, j_2\} \\ T(a_{ij}^k, b_{ij}^k, c_{ij}^k), & \text{if } j \in \{j_2+1, j_2+2, \ldots, j_3\} \\ I\left(\mu_{ij}^k, \upsilon_{ij}^k\right), & \text{if } j \in \{j_3+1, j_3+2, \ldots, j_4\} \\ H\{(h_{ij}^k)^1, \ldots, (h_{ij}^k)^{\#h_{ij}^k}\}, & \text{if } j \in \{j_4+1, j_4+2, \ldots, j_5\} \\ Tr(a_{ij}^k, b_{ij}^k, c_{ij}^k, d_{ij}^k), & \text{if } j \in \{j_5+1, j_5+2, \ldots, n\} \end{cases} \tag{6.6}$$

Thus, the heterogeneous MCGDM problem is expressed in the heterogeneous matrix format $\Re^k = (x_{ij}^k)_{m \times n}$ $(k \in \{1, 2, \ldots, g\})$. In the practical decision process, the criteria subset $\boldsymbol{C}_\xi$ can be further divided into two subsets $\boldsymbol{C}_\xi^I$ and $\boldsymbol{C}_\xi^{II}$, where $\boldsymbol{C}_\xi^I$ and $\boldsymbol{C}_\xi^{II}$ are the sets of benefit (the bigger the better) and cost (the smaller the better) criteria, respectively, $\boldsymbol{C}_\xi = \boldsymbol{C}_\xi^I \bigcup \boldsymbol{C}_\xi^{II}$ and $\boldsymbol{C}_\xi^I \bigcap \boldsymbol{C}_\xi^{II} = \varnothing$ $(\xi = 1, 2, \ldots, 6)$.

$$\begin{aligned} \boldsymbol{C}_1^I &= \{C_1, C_2, \ldots, C_{j_1^I}\}, & \boldsymbol{C}_1^{II} &= \{C_{j_1^I+1}, C_{j_1^I+2}, \ldots, C_{j_1}\}, \\ \boldsymbol{C}_2^I &= \{C_{j_1+1}, C_{j_1+2}, \ldots, C_{j_2^I}\}, & \boldsymbol{C}_2^{II} &= \{C_{j_2^I+1}, C_{j_2^I+2}, \ldots, C_{j_2}\}, \\ \boldsymbol{C}_3^I &= \{C_{j_2+1}, C_{j_2+2}, \ldots, C_{j_3^I}\}, & \boldsymbol{C}_3^{II} &= \{C_{j_3^I+1}, C_{j_3^I+2}, \ldots, C_{j_3}\}, \\ \boldsymbol{C}_4^I &= \{C_{j_3+1}, C_{j_3+2}, \ldots, C_{j_4^I}\}, & \boldsymbol{C}_4^{II} &= \{C_{j_4^I+1}, C_{j_4^I+2}, \ldots, C_{j_4}\}, \\ \boldsymbol{C}_5^I &= \{C_{j_4+1}, C_{j_4+2}, \ldots, C_{j_5^I}\}, & \boldsymbol{C}_5^{II} &= \{C_{j_5^I+1}, C_{j_5^I+2}, \ldots, C_{j_5}\}, \\ \boldsymbol{C}_6^I &= \{C_{j_5+1}, C_{j_5+2}, \ldots, C_{j_6^I}\}, & \boldsymbol{C}_6^{II} &= \{C_{j_6^I+1}, C_{j_6^I+2}, \ldots, C_n\}; \end{aligned} \tag{6.7}$$

where $1 \le j_1^I \le j_1 \le j_2^I \le j_2 \le j_3^I \le j_3 \le j_4^I \le j_4 \le j_5^I \le j_5 \le j_6^I \le n$.

Usually, the dimensions and measurements of criteria values are different because the natures of these criteria are different. Thus, the criteria values should be normalized in order to ensure their compatibility. According to the different natures of criteria, the criterion value x_{ij}^k can be normalized to $\bar{x}_{ij}^k$ in Eq. (6.8) (Zhang et al. 2015).

$$\bar{x}_{ij}^k = \begin{cases} a_{ij}^k / a_{\max j}^k \Leftrightarrow j = 1, 2, \ldots, j_1^l \\ 1 - (a_{ij}^k / a_{\max j}^k) \Leftrightarrow j = j_1^l + 1, j_1^l + 2, \ldots, j_1 \\ [\underline{a}_{ij}^k / \bar{a}_{\max j}^k, \bar{a}_{ij}^k / \bar{a}_{\max j}^k] \Leftrightarrow j = j_1 + 1, j_1 + 2, \ldots, j_2^l \\ [1 - \bar{a}_{ij}^k / \bar{a}_{\max j}^k, 1 - \underline{a}_{ij}^k / \bar{a}_{\max j}^k] \Leftrightarrow j = j_2^l + 1, j_2^l + 2, \ldots, j_2 \\ T(a_{ij}^k / c_{\max j}^k, b_{ij}^k / c_{\max j}^k, c_{ij}^k / c_{\max j}^k) \Leftrightarrow j = j_2 + 1, j_2 + 2, \ldots, j_3^l \\ T(1 - c_{ij}^k / c_{\max j}^k, 1 - b_{ij}^k / c_{\max j}^k, 1 - a_{ij}^k / c_{\max j}^k) \Leftrightarrow j = j_3^l + 1, j_3^l + 2, \ldots, j_3 \\ I(\mu_{ij}^k, v_{ij}^k) \Leftrightarrow j = j_3 + 1, j_3 + 2, \ldots, j_4^l \\ I(v_{ij}^k, \mu_{ij}^k) \Leftrightarrow j = j_4^l + 1, j_4^l + 2, \ldots, j_4 \\ H\{(\gamma_{ij}^k)^1, \ldots, (\gamma_{ij}^k)^{\#h_{ij}^k}\} \Leftrightarrow j = j_4 + 1, j_4 + 2, \ldots, j_5^l \\ H\{1 - (\gamma_{ij}^k)^{\#h_{ij}^k}, \ldots, 1 - (\gamma_{ij}^k)^1\} \Leftrightarrow j = j_5^l + 1, j_5^l + 2, \ldots, j_5 \\ Tr(a_{ij}^k / d_{\max j}^k, b_{ij}^k / d_{\max j}^k, c_{ij}^k / d_{\max j}^k, d_{ij}^k / d_{\max j}^k) \Leftrightarrow j = j_5 + 1, j_5 + 2, \ldots, j_6^l \\ Tr\begin{pmatrix} 1 - d_{ij}^k / d_{\max j}^k, 1 - c_{ij}^k / d_{\max j}^k, \\ 1 - b_{ij}^k / d_{\max j}^k, 1 - a_{ij}^k / d_{\max j}^k \end{pmatrix} \Leftrightarrow j = j_6^l + 1, j_6^l + 2, \ldots, n \end{cases} \tag{6.8}$$

where $a_{\max j}^k = \max_i \{a_{ij}^k\}$ $(j \in \{1, 2, \ldots, j_1\}, k \in \{1, 2, \ldots, g\})$; $\bar{a}_{\max j}^k = \max_i \{\bar{a}_{ij}^k\}$ $(j \in \{j_1 + 1, j_1 + 2, \ldots, j_2\}, k \in \{1, 2, \ldots, g\})$; $c_{\max j}^k = \max_i \{c_{ij}^k\}$ $(j \in \{j_2 + 1, j_2 + 2, \ldots, j_3\},\ k \in \{1, 2, \ldots, g\})$ and $d_{\max j}^k = \max_i \{d_{ij}^k\}$ $(j \in \{j_5 + 1, j_5 + 2, \ldots, n\},$ $k \in \{1, 2, \ldots, g\})$.

Thus, the decision matrices $\Re^k = (x_{ij}^k)_{m \times n}$ $(k \in \{1, 2, \ldots, g\})$ are transformed into the normalized decision matrices $\overline{\Re}^k = (\bar{x}_{ij}^k)_{m \times n}$ $(k \in \{1, 2, \ldots, g\})$ by using Eq. (6.8). It is easy to find that there exist three major difficulties and challenges for solving such a heterogeneous MCGDM problem (Zhang et al. 2015):

(1) How to effectively manage the heterogeneous decision information. It is usually difficult to unify them into one form because different forms of criteria values have different meanings and expression ways.
(2) How to derive objectively the weights of the experts and the criteria. These weights play an important role in the practical decision making process, but it is difficult for the managers and the experts to provide them precisely.
(3) How to determine the final ranking of the alternatives. Although several MCGDM methods have been developed to rank the alternatives, few of them

can be used to deal with the decision making problems with multiple forms of decision data, especially including the hesitant fuzzy decision data, such as HFEs and HFLTSs.

Bearing this fact in mind, Zhang et al. (2015) developed a deviation modeling method to deal with the heterogeneous MCGDM problems with incomplete weight information.

6.3 The Deviation Modeling Approach

In the developed deviation modelling approach, a maximizing deviation model is first established to determine the optimal weights of criteria for each expert. Then, a minimizing deviation model is developed to determine the weights of experts and identify the optimal alternative.

6.3.1 The Maximizing Deviation Model to Determine the Weights of Criteria

The estimation of the criteria weights plays an important role in the MCGDM process. Owing to the fact that the weights of criteria in the aforementioned heterogeneous MCGDM problem are completely unknown or partially known, in the following, we construct an optimal model based on the maximizing deviation method to derive the optimal weights of criteria for each expert. We first calculate the deviation value of each alternative to all other alternatives under a given criterion by using the different distance measures. Then, all deviation values for all the criteria under the expert can be obtained. According to the idea of Wang (1998), it is reasonable to identify the weights of criteria from the perspective of sorting the alternatives which maximizes all deviation values for all the criteria under the expert. Thus, we finally construct an optimal model that maximizes the overall deviation values for all criteria to determine the weights of criteria for each expert. The detail processes are presented as below:

For the expert e_k and the criterion C_j, the deviation values of the alternative A_ξ over all other alternatives can be defined as follows (Zhang et al. 2015):

$$D_{\xi j}^k(\boldsymbol{w}^k) = \sum_{\zeta=1}^{m} \left(w_j^k d_j(\bar{x}_{\xi j}^k, \bar{x}_{\zeta j}^k) \right) (\xi \in \{1,2,\ldots,m\}, \quad j \in \{1,2,\ldots,n\}), \quad (6.9)$$

where $\boldsymbol{w}^k = (w_1^k, w_2^k, \ldots, w_n^k)$ denotes the weighting vector of criteria for the expert $e_k \in \boldsymbol{E}$ which is unknown in advance and needs to be determined, and

$$d_j(\bar{x}_{\xi j}^k, \bar{x}_{\zeta j}^k) = \begin{cases} \left|a_{\xi j}^k / a_{\max j}^k - a_{\zeta j}^k / a_{\max j}^k\right| & \Leftrightarrow j = 1, 2, \ldots, j_1 \\ \frac{1}{2}\left(\left|\underline{a}_{\xi j}^k / \bar{a}_{\max j}^k - \underline{a}_{\zeta j}^k / \bar{a}_{\max j}^k\right| + \left|\bar{a}_{\xi j}^k / \bar{a}_{\max j}^k - \bar{a}_{\zeta j}^k / \bar{a}_{\max j}^k\right|\right) & \Leftrightarrow j = j_1 + 1, j_1 + 2, \ldots, j_2 \\ \frac{1}{3}\left(\begin{array}{l} \left|a_{\xi j}^k / c_{\max j}^k - a_{\zeta j}^k / c_{\max j}^k\right| + \left|b_{\xi j}^k / c_{\max j}^k - b_{\zeta j}^k / c_{\max j}^k\right| \\ + \left|c_{\xi j}^k / c_{\max j}^k - c_{\zeta j}^k / c_{\max j}^k\right| \end{array}\right) & \Leftrightarrow j = j_2 + 1, j_2 + 2, \ldots, j_3 \\ \frac{1}{2}\left(\left|u_{\xi j}^k - u_{\zeta j}^k\right| + \left|v_{\xi j}^k - v_{\zeta j}^k\right| + \left|\pi_{\xi j}^k - \pi_{\zeta j}^k\right|\right) & \Leftrightarrow j = j_3 + 1, j_3 + 2, \ldots, j_4 \\ \frac{1}{\#h} \sum_{\lambda=1}^{\#h} \left|(\gamma_{\xi j}^k)^\lambda - (\gamma_{\zeta j}^k)^\lambda\right| & \Leftrightarrow j = j_4 + 1, j_4 + 2, \ldots, j_5 \\ \frac{1}{4}\left(\begin{array}{l} \left|a_{\xi j}^k / d_{\max j}^k - a_{\zeta j}^k / d_{\max j}^k\right| + \left|b_{\xi j}^k / d_{\max j}^k - b_{\zeta j}^k / d_{\max j}^k\right| \\ + \left|c_{\xi j}^k / d_{\max j}^k - c_{\zeta j}^k / d_{\max j}^k\right| + \left|d_{\xi j}^k / d_{\max j}^k - d_{\zeta j}^k / d_{\max j}^k\right| \end{array}\right) & \Leftrightarrow j = j_5 + 1, j_5 + 2, \ldots, n \end{cases} \tag{6.10}$$

The deviation values of all alternatives to the other alternatives for the criterion $C_j \in \boldsymbol{C}$ and the expert $e_k \in \boldsymbol{E}$ can be expressed as:

$$D_j(\boldsymbol{w}^k) = \sum_{\xi=1}^{m} D_{\xi j}(\boldsymbol{w}^k) = \sum_{\xi=1}^{m}\sum_{\zeta=1}^{m}\left(w_j^k d_j(\bar{x}_{\xi j}^k, \bar{x}_{\zeta j}^k)\right), \quad j \in \{1, 2, \ldots, n\}. \tag{6.11}$$

Based on the above analysis, it is reasonable to select the weight vector $\boldsymbol{w}^k$ to maximize all deviation values for all criteria under the expert $e_k \in \boldsymbol{E}$. Thus, a maximization deviation model (MOD-6.1) is constructed as (Zhang et al. 2015):

$$\begin{cases} \max \quad D(\boldsymbol{w}^k) = \sum_{j=1}^{n}\sum_{\xi=1}^{m}\sum_{\zeta=1}^{m}\left(w_j^k d_j(\bar{x}_{\xi j}^k, \bar{x}_{\zeta j}^k)\right) \\ s.t. \quad \sum_{j=1}^{n}(w_j^k)^2 = 1, \quad w_j^k \geq 0, \quad j \in \{1, 2, \ldots, n\} \end{cases} \tag{MOD-6.1}$$

To solve the above model, let

$$Q(\boldsymbol{w}^k, \eta) = \sum_{j=1}^{n}\sum_{\xi=1}^{m}\sum_{\zeta=1}^{m}\left(w_j^k d_j(\bar{x}_{\xi j}^k, \bar{x}_{\zeta j}^k)\right) + \frac{\eta}{2}\left(\sum_{j=1}^{n}\left(w_j^k\right)^2 - 1\right), \tag{6.12}$$

which expresses the Lagrange function of the optimization problem (MOD-6.1), where η denoting Lagrange multiplier variable is a real number. Then, the partial derivatives of Q are computed as:

$$\begin{cases} \frac{\partial Q}{\partial w_j^k} = \sum_{\xi=1}^{m}\sum_{\zeta=1}^{m}\left(d_j(\bar{x}_{\xi j}^k, \bar{x}_{\zeta j}^k)\right) + \eta w_j^k = 0 \\ \frac{\partial Q}{\partial \eta} = \frac{1}{2}\left(\sum_{j=1}^{n}\left(w_j^k\right)^2 - 1\right) = 0 \end{cases} \tag{6.13}$$

By solving Eq. (6.13), we can get

$$w_j^k = \frac{\sum_{\xi=1}^{m}\sum_{\zeta=1}^{m}\left(d_j(\bar{x}_{\xi j}^k, \bar{x}_{\zeta j}^k)\right)}{\sqrt{\sum_{j=1}^{n}\left(\sum_{\xi=1}^{m}\sum_{\zeta=1}^{m}\left(d_j(\bar{x}_{\xi j}^k, \bar{x}_{\zeta j}^k)\right)\right)^2}}. \tag{6.14}$$

For the sake of simplicity, let

$$\aleph_j^k = \sum_{\xi=1}^{m}\sum_{\zeta=1}^{m}\left(d_j(\bar{x}_{\xi j}^k, \bar{x}_{\zeta j}^k)\right),$$

and Eq. (6.14) is rewritten as:

$$w_j^k = \frac{\aleph_j^k}{\sqrt{\sum_{j=1}^{n}(\aleph_j^k)^2}} \tag{6.15}$$

It can be easily verified from Eq. (6.15) that all $w_j^k (j \in \{1, 2, \ldots, n\})$ are positive such that they do satisfy the constrained conditions in the model (MOD-6.1) and the solution is unique. In order to normalize the optimal weights, we normalize the sum of w_j^k into a unit, then

$$\bar{w}_j^k = \frac{w_j^k}{\sum_{j=1}^{n} w_j^k} = \frac{\aleph_j^k}{\sum_{j=1}^{n}\aleph_j^k} = \frac{\sum_{\xi=1}^{m}\sum_{\zeta=1}^{m}\left(d_j(\bar{x}_{\xi j}^k, \bar{x}_{\zeta j}^k)\right)}{\sum_{j=1}^{n}\sum_{\xi=1}^{m}\sum_{\zeta=1}^{m}\left(d_j(\bar{x}_{\xi j}^k, \bar{x}_{\zeta j}^k)\right)} \tag{6.16}$$

Usually, there are actual decision situations that the weights of the criteria are not completely unknown but partially known. For these cases, based on the set of the known weight information $\mathbf{\Delta}$, we can construct the following constrained optimization model (Zhang et al. 2015):

$$\begin{cases} \max & D(\mathbf{w}^k) = \sum_{j=1}^{n}\sum_{\xi=1}^{m}\sum_{\zeta=1}^{m}\left(w_j^k d_j(\bar{x}_{\xi j}^k, \bar{x}_{\zeta j}^k))\right) \\ s.t & (w_1^k, w_2^k, \ldots, w_n^k) \in \mathbf{\Delta} \end{cases} \tag{MOD-6.2}$$

where $\mathbf{\Delta}$ is a set of constraint conditions described in Chap. 1 that the weight vector $\mathbf{w}^k = (w_1^k, w_2^k, \ldots, w_n^k)$ $(k \in \{1, 2, \ldots, g\})$ should satisfy according to the requirements in real situations.

This model (MOD-6.2) is a linear programming model that can be executed using the Lingo 11.0 software package. By solving this model, we get the optimal solution, i.e., the optimal weight vector of criteria for each expert, $\mathbf{w}^{k*} = (w_1^{k*}, w_2^{k*}, \ldots, w_n^{k*})^T$ $(k \in \{1, 2, \ldots, g\})$.

6.3.2 The Minimizing Deviation Model to Solve the Heterogeneous MCGDM Problems

In general, after obtaining the weights of criteria based on the maximizing deviation method, analogous to the literature (Wu and Chen 2007; Wei 2008; Xu 2010) we should aggregate the given decision information so as to get the overall preference value of each alternative and then to rank the alternatives. However, there exists no aggregation operator to aggregate the heterogeneous decision data introduced in Sect. 6.2. Borrowing the idea of the TOPSIS method (Hwang and Yoon 1981), we calculate the distances of the alternatives to the ideal solutions under each criterion for each expert, aiming at choosing the alternative with the shortest distance from the heterogeneous PIS and the farthest distance from the heterogeneous NIS.

For the expert $e_k \in \boldsymbol{E}$, we denoted the heterogeneous PIS by $A^{k+} = ((\bar{x}_1^k)^+, (\bar{x}_2^k)^+, \ldots, (\bar{x}_n^k)^+)$ and the heterogeneous NIS by $A^{k-} = ((\bar{x}_1^k)^-, (\bar{x}_2^k)^-, \ldots, (\bar{x}_n^k)^-)$, and they can be determined by the following equations (Zhang et al. 2015):

$$\left(\bar{x}_j^k\right)^+ = \begin{cases} \max_{i=1}^m \left\{a_{ij}^k / a_{\max j}^k\right\} \Leftarrow j = 1, 2, \ldots, j_1^I \\ \max_{i=1}^m \left\{1 - a_{ij}^k / a_{\max j}^k\right\} \Leftarrow j = j_1^I + 1, j_1^I + 2, \ldots, j_1 \\ \max_{i=1}^m \left\{[\underline{a}_{ij}^k / \bar{a}_{\max j}^k, \bar{a}_{ij}^k / \bar{a}_{\max j}^k]\right\} \Leftarrow j = j_1 + 1, j_1 + 2, \ldots, j_2^I \\ \max_{i=1}^m \left\{[1 - \bar{a}_{ij}^k / \bar{a}_{\max j}^k, 1 - \underline{a}_{ij}^k / \bar{a}_{\max j}^k]\right\} \Leftarrow j = j_2^I + 1, j_2^I + 2, \ldots, j_2 \\ \max_{i=1}^m \left\{T(a_{ij}^k / c_{\max j}^k, b_{ij}^k / c_{\max j}^k, c_{ij}^k / c_{\max j}^k)\right\} \Leftarrow j = j_2 + 1, j_2 + 2, \ldots, j_3^I \\ \max_{i=1}^m \left\{\begin{array}{l} T(1 - c_{ij}^k / c_{\max j}^k, \\ 1 - b_{ij}^k / c_{\max j}^k, 1 - a_{ij}^k / c_{\max j}^k) \end{array}\right\} \Leftarrow j = j_3^I + 1, j_3^I + 2, \ldots, j_3 \\ \max_{i=1}^m \left\{I(\mu_{ij}^k, v_{ij}^k)\right\} \Leftarrow j = j_3 + 1, j_3 + 2, \ldots, j_4^I \\ \max_{i=1}^m \left\{I(v_{ij}^k, \mu_{ij}^k)\right\} \Leftarrow j = j_4^I + 1, j_4^I + 2, \ldots, j_4 \\ \max_{i=1}^m \left\{H\{(h_{ij}^k)^1, \ldots, (h_{ij}^k)^{\# h_{ij}^k}\}\right\} \Leftarrow j = j_4 + 1, j_4 + 2, \ldots, j_5^I \\ \max_{i=1}^m \left\{H\{1 - (h_{ij}^k)^{\# h_{ij}^k}, \ldots, 1 - (h_{ij}^k)^1\}\right\} \Leftarrow j = j_5^I + 1, j_5^I + 2, \ldots, j_5 \\ \max_{i=1}^m \left\{\begin{array}{l} Tr(a_{ij}^k / d_{\max j}^k, b_{ij}^k / d_{\max j}^k, \\ c_{ij}^k / d_{\max j}^k, d_{ij}^k / d_{\max j}^k) \end{array}\right\} \Leftarrow j = j_5 + 1, j_5 + 2, \ldots, j_6^I \\ \max_{i=1}^m \left\{\begin{array}{l} Tr(1 - d_{ij}^k / d_{\max j}^k, 1 - c_{ij}^k / d_{\max j}^k, \\ 1 - b_{ij}^k / d_{\max j}^k, 1 - a_{ij}^k / d_{\max j}^k) \end{array}\right\} \Leftarrow j = j_6^I + 1, j_6^I + 2, \ldots, n \end{cases} \tag{6.17}$$

and

$$
\left(\bar{x}_j^k\right)^- = \begin{cases}
\min_{i=1}^m \left\{a_{ij}^k/a_{\max j}^k\right\} \Leftarrow j = 1,2,\ldots,j_1^I \\
\min_{i=1}^m \left\{1 - a_{ij}^k/a_{\max j}^k\right\} \Leftarrow j = j_1^I + 1, j_1^I + 2,\ldots,j_1 \\
\min_{i=1}^m \left\{\left[\underline{a}_{ij}^k/\bar{a}_{\max j}^k, \bar{a}_{ij}^k/\bar{a}_{\max j}^k\right]\right\} \Leftarrow j = j_1 + 1, j_1 + 2,\ldots,j_2^I \\
\min_{i=1}^m \left\{\left[1 - \bar{a}_{ij}^k/\bar{a}_{\max j}^k, 1 - \underline{a}_{ij}^k/\bar{a}_{\max j}^k\right]\right\} \Leftarrow j = j_2^I + 1, j_2^I + 2,\ldots,j_2 \\
\min_{i=1}^m \left\{T\left(a_{ij}^k/c_{\max j}^k, b_{ij}^k/c_{\max j}^k, c_{ij}^k/c_{\max j}^k\right)\right\} \Leftarrow j = j_2 + 1, j_2 + 2,\ldots,j_3^I \\
\min_{i=1}^m \left\{\begin{matrix} T(1 - c_{ij}^k/c_{\max j}^k, \\ 1 - b_{ij}^k/c_{\max j}^k, 1 - a_{ij}^k/c_{\max j}^k) \end{matrix}\right\} \Leftarrow j = j_3^I + 1, j_3^I + 2,\ldots,j_3 \\
\min_{i=1}^m \left\{I\left(\mu_{ij}^k, v_{ij}^k\right)\right\} \Leftarrow j = j_3 + 1, j_3 + 2,\ldots,j_4^I \\
\min_{i=1}^m \left\{I\left(v_{ij}^k, \mu_{ij}^k\right)\right\} \Leftarrow j = j_4^I + 1, j_4^I + 2,\ldots,j_4 \\
\min_{i=1}^m \left\{H\{(h_{ij}^k)^1,\ldots,(h_{ij}^k)^{\# h_{ij}^k}\}\right\} \Leftarrow j = j_4 + 1, j_4 + 2,\ldots,j_5^I \\
\min_{i=1}^m \left\{H\{1 - (h_{ij}^k)^{\# h_{ij}^k},\ldots,1 - (h_{ij}^k)^1\}\right\} \Leftarrow j = j_5^I + 1, j_5^I + 2,\ldots,j_5 \\
\min_{i=1}^m \left\{\begin{matrix} Tr(a_{ij}^k/d_{\max j}^k, b_{ij}^k/d_{\max j}^k, \\ c_{ij}^k/d_{\max j}^k, d_{ij}^k/d_{\max j}^k) \end{matrix}\right\} \Leftarrow j = j_5 + 1, j_5 + 2,\ldots,j_6^I \\
\min_{i=1}^m \left\{\begin{matrix} Tr(1 - d_{ij}^k/d_{\max j}^k, 1 - c_{ij}^k/d_{\max j}^k, \\ 1 - b_{ij}^k/d_{\max j}^k, 1 - a_{ij}^k/d_{\max j}^k) \end{matrix}\right\} \Leftarrow j = j_6^I + 1, j_6^I + 2,\ldots,n
\end{cases} \quad (6.18)
$$

For the expert $e_k \in \boldsymbol{E}$, the distances between the alternative A_i and the heterogeneous PIS A^{k+} as well as the heterogeneous NIS A^{k-} can be defined, respectively, as follows:

$$
D_i^{k+} = \sum_{j=1}^n \left(w_j^k d_j\left(\bar{x}_{ij}^k, \left(\bar{x}_j^k\right)^+\right)\right) (i \in \{1,2,\ldots,m\}, k \in \{1,2,\ldots,g\}) \quad (6.19)
$$

$$
D_i^{k-} = \sum_{j=1}^n \left(w_j^k d_j\left(\bar{x}_{ij}^k, \left(\bar{x}_j^k\right)^-\right)\right) (i \in \{1,2,\ldots,m\}, k \in \{1,2,\ldots,g\}) \quad (6.20)
$$

where $d_j(\cdot)$ is the distance measure introduced in Eq. (6.10).

Then, for the expert $e_k \in \boldsymbol{E}$, the relative closeness index of the alternative A_i to the heterogeneous PIS A^{k+} is defined as:

$$CI_i^k = \frac{D_i^{k-}}{D_i^{k+} + D_i^{k-}} \quad (i \in \{1,2,\ldots,m\}, k \in \{1,2,\ldots,g\}) \tag{6.21}$$

The matrix format of the relative closeness indices of all alternatives for all experts can be expressed as follows:

$$\boldsymbol{CI} = (CI_i^k)_{m\times g} = \begin{array}{c} \\ A_1 \\ A_2 \\ \vdots \\ A_m \end{array} \begin{array}{c} \begin{array}{cccc} e_1 & e_2 & \ldots & e_g \end{array} \\ \begin{pmatrix} CI_1^1 & CI_1^2 & \ldots & CI_1^g \\ CI_2^1 & CI_2^2 & \ldots & CI_2^g \\ \vdots & \vdots & \vdots & \vdots \\ CI_m^1 & CI_m^2 & \ldots & CI_m^g \end{pmatrix} \end{array} \tag{6.22}$$

In the practical group decision process, the experts usually come from various research areas and may have many differences in knowledge structures, express abilities, evaluation levels and individual preferences as well as practical experience, and thus they have a variety of views for the same group decision problem and may provide different assessments. The relative closeness indices of each alternative derived from these different original assessments provided by different experts may be different. Ideally, the final decision result (i.e., the opinion of the group) should be consistent with the individual expert's opinion to the greatest extent. In other words, the deviation values between the individual expert's opinion and the group's opinion should be minimized.

Let the relative closeness indices of the alternatives for the group be $\boldsymbol{CI}^* = (CI_1^*, CI_2^*, \ldots, CI_m^*)$ which need to be determined. The deviation value between the relative closeness indices of the alternative A_i for the individual expert and for the group is defined by using a l^p-distance (Yu 1973) as:

$$Dev_i = \sum_{k=1}^{g} \varpi_k \left| CI_i^* - CI_i^k \right|^p \quad (i \in \{1,2,\ldots,m\}) \tag{6.23}$$

where the parameter $p \in [1, \infty)$.

In the practical group decision process, the deviation value between the relative closeness indices of all alternative for all the individual experts and for the group should be minimized. Thus, a minimization deviation model is constructed to obtain the $\boldsymbol{CI}^*$ as (Zhang et al. 2015):

$$\begin{cases} \min \quad Z = \sum_{i=1}^{m} \sum_{k=1}^{g} \varpi_k \left| CI_i^* - CI_i^k \right|^p \\ s.t. \quad \sum_{k=1}^{g} \varpi_k = 1, \varpi_k \geq 0 \, (k \in \{1,2,\ldots,g\}) \end{cases} \tag{MOD-6.3}$$

It is noted that if the parameter $p = 1$, the model (MOD-6.3) turns into the following one:

$$\begin{cases} \min & Z = \sum_{i=1}^{m} \sum_{k=1}^{g} \varpi_k \left| CI_i^* - CI_i^k \right| \\ s.t. & \sum_{k=1}^{g} \varpi_k = 1, \varpi_k \geq 0 \, (k \in \{1, 2, \ldots, g\}) \end{cases} \tag{MOD-6.4}$$

while if the parameter $p = \infty$, then the model (MOD-6.3) turns into the following one:

$$\begin{cases} \min & Z = \sum_{i=1}^{m} \left(\max_{k=1,2,\ldots,g} \varpi_k \left| CI_i^* - CI_i^k \right| \right) \\ s.t. & \sum_{k=1}^{g} \varpi_k = 1, \varpi_k \geq 0 \, (k \in \{1, 2, \ldots, g\}) \end{cases} \tag{MOD-6.5}$$

It is easily seen that in the model (MOD-6.4) the obtained CI_i^* represents the best solution based on the majority principle. This is the solution for which the weighted sum of the deviation values between the relative closeness indices of all alternative for all the individual experts and for the group should be minimized. While in the model (MOD-6.5) the obtained CI_i^* represents the best solution based on the minority principle. Namely, this is the solution for which the sum of the maximization deviations between the relative closeness indices of all alternative for all the individual experts and for the group should be minimized. Usually, as p increases, more importance is given to the largest deviations (Yu 1973). Thus, the case $p = 1$ leads to a more robust consensus, while the case $p = \infty$ is more sensitive to the extreme deviation values (González-Pachón et al. 2014).

To comprehensively consider the majority and minority principle, we employ a control parameter $\theta (\theta \in [0, 1])$ to construct a convex combination of the cases $p = 1$ and $p = \infty$. To do so, the model (MOD-6.3) is transformed into the following model (Zhang et al. 2015):

$$\begin{cases} \min & Z = \sum_{i=1}^{m} \begin{pmatrix} (1-\theta) \left(\max_{k=1,2,\ldots,g} \varpi_k \left| CI_i^* - CI_i^k \right| \right) \\ + \theta \sum_{k=1}^{g} \varpi_k \left| CI_i^* - CI_i^k \right| \end{pmatrix} \\ s.t. & \sum_{k=1}^{g} \varpi_k = 1, \varpi_k \geq 0 \, (k \in \{1, 2, \ldots, g\}) \end{cases} \tag{MOD-6.6}$$

The model (MOD-6.6) is an appealing one because it not only encompasses as particular cases the model (MOD-6.4) with $\theta = 1$ and the model (MOD-6.5) with $\theta = 0$, but also yields the compromise solutions (if they exist) which trade-off majority versus minority principles by changing the values of the control parameter

$\theta \in (0,1)$. To solve the model (MOD-6.6), we adapt a change of variables proposed by Charnes and Cooper (1977) within a goal programming context. Let

$$\begin{cases} \varphi_i^k = \frac{1}{2}\left(\left|CI_i^* - CI_i^k\right| + \left(CI_i^* - CI_i^k\right)\right) \\ \psi_i^k = \frac{1}{2}\left(\left|CI_i^* - CI_i^k\right| - \left(CI_i^* - CI_i^k\right)\right) \end{cases} \tag{6.24}$$

Then, we can get

$$\varphi_i^k + \psi_i^k = \left|CI_i^* - CI_i^k\right|, \quad \varphi_i^k - \psi_i^k = CI_i^* - CI_i^k,$$

and the object function of the model (MOD-6.6) can be transformed as:

$$Z = \sum_{i=1}^{m}\left[(1-\theta)\left(\max_{k=1,2,\ldots,g} \varpi_k(\varphi_i^k + \psi_i^k)\right) + \theta\sum_{k=1}^{g}\varpi_k(\varphi_i^k + \psi_i^k)\right].$$

Therefore, the model (MOD-6.6) is converted into the following one (Zhang et al. 2015):

$$\begin{cases} \min \quad Z = \sum_{i=1}^{m}\begin{pmatrix} (1-\theta)\left(\max_{k=1,2,\ldots,g} \varpi_k(\varphi_i^k + \psi_i^k)\right) \\ +\theta\sum_{k=1}^{g}\varpi_k(\varphi_i^k + \psi_i^k) \end{pmatrix} \\ s.t. \quad \varphi_i^k - \psi_i^k + CI_i^k = CI_i^*, \quad \forall i,k \\ \qquad \varphi_i^k, \psi_i^k \geq 0, \quad \forall i,k \\ \qquad \sum_{k=1}^{g}\varpi_k = 1, \varpi_k \geq 0\,(k \in \{1,2,\ldots,g\}) \end{cases} \tag{MOD-6.7}$$

Moreover, let

$$\wp_i = \max_{k=1,2,\ldots,g} \varpi_k(\varphi_i^k + \psi_i^k),$$

then we obtain

$$\varpi_k\left(\varphi_i^k + \psi_i^k\right) \leq \wp_i\,(i \in \{1,2,\ldots,m\}),$$

$$Z = \sum_{i=1}^{m}[(1-\theta)\wp_i + \theta\sum_{k=1}^{g}\varpi_k(\varphi_i^k + \psi_i^k)].$$

It is easily seen that the object function Z can be formally rewritten as:

$$Z = (1-\theta)\sum_{i=1}^{m} \wp_i + \theta \sum_{i=1}^{m}\sum_{k=1}^{g} \varpi_k(\varphi_i^k + \psi_i^k)$$

Thus, the model (MOD-6.7) is equivalent to the following linear programming model (Zhang et al. 2015):

$$\begin{cases} \min \quad Z = (1-\theta)\sum\limits_{i=1}^{m} \wp_i + \theta \sum\limits_{i=1}^{m}\sum\limits_{k=1}^{g} \varpi_k\left(\varphi_i^k + \psi_i^k\right) \\ \quad\quad\ \varphi_i^k - \psi_i^k + CI_i^k = CI_i^*, \quad k \in \{1,2,\ldots,g\}, i \in \{1,2,\ldots,m\} \\ s.t. \quad \varpi_k\left(\varphi_i^k + \psi_i^k\right) \le \wp_i, \quad k \in \{1,2,\ldots,g\}, i \in \{1,2,\ldots,m\} \\ \quad\quad\ \varphi_i^k,\ \psi_i^k \ge 0, \quad k \in \{1,2,\ldots,g\}, i \in \{1,2,\ldots,m\} \\ \quad\quad\ \sum\limits_{k=1}^{g} \varpi_k = 1, \varpi_k \ge 0\,(k \in \{1,2,\ldots,g\}) \end{cases} \quad \text{(MOD-6.8)}$$

It is easily seen that the model (MOD-6.8) is a parametric linear programming model which can be solved easily by using the Lingo 11.0 software package or MATLAB 7.4.0 mathematics software package. As aforementioned previously, there are actual situations that the weights of experts are not completely unknown but partially known. For this situation, based on the set of the given weights of the experts $\boldsymbol{\Gamma}$, we can construct the following constrained optimization model (Zhang et al. 2015):

$$\begin{cases} \min \quad Z = (1-\theta)\sum\limits_{i=1}^{m} \wp_i + \theta \sum\limits_{i=1}^{m}\sum\limits_{k=1}^{g} \varpi_k\left(\varphi_i^k + \psi_i^k\right) \\ \quad\quad\ \varphi_i^k - \psi_i^k + CI_i^k = CI_i^*, \quad k \in \{1,2,\ldots,g\}, i \in \{1,2,\ldots,m\} \\ s.t. \quad \varpi_k\left(\varphi_i^k + \psi_i^k\right) \le \wp_i, \quad k \in \{1,2,\ldots,g\}, i \in \{1,2,\ldots,m\} \\ \quad\quad\ \varphi_i^k,\ \psi_i^k \ge 0, \quad k \in \{1,2,\ldots,g\}, i \in \{1,2,\ldots,m\} \\ \quad\quad\ (\varpi_1, \varpi_2, \ldots, \varpi_g) \in \boldsymbol{\Gamma}, \end{cases} \quad \text{(MOD-6.9)}$$

where $\boldsymbol{\Gamma}$ is also a set of constraint conditions described in Sect. 2.5 that the weight value ϖ^k should satisfy according to the requirements in real situations.

By solving the model (MOD-6.8) or (MOD-6.9), we get the relative closeness indices of alternatives for the group $\boldsymbol{CI}^* = (CI_1^*, CI_2^*, \ldots, CI_m^*)$ and the optimal weights of the experts $\boldsymbol{\varpi} = (\varpi_1, \varpi_2, \ldots, \varpi_g)$. At length, we can obtain the ranking of the alternatives according to the relative closeness indices $\boldsymbol{CI}^* = (CI_1^*, CI_2^*, \ldots, CI_m^*)$.

6.3.3 The Proposed Algorithm

Based on the above models and analysis, a practical algorithm of the proposed method is introduced for solving the heterogeneous MCGDM problems in which both the weights of criteria and the weights of experts are partially known or completely unknown, and the decision data take the forms of real numbers, interval numbers, linguistic variables, IFNs, HFEs and HFLTSs. The proposed algorithm (Algorithm 6.1) involves the following steps (Zhang et al. 2015):

Step 1. For a heterogeneous MCGDM problem, we construct the decision making matrix $\Re^k = (x_{ij}^k)_{m\times n}$ $(k \in \{1,2,\ldots,g\})$ where the element x_{ij}^k is the assessment of the alternative $A_i \in \boldsymbol{A}$ with respect to the criterion $C_j \in \boldsymbol{C}$ given by the expert $e_k \in \boldsymbol{E}$. Then, the heterogeneous decision matrices $\Re^k = (x_{ij}^k)_{m\times n}$ $(k \in \{1,2,\ldots,g\})$ are transformed into the normalized decision making matrices $\bar{\Re}^k = (\bar{x}_{ij}^k)_{m\times n}$ $(k \in \{1,2,\ldots,g\})$ by using Eq. (6.8).

Step 2. If the weights of the criteria for the expert $e_k \in \boldsymbol{E}$ are completely unknown, then we employ Eq. (6.16) to obtain the optimal weights of the criteria for the expert $e_k \in \boldsymbol{E}$, while if the weights of the criteria for the expert $e_k \in \boldsymbol{E}$ are partially known, then we employ the model (MOD-6.2) to obtain the optimal weights of the criteria for the expert $e_k \in \boldsymbol{E}$.

Step 3. Use Eqs. (6.17) and (6.18) to determine the heterogeneous PIS $A^{k+} = ((\bar{x}_1^k)^+, (\bar{x}_2^k)^+, \ldots, (\bar{x}_n^k)^+)$ and the heterogeneous NIS $A^{k-} = ((\bar{x}_1^k)^-, (\bar{x}_2^k)^-, \ldots, (\bar{x}_n^k)^-)$ for the expert $e_k \in \boldsymbol{E}$, respectively.

Step 4. Employ Eqs. (6.19) and (6.20) to calculate the distances between the alternative $A_i \in \boldsymbol{A}$ and the PIS A^{k+} as well as the NIS A^{k-}, respectively.

Step 5. Utilize Eq. (6.21) to calculate the relative closeness index CI_i^k of the alternative $A_i \in \boldsymbol{A}$ to the heterogeneous PIS A^{k+} for the expert $e_k \in \boldsymbol{E}$.

Step 6. If the weights of the experts are completely unknown, then we employ the model (MOD-6.8) to obtain the optimal weights of the experts and the group closeness index CI_i^* of each alternative A_i, while if the weights of the criteria for the experts $e_k \in \boldsymbol{E}$ are partially known, then we use the model (MOD-6.9) to obtain the optimal weights of the experts and the group closeness index CI_i^* of the alternative $A_i \in \boldsymbol{A}$.

Step 7. Determine the optimal ranking order of the alternatives and identify the optimal alternative. On the basis of the group closeness index CI_i^* obtained on Step 6, we put the alternatives into orders with respect to the decreasing values of $CI_i^* (i \in \{1,2,\ldots,m\})$, and the alternative with the maximal relative closeness index for the common will of group is the best alternative, namely,

$$A^* = \left\{ A_i : \left\langle i : CI_i^* = \max_{1\leq\xi\leq m} CI_\xi^* \right\rangle \right\} \tag{6.25}$$

6.4 Empirical Analysis for the Selection of Strategic Freight Forwarder

Zhang et al. (2015) presented an empirical case (adapted from Feng and Lai 2014) concerned with the selection of Strategic Freight Forwarder to demonstrate the applicability and the implementation process of the proposed method. Moreover, a comparative analysis with the similar method is conducted.

6.4.1 Description of the Decision Making Problem

China Southern Airlines (CSA) is one of the largest transportation firms in the world (the fourth worldwide and the first in Asia). In CSA, the Air Cargo Transportation (ACT) Department provides cargo transportation services using dedicated freighters and the belly space of passenger planes. It provides freight forwarders and shippers with one-stop services, including such as consulting, capacity reservation, pickup, receiving, packaging, sorting, loading, transportation, delivering, and cargo tracking and tracing. A large number of freight forwarders cooperate with the ACT Department. With the increasing competition in the air cargo market, the ACT Department had to develop some strategic partners, with whom the ACT Department intended to build a long-term collaborative relationship. The ACT Department was planning to select the best freight forwarder from four candidates, as its strategic partner. To achieve this strategic objective, a committee consisting of four experts was formed for developing the decision criteria. The weights of the experts are partially known and are given as follows (Zhang et al. 2015):

$$\Gamma = \{\varpi_4 \geq \varpi_1, 0.15 \leq \varpi_2 - \varpi_3 \leq 0.25, \varpi_4 - \varpi_3 \geq \varpi_2 - \varpi_1,$$
$$\varpi_2 \geq 2\varpi_1, 0.15 \leq \varpi_4 \leq 0.35, \sum_{k=1}^{4} \varpi_k = 1, \varpi_k > 0, k = 1, 2, 3, 4\}$$

The committee finalized the criteria for the strategic forwarders selection after a brainstorming panel meeting and an in-depth discussion. The seven decision criteria identified are supply quantity (C_1), supply stability (C_2), delay time (C_3), supply ratio of low-demand routes (C_4), IT support (C_5), payment reputation (C_6) and switching risk (C_7). The descriptions of the decision criteria are listed in Table 6.1 [more details about criteria refer to Feng and Lai (2014)].

Supply quantity (C_1), supply stability (C_2), delay time (C_3) and supply ratio of low-demand routes (C_4) are all quantitative criteria. The assessments for supply quantity (C_1) can be represented by real numbers. The assessments for supply stability (C_2) are divided into two parts: stability degree and disstability degree, which just consist in the membership degree and non-membership degree of IFN.

Table 6.1 Criteria used for strategic freight forwarder of CSA (Feng and Lai 2014)

Evaluation criteria	Descriptions of criteria
Supply quantity (C_1)	The quantity of cargos provided by the freight forwarder
Supply stability (C_2)	The stability of the supply quantity of the freight forwarder
Delay time (C_3)	The time between the actual delivery time and the scheduled delivery time of consignments
Supply ratio of low-demand routes (C_4)	The quantity of cargos from the freight forwarder, which is transported on low-demand routes
IT supports (C_5)	The online service system available at the freight forwarder
Payment reputation (C_6)	The reputation of the candidate in terms of the length of payment period, days of delayed payment, amount of deferred payable balance, and payment methods when a contract is fulfilled
Switching risk (C_7)	The risk level that a strategic partner would switch its business to other airlines

Table 6.2 The linguistic scales and their corresponding TFNs

Linguistic variables and semantic	Abbreviation	TFNs
l_0: *Definitely low*	*DL*	T(0.0, 0.0, 0.17)
l_1: *Very low*	*VL*	T(0.0, 0.17, 0.33)
l_2: *Low*	*L*	T(0.17, 0.33, 0.5)
l_3: *Medium*	*M*	T(0.33, 0.5, 0.67)
l_4: *High*	*H*	T(0.5, 0.67, 0.83)
l_5: *Very high*	*VH*	T(0.67, 0.83, 1.0)
l_6: *Definitely high*	*DH*	T(0.83, 1.0, 1.0)

Thus, IFNs are used to express the assessment information for supply stability (C_2). Due to the uncertainty of transportation process, it is better to use the interval numbers to represent the delay time (C_3). The assessments for supply ratio of low-demand routes (C_4) can be represented by HFEs because the experts often hesitate among several values to provide their preference for C_4. While IT support (C_5), payment reputation (C_6) and switching risk (C_7) are all qualitative criteria. The assessments for IT support (C_5) and payment reputation (C_6) are represented by linguistic terms, and the assessments for switching risk (C_7) are represented by comparative linguistic terms. The linguistic terms used here and their corresponding TFNs are shown in Table 6.2. It is noted that the criteria C_3 and C_7 are the cost criteria, and others are the benefit criteria. The weights of all criteria are completely unknown. The decision data used in the decision making process of the Strategic Freight Forwarder Selection is presented in Table 6.3 (Zhang et al. 2015).

6.4.2 Decision Making Process

In the following, we utilize the proposed deviation modeling approach to solve the heterogeneous MCGDM problems mentioned in Sect. 6.4.1. The solution process and the computation results are summarized as follows.

Firstly, we use Eq. (6.6) to normalize the heterogeneous decision matrix of four experts presented in Table 6.3. The normalized results are shown in Table 6.4 (Zhang et al. 2015). Secondly, we employ Eq. (6.16) to get the optimal weights of all criteria of each expert:

$$\begin{aligned} w^1 &= (w_1^1, w_2^1, w_3^1, w_4^1, w_5^1, w_6^1, w_7^1) \\ &= (0.2043, 0.1626, 0.0625, 0.0576, 0.1521, 0.1464, 0.2145), \\ w^2 &= (w_1^2, w_2^2, w_3^2, w_4^2, w_5^2, w_6^2, w_7^2) \\ &= (0.1844, 0.1567, 0.0438, 0.1014, 0.1537, 0.1226, 0.1507), \\ w^3 &= (w_1^3, w_2^3, w_3^3, w_4^3, w_5^3, w_6^3, w_7^3) \\ &= (0.1847, 0.0920, 0.2337, 0.0869, 0.0764, 0.2299, 0.0966), \\ w^4 &= (w_1^4, w_2^4, w_3^4, w_4^4, w_5^4, w_6^4, w_7^4) \\ &= (0.1862, 0.1788, 0.0742, 0.0806, 0.1052, 0.1928, 0.1822). \end{aligned}$$

Thirdly, we utilize Eqs. (6.17) and (6.18) to determine the heterogeneous PIS $\boldsymbol{A}^{k+} = ((\bar{x}_1^k)^+, (\bar{x}_2^k)^+, (\bar{x}_3^k)^+, (\bar{x}_4^k)^+)$ and the heterogeneous NIS $\boldsymbol{A}^{k-} = ((\bar{x}_1^k)^-, (\bar{x}_2^k)^-, (\bar{x}_3^k)^-, (\bar{x}_4^k)^-)$ for four experts, respectively, and the obtained results are listed in Table 6.4. Then, we use Eqs. (6.19) and (6.20) to calculate the distances for the expert e_k between the alternative A_i and the heterogeneous PIS A^{k+} as well as the heterogeneous NIS A^{k-}, respectively. The obtained results are displayed in Table 6.5 (Zhang et al. 2015).

Furthermore, we calculate the relative closeness index CI_i^k of the alternative A_i to the heterogeneous PIS A^{k+} for the expert e_k by using Eq. (6.21). The relative closeness index $CI_i^k(i, k = 1, 2, 3, 4)$ are also presented in Table 6.5.

Fourthly, based on the data presented in Table 6.5, we utilize the model (MOD-6.9) to construct a parametric linear programming model as follows (Zhang et al. 2015):

Table 6.3 The decision matrices of four experts

Experts	Alternatives	Criteria			
		C_1	C_2	C_3	C_4
e_1	A_1	2.35	I(0.5, 0.3)	[15.5, 16.8]	H{0.4, 0.5}
	A_2	2.48	I(0.6, 0.2)	[16.6, 18.35]	H{0.4, 0.5, 0.7}
	A_3	1.53	I(0.4, 0.4)	[14.34, 16.42]	H{0.5, 0.6}
	A_4	4.50	I(0.3, 0.6)	[12.67, 15.32]	H{0.4}
e_2	A_1	4.35	I(0.4, 0.5)	[12.5, 15.8]	H{0.3, 0.5}
	A_2	2.28	I(0.7, 0.2)	[11.6, 14.65]	H{0.6, 0.7}
	A_3	1.53	I(0.6, 0.4)	[13.34, 16.42]	H{0.3, 0.5, 0.6}
	A_4	2.50	I(0.3, 0.7)	[14.67, 15.32]	H{0.2, 0.5}
e_3	A_1	1.55	I(0.4, 0.3)	[13.25, 15.48]	H{0.3, 0.5}
	A_2	4.48	I(0.6,0.2)	[11.6, 15.45]	H{0.4, 0.5, 0.7}
	A_3	1.53	I(0.8, 0.1)	[13.34, 15.42]	H{0.6}
	A_4	3.50	I(0.6, 0.2)	[16.67, 18.32]	H{0.2, 0.5}
e_4	A_1	2.35	I(0.7, 0.3)	[12.5, 14.85]	H{0.4, 0.5, 0.6}
	A_2	1.48	I(0.6, 0.2)	[16.6, 18.62]	H{0.2, 0.5}
	A_3	3.53	I(0.5, 0.2)	[14.34, 16.42]	H{0.5, 0.6}
	A_4	2.56	I(0.4, 0.6)	[12.67, 15.32]	H{0.5}

Experts	Alternatives	Criteria		
		C_5	C_6	C_7
e_1	A_1	*M*	*M*	*Between M and VH*
	A_2	*H*	*DL*	*Atleast H*
	A_3	*VH*	*DL*	*Between M and H*
	A_4	*DH*	*VL*	*Atmost L*
e_2	A_1	*H*	*M*	*Between L and M*
	A_2	*L*	*L*	*Atleast VH*
	A_3	*VH*	*L*	*Between H and VH*
	A_4	*M*	*DL*	*Between DL and L*
e_3	A_1	*H*	*M*	*Between M and H*
	A_2	*H*	*M*	*Atleast H*
	A_3	*VH*	*VL*	*Between M and H*
	A_4	*DH*	*H*	*Between L and M*
e_4	A_1	*L*	*H*	*Atleast H*
	A_2	*H*	*M*	*Atleast VH*
	A_3	*M*	*VL*	*Between M and H*
	A_4	*M*	*VL*	*Between L and M*

Table 6.4 The normalized decision matrix and the heterogeneous NIS as well as the heterogeneous PIS for four experts

Experts	Alternatives	Criteria			
		C_1	C_2	C_3	C_4
e_1	A_1	0.52	I(0.5, 0.3)	[0.08, 0.16]	H{0.4, 0.4, 0.5}
	A_2	0.55	I(0.6, 0.2)	[0.0, 0.1]	H{0.4, 0.5, 0.7}
	A_3	0.34	I(0.4, 0.4)	[0.11, 0.22]	H{0.5, 0.5, 0.6}
	A_4	1.00	I(0.3, 0.6)	[0.17, 0.31]	H{0.4, 0.4, 0.4}
	A^{1+}	1.00	I(0.6, 0.2)	[0.17, 0.31]	H{0.5, 0.5, 0.6}
	A^{1-}	0.34	I(0.3, 0.6)	[0.0, 0.1]	H{0.4, 0.4, 0.4}
e_2	A_1	1.00	I(0.4, 0.5)	[0.04, 0.24]	H{0.3, 0.3, 0.5}
	A_2	0.52	I(0.7, 0.2)	[0.11, 0.29]	H{0.6, 0.6, 0.7}
	A_3	0.35	I(0.6, 0.4)	[0.0, 0.19]	H{0.3, 0.5, 0.6}
	A_4	0.57	I(0.3, 0.7)	[0.07, 0.11]	H{0.2, 0.2, 0.5}
	A^{2+}	1.00	I(0.7, 0.2)	[0.11, 0.29]	H{0.6, 0.6, 0.7}
	A^{2-}	0.35	I(0.3, 0.7)	[0.0, 0.19]	H{0.2, 0.2, 0.5}
e_3	A_1	0.35	I(0.4, 0.3)	[0.15, 0.28]	H{0.3, 0.3, 0.5}
	A_2	1.00	I(0.6, 0.2)	[0.16, 0.37]	H{0.4, 0.5, 0.7}
	A_3	0.34	I(0.8, 0.1)	[0.16, 0.27]	H{0.6, 0.6, 0.6}
	A_4	0.78	I(0.6, 0.2)	[0.0, 0.09]	H{0.2, 0.2, 0.5}
	A^{3+}	1.00	I(0.8, 0.1)	[0.16, 0.27]	H{0.6, 0.6, 0.6}
	A^{3-}	0.34	I(0.4, 0.3)	[0.0, 0.09]	H{0.2, 0.2, 0.5}
e_4	A_1	0.67	I(0.7, 0.3)	[0.2, 0.33]	H{0.4, 0.5, 0.6}
	A_2	0.42	I(0.6, 0.2)	[0.0, 0.11]	H{0.2, 0.2, 0.5}
	A_3	1.00	I(0.5, 0.2)	[0.12, 0.23]	H{0.5, 0.5, 0.6}
	A_4	0.73	I(0.4, 0.6)	[0.18, 0.32]	H{0.5, 0.5, 0.5}
	A^{4+}	1.00	I(0.7, 0.3)	[0.2, 0.33]	H{0.5, 0.5, 0.6}
	A^{4-}	0.42	I(0.4, 0.6)	[0.0, 0.11]	H{0.2, 0.2, 0.5}

Experts	Alternatives	Criteria		
		C_5	C_6	C_7
e_1	A_1	T(0.33, 0.5, 0.67)	T(0.33, 0.5, 0.67)	Tr(0.0, 0.3, 0.36, 0.67)
	A_2	T(0.5, 0.67, 0.83)	T(0.0, 0.0, 0.17)	Tr(0.0, 0.0, 0.15, 0.5)
	A_3	T(0.67, 0.83, 1.0)	T(0.0, 0.0, 0.17)	Tr(0.17, 0.33, 0.5, 0.67)
	A_4	T(0.83, 1.0, 1.0)	T(0.0, 0.17, 0.33)	Tr(0.5, 0.85, 1.0, 1.0)
	A^{1+}	T(0.83, 1.0, 1.0)	T(0.33, 0.5, 0.67)	Tr(0.0, 0.0, 0.15, 0.5)
	A^{1-}	T(0.33, 0.5, 0.67)	T(0.0, 0.0, 0.17)	Tr(0.5, 0.85, 1.0, 1.0)
e_2	A_1	T(0.5, 0.67, 0.83)	T(0.33, 0.5, 0.67)	Tr(0.33, 0.5, 0.67, 0.83)
	A_2	T(0.17, 0.33, 0.5)	T(0.17, 0.33, 0.5)	Tr(0.0, 0.0, 0.03, 0.33)
	A_3	T(0.67, 0.83, 1.0)	T(0.17, 0.33, 0.5)	Tr(0.0, 0.17, 0.33, 0.5)
	A_4	T(0.33, 0.5, 0.67)	T(0.0, 0.0, 0.17)	Tr(0.5, 0.85, 1.0, 1.0)
	A^{2+}	T(0.67, 0.83, 1.0)	T(0.33, 0.5, 0.67)	Tr(0.5, 0.85, 1.0, 1.0)
	A^{2-}	T(0.17, 0.33, 0.5)	T(0.0, 0.0, 0.17)	Tr(0.0, 0.0, 0.03, 0.33)

(continued)

Table 6.4 (continued)

Experts	Alternatives	Criteria		
		C_5	C_6	C_7
e_3	A_1	T(0.5, 0.67, 0.83)	T(0.33, 0.5, 0.67)	Tr(0.17, 0.33, 0.5, 0.67)
	A_2	T(0.5, 0.67, 0.83)	T(0.33, 0.5, 0.67)	Tr(0.0, 0.0, 0.15, 0.5)
	A_3	T(0.67, 0.83, 1.0)	T(0.0, 0.17, 0.33)	Tr(0.17, 0.33, 0.5, 0.67)
	A_4	T(0.83, 1.0, 1.0)	T(0.5, 0.67, 0.83)	Tr(0.33, 0.5, 0.67, 0.83)
	A^{3+}	T(0.83, 1.0, 1.0)	T(0.5, 0.67, 0.83)	Tr(0.33, 0.5, 0.67, 0.83)
	A^{3-}	T(0.5, 0.67, 0.83)	T(0.0, 0.17, 0.33)	Tr(0.0, 0.0, 0.15, 0.5)
e_4	A_1	T(0.17, 0.33, 0.5)	T(0.5, 0.67, 0.83)	Tr(0.0, 0.0, 0.15, 0.5)
	A_2	T(0.5, 0.67, 0.83)	T(0.33, 0.5, 0.67)	Tr(0.0, 0.0, 0.03, 0.33)
	A_3	T(0.33, 0.5, 0.67)	T(0.0, 0.17, 0.33)	Tr(0.17, 0.33, 0.5, 0.67)
	A_4	T(0.33, 0.5, 0.67)	T(0.0, 0.17, 0.33)	Tr(0.33, 0.5, 0.67, 0.83)
	A^{4+}	T(0.5, 0.67, 0.83)	T(0.5, 0.67, 0.83)	Tr(0.33, 0.5, 0.67, 0.83)
	A^{4-}	T(0.17, 0.33, 0.5)	T(0.0, 0.17, 0.33)	Tr(0.0, 0.0, 0.15, 0.5)

Table 6.5 The distances of each alternative from PIS and NIS, and the relative closeness indices

Experts	Alternatives	D_i^{k+}	D_i^{k-}	CI_i^{k+}
e_1	A_1	0.2315	0.2651	0.5338
	A_2	0.2146	0.2858	0.5711
	A_3	0.3084	0.1882	0.3790
	A_4	0.2663	0.2302	0.4637
e_2	A_1	0.1407	0.3397	0.7071
	A_2	0.2984	0.1820	0.3788
	A_3	0.2817	0.1988	0.4138
	A_4	0.3018	0.1821	0.3763
e_3	A_1	0.2549	0.1486	0.3684
	A_2	0.1417	0.2886	0.6707
	A_3	0.2612	0.1400	0.3489
	A_4	0.1248	0.2763	0.6888
e_4	A_1	0.1757	0.2283	0.5651
	A_2	0.3000	0.1841	0.3802
	A_3	0.2043	0.2712	0.5703
	A_4	0.2216	0.1824	0.4514

$$\min \quad Z = (1-\theta)\sum_{i=1}^{4}\wp_i + \theta\sum_{i=1}^{4}\sum_{k=1}^{4}\varpi_k\left(\varphi_i^k + \psi_i^k\right)$$

$$\text{s.t.} \begin{cases} \varphi_1^1 - \psi_1^1 + 0.5338 = CI_1^*, \varphi_1^2 - \psi_1^2 + 0.7071 = CI_1^*, \varphi_1^3 - \psi_1^3 + 0.3684 = CI_1^*, \\ \varphi_1^4 - \psi_1^4 + 0.5651 = CI_1^*, \varphi_2^1 - \psi_2^1 + 0.5711 = CI_2^*, \varphi_2^2 - \psi_2^2 + 0.3788 = CI_2^*, \\ \varphi_2^3 - \psi_2^3 + 0.6707 = CI_2^*, \varphi_2^4 - \psi_2^4 + 0.3802 = CI_2^*, \varphi_3^1 - \psi_3^1 + 0.3790 = CI_3^*, \\ \varphi_3^2 - \psi_3^2 + 0.4138 = CI_3^*, \varphi_3^3 - \psi_3^3 + 0.3489 = CI_3^*, \varphi_3^4 - \psi_3^4 + 0.5703 = CI_3^*, \\ \varphi_4^1 - \psi_4^1 + 0.4637 = CI_4^*, \varphi_4^2 - \psi_4^2 + 0.3763 = CI_4^*, \\ \varphi_4^3 - \psi_4^3 + 0.6888 = CI_4^*, \varphi_4^4 - \psi_4^4 + 0.4514 = CI_4^*, \\ \varpi_1\left(\varphi_1^1 + \psi_1^1\right) \le \wp_1, \varpi_2\left(\varphi_1^2 + \psi_1^2\right) \le \wp_1, \varpi_3\left(\varphi_1^3 + \psi_1^3\right) \le \wp_1, \varpi_4\left(\varphi_1^4 + \psi_1^4\right) \le \wp_1 \\ \varpi_1\left(\varphi_2^1 + \psi_2^1\right) \le \wp_2, \varpi_2\left(\varphi_2^2 + \psi_2^2\right) \le \wp_2, \varpi_3\left(\varphi_2^3 + \psi_2^3\right) \le \wp_2, \varpi_4\left(\varphi_2^4 + \psi_2^4\right) \le \wp_2 \\ \varpi_1\left(\varphi_3^1 + \psi_3^1\right) \le \wp_3, \varpi_2\left(\varphi_3^2 + \psi_3^2\right) \le \wp_3, \varpi_3\left(\varphi_3^3 + \psi_3^3\right) \le \wp_3, \varpi_4\left(\varphi_3^4 + \psi_3^4\right) \le \wp_3 \\ \varpi_1\left(\varphi_4^1 + \psi_4^1\right) \le \wp_4, \varpi_2\left(\varphi_4^2 + \psi_4^2\right) \le \wp_4, \varpi_3\left(\varphi_4^3 + \psi_4^3\right) \le \wp_4, \varpi_4\left(\varphi_4^4 + \psi_4^4\right) \le \wp_4 \\ \varpi_4 \ge \varpi_1, 0.15 \le \varpi_2 - \varpi_3 \le 0.25, \ \varpi_4 - \varpi_3 \ge \varpi_2 - \varpi_1, \ \varpi_2 \ge 2\varpi_1, \\ 0.15 \le \varpi_4 \le 0.35, \varpi_1 + \varpi_2 + \varpi_3 + \varpi_4 = 1 \\ \varphi_i^k, \psi_i^k, \varpi_k \ge 0, \forall i, k \in \{1,2,3,4\} \end{cases} \tag{MOD-6.10}$$

The above parametric linear programming model (MOD-6.10) can be solved with the help of the Lingo 11.0 software for different values of the control parametric θ. The obtained solutions and the final ranking orders of alternatives are shown in Table 6.6 (Zhang et al. 2015).

From Table 6.6, it is clearly seen that there are three vectors of the relative closeness indices for the group. The first one is the best solution based on the minority principle, the third one is the best solution on the basis of the majority principle and the second one is the best solution based on a compromise between the two previous principles. Obviously, the second solution trade-offs majority versus minority principles, which is the comprehensive solution. Thus, the final ranking order of all the alternatives is $A_1 \succ A_4 \succ A_3 \succ A_2$, and the best alternative is A_1.

Table 6.6 The calculation results based on different values of the parametric θ

	CI_1^*	CI_2^*	CI_3^*	CI_4^*	ϖ_1
$\theta \in [0.0, 0.54)$	0.6278	0.3802	0.4909	0.4494	0.18
$\theta \in [0.53, 0.56]$	0.6278	0.3802	0.4138	0.4494	0.18
$\theta \in (0.56, 1.0]$	0.5651	0.3802	0.4138	0.4514	0.18
	ϖ_2	ϖ_3	ϖ_4	*Rankings of alternatives*	
$\theta \in [0.0, 0.54)$	0.36	0.11	0.35	$A_1 \succ A_3 \succ A_4 \succ A_2$	
$\theta \in [0.53, 0.56]$	0.36	0.11	0.35	$A_1 \succ A_4 \succ A_3 \succ A_2$	
$\theta \in (0.56, 1.0]$	0.36	0.11	0.35	$A_1 \succ A_4 \succ A_3 \succ A_2$	

6.4.3 Comparative Analysis

To show the advantage of the proposed approach, Zhang et al. (2015) made a comparative analysis with the similar method proposed by Li et al. (2010) which is denoted by Li et al. (2010)'s method for convenience. It is worthwhile to point out that Li et al. (2010)'s method is just suitable to handle the heterogeneous MCGDM problem whose decision information takes the forms of linguistic variables, TFNs, interval numbers and crisp numbers, but fails to deal with the heterogeneous MCGDM problem described in Sect. 6.4.1 where the decision data also contains IFNs, HFEs and HFLTSs. Therefore, Zhang et al. (2015) extended Li et al. (2010)'s method to tackle appropriately the practical decision making problem presented in Sect. 6.4.1. The algorithm (Algorithm 6.2) of the extension of Li et al. (2010)'s method is summarized as follows (Zhang et al. 2015):

Step 1. Normalize the heterogeneous decision matrix, and determine the heterogeneous PIS $A^{k+} = \left((\bar{x}_1^k)^+, (\bar{x}_2^k)^+, \ldots, (\bar{x}_n^k)^+\right)$ and the heterogeneous NIS $A^{k-} = \left((\bar{x}_1^k)^-, (\bar{x}_2^k)^-, \ldots, (\bar{x}_n^k)^-\right)$ for the expert $e_k \in \boldsymbol{E}$, please refer to Step 1 and Step 3 presented in Algorithm 6.1.

Step 2. Based on the given weights of criteria for each expert, an optimal model is constructed to determine the weights of the criteria for the group:

$$\begin{cases} \min \quad Z^w = \sum_{k=1}^{g}\sum_{j=1}^{n}\left(w_j^* - w_j^k\right)^2 \\ s.t. \quad w_j^* \geq 0\,(j \in \{1,2,\ldots,n\}), \sum_{j=1}^{n} w_j^* = 1 \end{cases} \tag{MOD-6.11}$$

where w_j^* denotes the weight of the criterion C_j for the group. By solving the model (MOD-6.11), we can obtain

$$w_j^* = \frac{\sum_{k=1}^{g} w_j^k}{g} \quad (j \in \{1,2,\ldots,n\}).$$

Step 3. Calculate the distances between the alternative A_i and the heterogeneous PIS A^{k+} as well as the heterogeneous NIS A^{k-}, respectively. Please refer to Step 4 presented in Algorithm 6.1.

Step 4. Calculate the relative closeness index CI_i^k of the alternative A_i to the heterogeneous PIS A^{k+} for the expert $e_k \in \boldsymbol{E}$ using the following equation:

$$CI_i^k = \varepsilon^k \frac{D_{\max i}^{k+} - D_i^{k+}}{D_{\max i}^{k+} - D_{\min i}^{k+}} + \left(1 - \varepsilon^k\right) \frac{D_i^{k-} - D_{\min i}^{k-}}{D_{\max i}^{k-} - D_{\min i}^{k-}} \tag{6.26}$$

where the parameters $\varepsilon^k \in [0, 1]$ and

$$D^{k+}_{\max i} = \max_{i=1,2,\ldots,m}\{D^{k+}_i\}, D^{k+}_{\min i} = \min_{i=1,2,\ldots,m}\{D^{k+}_i\},$$
$$D^{k-}_{\max i} = \max_{i=1,2,\ldots,m}\{D^{k-}_i\}, D^{k-}_{\min i} = \min_{i=1,2,\ldots,m}\{D^{k-}_i\}.$$

Step 5. Based on the obtained weights of criteria for the group and the given weights of criteria by each expert, a nonlinear programming model is established to determine the optimal weights of the experts as follows:

$$\begin{cases} \min \quad Z^{\varpi} = \sum_{k=1}^{g}\sum_{j=1}^{n}\left(\varpi_k(w^*_j - w^k_j)\right)^2 \\ s.t. \quad \sum_{k=1}^{g}\varpi_k = 1, \varpi_k \geq 0, k \in \{1, 2, \ldots, g\} \end{cases} \tag{MOD-6.12}$$

By solving the model (MOD-6.12), we can obtain

$$\varpi_k = \left\{\sum_{\xi=1}^{g}\left(\frac{\sum_{j=1}^{n}\left(w^*_j - w^k_j\right)^2}{\sum_{j=1}^{n}\left(w^*_j - w^{\xi}_j\right)^2}\right)\right\}^{-1} \quad (k \in \{1, 2, \ldots, g\}) \tag{6.27}$$

Step 6. Calculate the relative closeness index CI^*_i of the alternative A_i to the heterogeneous PIS A^{k+} for the group. The matrix format of the derived relative closeness indices of all alternatives for all experts is also expressed as $\boldsymbol{CI} = (CI^k_i)_{m\times g}$. Li et al. (2010) pointed out that the experts in the group can be regarded as "*the criteria*". Similar to the method to determine the PIS and NIS for the individual expert, we can identify the relative closeness index-based PIS $A^{CI+} = (CI^{1+}, CI^{2+}, \ldots, CI^{g+})$ and the relative closeness index-based NIS $A^{CI-} = (CI^{1-}, CI^{2-}, \ldots, CI^{g-})$ for the group, where

$$\begin{aligned} CI^{k+} &= \max_{i=1,2,\ldots,m}\{CI^k_i\}, \\ CI^{k-} &= \min_{i=1,2,\ldots,m}\{CI^k_i\} \end{aligned}, \tag{6.28}$$

The distances between the relative closeness index of the alternative A_i and the relative closeness index-based PIS A^{CI+} as well as NIS A^{CI-} for the group, respectively, can be calculated by using the following equations:

$$\begin{aligned} DCI_i^+ &= \sum_{k=1}^{g} \varpi_k \left| CI^{k+} - CI_i^k \right|, \\ DCI_i^- &= \sum_{k=1}^{g} \varpi_k \left| CI^{k-} - CI_i^k \right| \end{aligned}, \tag{6.29}$$

Then, the group closeness index DCI_i^* of the alternative A_i to the relative closeness index-based PIS A^{CI+} can be defined as:

$$DCI_i^* = \varepsilon \frac{DCI_{\max i}^+ - DCI_i^+}{DCI_{\max i}^+ - DCI_{\min i}^+} + (1-\varepsilon) \frac{DCI_i^- - DCI_{\min i}^-}{DCI_{\max i}^- - DCI_{\min i}^-} \tag{6.30}$$

where the parameters $\varepsilon \in [0, 1]$ and

$$\begin{aligned} DCI_{\max i}^+ &= \max_{i=1,2,\ldots,m} \{DCI_i^+\}, DCI_{\min i}^+ = \min_{i=1,2,\ldots,m} \{DCI_i^+\}, \\ DCI_{\max i}^- &= \max_{i=1,2,\ldots,m} \{DCI_i^-\}, DCI_{\min i}^- = \min_{i=1,2,\ldots,m} \{DCI_i^-\}. \end{aligned}$$

On the basis of the group closeness indices DCI_i^* $(i \in \{1, 2, \ldots, m\})$, it is easy to obtain the ranking of alternatives.

In what follows, we utilize the extension of Li et al. (2010)'s method to solve the heterogeneous MCGDM problem mentioned in Sect. 6.4.1. Firstly, we calculate the normalized decision matrix and the NIS as well as the PIS for four experts, which are shown in Table 6.4. Due to the fact that the extension of Li et al. (2010)'s method needs the experts to provide the weights of criteria in advance, to make this comparison more validly we assume that the weights of criteria provided by the experts in advance are the same as those obtained by the maximizing deviation model proposed by Zhang et al. (2015), namely,

$$\begin{aligned} \boldsymbol{w}^1 &= (w_1^1, w_2^1, w_3^1, w_4^1, w_5^1, w_6^1, w_7^1) \\ &= (0.2043, 0.1626, 0.0625, 0.0576, 0.1521, 0.1464, 0.2145), \\ \boldsymbol{w}^2 &= (w_1^2, w_2^2, w_3^2, w_4^2, w_5^2, w_6^2, w_7^2) \\ &= (0.1844, 0.1567, 0.0438, 0.1014, 0.1537, 0.1226, 0.1507), \\ \boldsymbol{w}^3 &= (w_1^3, w_2^3, w_3^3, w_4^3, w_5^3, w_6^3, w_7^3) \\ &= (0.1847, 0.0920, 0.2337, 0.0869, 0.0764, 0.2299, 0.0966), \\ \boldsymbol{w}^4 &= (w_1^4, w_2^4, w_3^4, w_4^4, w_5^4, w_6^4, w_7^4) \\ &= (0.1862, 0.1788, 0.0742, 0.0806, 0.1052, 0.1928, 0.1822). \end{aligned}$$

Table 6.7 The distances of alternatives from the PIS and NIS, and the derived closeness indices

Experts	Alternatives	D_i^{k+}	D_i^{k-}	CI_i^{k+}
e_1	A_1	0.2079	0.2464	1.0000
	A_2	0.2206	0.2388	0.8795
	A_3	0.2937	0.1605	0.0000
	A_4	0.2362	0.2181	0.6700
e_2	A_1	0.1336	0.3596	1.0000
	A_2	0.3012	0.1920	0.0335
	A_3	0.3008	0.1924	0.0359
	A_4	0.3113	0.1902	0.0000
e_3	A_1	0.2916	0.1236	0.0000
	A_2	0.1757	0.2543	0.6938
	A_3	0.2518	0.1625	0.2213
	A_4	0.1134	0.3009	1.0000
e_4	A_1	0.1736	0.2163	0.7543
	A_2	0.2886	0.1689	0.0000
	A_3	0.1869	0.2620	0.9422
	A_4	0.2066	0.1833	0.4339

Using the model (MOD-6.11), we obtain the weights of criteria for the group as follows:

$$\begin{aligned} \boldsymbol{w} &= (w_1, w_2, w_3, w_4, w_5, w_6, w_7) \\ &= (0.1899, 0.1475, 0.1035, 0.0816, 0.1219, 0.1729, 0.1610). \end{aligned}$$

Then, we use Eqs. (6.19) and (6.20) to calculate the distances between the alternative A_i for the expert e_k and the heterogeneous PIS A^{k+} as well as the heterogeneous NIS A^{k-}, respectively. The obtained results are displayed in Table 6.7. Furthermore, the relative closeness index CI_i^k of each alternative A_i to the heterogeneous PIS A^{k+} for the expert $e_k \in \boldsymbol{E}$ can be obtained by using Eq. (6.26) with $\varepsilon^k = 0.5(k = 1, 2, 3, 4)$, and the results are also presented in Table 6.7 (Zhang et al. 2015).

Using Eqs. (6.28)–(6.30), we finally obtain the relative closeness indices of all the alternatives for the group as follows:

$$DCI_1^* = 1.0, \quad DCI_2^* = 0.0, \quad DCI_3^* = 0.8, \quad DCI_4^* = 0.3649$$

Thus, the ranking of all alternatives is $A_1 \succ A_3 \succ A_4 \succ A_2$, which is the same as that obtained by the proposed method (we also call it Zhang et al. 2015's method) based on the minority principle, but is different from that derived from the proposed method on the basis of a compromise between the majority and minority principles.

Table 6.8 Problem requirements of different heterogeneous MCGDM methods

Problems with different characteristics and the decision results	Heterogeneous MCGDM methods		
	Zhang et al. (2015)'s method	The extension of Li et al. (2010)'s method	Li et al. (2010)'s method
Performance ratings	HFEs, HFLTSs, IFNs, linguistic variables, TFNs, interval numbers and real numbers	HFEs, HFLTSs, IFNs, linguistic variables, TFNs, interval numbers and real numbers	linguistic variables, TFNs, interval numbers and real numbers
Criteria weights for each expert	Completely unknown or partially known	Completely known	Completely known
Weights of the experts	Completely unknown or partially known	Completely unknown	Completely unknown
The final decision results	The consistency between the individual expert and the group has been taken into account	The consistency between the individual expert and the group has not been considered	The consistency between the individual expert and the group has not been considered

Based on the step by step demonstration of Zhang et al. (2015)'s method and the extension of Li et al. (2010)'s method and the final decision results derived from these two methods on the evaluation of Strategic Freight Forwarder of CSA, it is not hard to see that Zhang et al. (2015)'s method has some desirable advantages over the extension of Li et al. (2010)'s method.

As shown in Table 6.8, we can conclude that (Zhang et al., 2015):

(1) Besides the decision data expressed in the forms of real numbers, interval numbers, linguistic variables and fuzzy numbers, Zhang et al. (2015)'s method can accommodate more complicated decision data, including IFNs, HFEs, and HFLTSs. While Li's method (Li et al. 2010) just is suitable to solve the heterogeneous MCGDM problems in which the decision data are expressed as real numbers, interval numbers, linguistic variables and TFNs, which fails to deal with more complex problems such as the practical decision making problem described in Sect. 6.4.1.

(2) Zhang et al. (2015)'s method does not require the weights of the criteria provided subjectively by the experts in advance, but establishes the maximization deviation model to determine objectively the weights of the criteria for each expert, which is more reasonable; while the extension of Li et al. (2010)'s method needs the experts to provide the weights of criteria in advance, which is subjective and sometimes cannot yield the persuasive results.

(3) Zhang et al. (2015)'s method constructs a minimization deviation model to determine the ranking of the alternatives, which takes fully into account the consistency between the opinions of the individual expert and the group, and thus the final decision results derived by this method are more persuasive. While the extension of Li et al. (2010)'s method regards the expert as the "*criterion*" and then determines the final rankings of the alternatives on the basis of the distances to the ideal solutions, which fails to consider the consistency between the individual experts and the group.

6.5 Conclusions

It is common that, in the practical heterogeneous MCGDM problems, the decision data provided by the experts take multiple distinct forms, such as crisp numbers, interval numbers, linguistic variables, IFNs, HFEs and HFLTSs. This kind of heterogeneous MCGDM problems has broad applications in the fields of natural science, social science, economy and management, etc. This chapter introduces a deviation modeling method proposed by Zhang et al. (2015) to deal with such heterogeneous MCGDM problems in which the weights of both experts and criteria are partially known or completely unknown. The most characteristic of the deviation modeling method is that it does not unify the heterogeneous information but directly calculates the distances to the PIS and NIS. In the deviation modeling method, an optimal model based on the maximizing deviation method is established to determine the optimal weights of criteria for each expert, and a minimizing deviation model on the basis of the idea that the opinions of the individual expert should be consistent with those of the group is proposed to determine the weights of experts and identify the optimal alternative.

The deviation modeling method is finally illustrated by using the selection problem of Strategic Freight Forwarder of China Southern Airlines. The comparison analysis with Li et al. (2010)'s method shows that (1) the deviation modeling method can accommodate more complicated decision data, including IFNs, HFEs and HFLTSs; (2) the deviation modeling method utilizes the maximization deviation model to determine objectively the weights of the criteria for each expert, which avoids the subjective randomness of selecting the weights; (3) the deviation modeling method considers fully the consistency between the opinions of the individual experts and the group, thus the final decision results derived by the proposed method are more persuasive. Additionally, the deviation modeling method can be also expected to be applicable to other similar decision making problems, such as risk investment, supply chain management, etc.

References

Bordogna, G., & Pasi, G. (1993). A fuzzy linguistic approach generalizing boolean information retrieval: A model and its evaluation. *Journal of the American Society for Information Science, 44*, 70–82.

Charnes, A., & Cooper, W. W. (1977). Goal programming and multiple objective optimizations: Part 1. *European Journal of Operational Research, 1*, 39–54.

Chen, C. T. (2000). Extensions of the TOPSIS for group decision-making under fuzzy environment. *Fuzzy Sets and Systems, 114*, 1–9.

Cheng, S., Chan, C. W., & Huang, G. H. (2003). An integrated multi-criteria decision analysis and inexact mixed integer linear programming approach for solid waste management. *Engineering Applications of Artificial Intelligence, 16*, 543–554.

Delgado, M., Vila, M., & Voxman, W. (1998). A fuzziness measure for fuzzy numbers: Applications. *Fuzzy Sets and Systems, 94*, 205–216.

Feng, B., & Lai, F. J. (2014). Multi-attribute group decision making with aspirations: A case study. *Omega, 44*, 136–147.

González-Pachón, J., Diaz-Balteiro, L., & Romero, C. (2014). How to combine inconsistent ordinal and cardinal preferences: A satisficing modelling approach. *Computers & Industrial Engineering, 67*, 168–172.

Herrera, F., Martinez, L., & Sánchez, P. J. (2005). Managing non-homogeneous information in group decision making. *European Journal of Operational Research, 166*, 115–132.

Hwang, C. L., & Yoon, K. (1981). *Multiple attribute decision making methods and applications*. Berlin, Heidelberg: Springer.

Li, D. F., Huang, Z. G., & Chen, G. H. (2010). A systematic approach to heterogeneous multiattribute group decision making. *Computers & Industrial Engineering, 59*, 561–572.

Liu, H. B., & Rodríguez, R. M. (2014). A fuzzy envelope for hesitant fuzzy linguistic term set and its application to multicriteria decision making. *Information Sciences, 258*, 220–238.

Ma, J., Lu, J., & Zhang, G. Q. (2010). Decider: A fuzzy multi-criteria group decision support system. *Knowledge-Based Systems, 23*, 23–31.

Martínez, L., Liu, J., Ruan, D., & Yang, J. B. (2007). Dealing with heterogeneous information in engineering evaluation processes. *Information Sciences, 177*, 1533–1542.

Rodriguez, R. M., Martinez, L., & Herrera, F. (2012). Hesitant fuzzy linguistic term sets for decision making. *IEEE Transactions on Fuzzy Systems, 20*, 109–119.

Rodríguez, R. M., Martínez, L., & Herrera, F. (2013). A group decision making model dealing with comparative linguistic expressions based on hesitant fuzzy linguistic term sets. *Information Sciences, 241*, 28–42.

Wan, S. P., & Li, D. F. (2014). Atanassov's intuitionistic fuzzy programming method for heterogeneous multiattribute group decision making with Atanassov's intuitionistic fuzzy truth degrees. *IEEE Transactions on Fuzzy Systems, 22*, 300–312.

Wang, Y. M. (1998). Using the method of maximizing deviations to make decision for multi-indices. *System Engineering and Electronics, 7*, 31.

Wei, G. W. (2008). Maximizing deviation method for multiple attribute decision making in intuitionistic fuzzy setting. *Knowledge-Based Systems, 21*, 833–836.

Wu, Z. B., & Chen, Y. H. (2007). The maximizing deviation method for group multiple attribute decision making under linguistic environment. *Fuzzy Sets and Systems, 158*, 1608–1617.

Xu, Z. S. (2010). A deviation-based approach to intuitionistic fuzzy multiple attribute group decision making. *Group Decision and Negotiation, 19*, 57–76.

Yager, R. R. (1995). An approach to ordinal decision making. *International Journal of Approximate Reasoning, 12*, 237–261.

Yu, P. L. (1973). A class of solutions for group decision problems. *Management Science, 19*, 936–946.

Zadeh, L. A. (1975). The concept of a linguistic variable and its application to approximate reasoning. *Information Sciences, Part I, II, III, 8, 9*, 199–249, 301–357, 143–180.

Zhang, G. Q., & Lu, J. (2003). An integrated group decision-making method dealing with fuzzy preferences for alternatives and individual judgments for selection criteria. *Group Decision and Negotiation, 12*, 501–515.

Zhang, X. L., Xu, Z. S., & Wang, H. (2015). Heterogeneous multiple criteria group decision making with incomplete weight information: A deviation modeling approach. *Information Fusion, 25*, 25–62.

Printed in the United States
By Bookmasters